Master of Modern Physics

PRINCETON SERIES IN PHYSICS
Edited by Sam B. Treiman (published since 1976)

Studies in Mathematical Physics: Essays in Honor of Valentine Bargmann *edited by Elliot H. Lieb, B. Simon, and A. S. Wightman*

Convexity in the Theory of Lattice Gases *by Robert B. Israel*

Works on the Foundations of Statistical Physics *by N. S. Krylov*

Surprises in Theoretical Physics *by Rudolf Peierls*

The Large-Scale Structure of the Universe *by P.J.E. Peebles*

Statistical Physics and the Atomic Theory of Matter, From Boyle and Newton to Landau and Onsager *by Stephen G. Brush*

Quantum Theory and Measurement *edited by John Archibald Wheeler and Wojciech Hubert Zurek*

Current Algebra and Anomalies *by Sam B. Treiman, Roman Jackiw, Bruno Zumino, and Edward Witten*

Quantum Fluctuations *by E. Nelson*

Spin Glasses and Other Frustrated Systems *by Debashish Chowdhury*
(*Spin Glasses and Other Frustrated Systems* is published in co-operation with World Scientific Publishing Co. Pte. Ltd., Singapore.)

Weak Interactions in Nuclei *by Barry R. Holstein*

Large-Scale Motions in the Universe: A Vatican Study Week *edited by Vera C. Rubin and George V. Coyne, S.J.*

Instabilities and Fronts in Extended Systems *by Pierre Collet and Jean-Pierre Eckmann*

More Surprises in Theoretical Physics *by Rudolf Peierls*

From Perturbative to Constructive Renormalization *by Vincent Rivasseau*

Supersymmetry and Supergravity (2nd ed.) *by Julius Wess and Jonathan Bagger*

Maxwell's Demon: Entropy, Information, Computing *edited by Harvey S. Leff and Andrew F. Rex*

Introduction to Algebraic and Constructive Quantum Field Theory *by John C. Baez, Irving E. Segal, and Zhengfang Zhou*

Principles of Physical Cosmology *by P.J.E. Peebles*

Scattering in Quantum Field Theories: The Axiomatic and Constructive Approaches *by Daniel Iagolnitzer*

QED and the Men Who Made It: Dyson, Feynman, Schwinger, and Tomonaga *by Silvan S. Schweber*

The Interpretation of Quantum Mechanics *by Roland Omnès*

Gravitation and Inertia *by Ignazio Ciufolini and John Archibald Wheeler*

The Dawning of Gauge Theory *by Lochlainn O'Raifeartaigh*

The Theory of Superconductivity in the High-T_c Cuprates *by P. W. Anderson*

Master of Modern Physics: The *Scientific* Contributions of H. A. Kramers *by D. ter Haar*

Master of Modern Physics

The *Scientific* Contributions

of H. A. Kramers

D. TER HAAR

Princeton Series in Physics

PRINCETON UNIVERSITY PRESS · PRINCETON, NEW JERSEY

Copyright © 1998 by Princeton University Press
Published by Princeton University Press, 41 William Street,
Princeton, New Jersey 08540
In the United Kingdom: Princeton University Press, Chichester, West Sussex

All Rights Reserved

Library of Congress Cataloging-in-Publication Data

Haar, D. ter.
Master of modern physics : the scientific contributions of H.A. Kramers / Dirk ter Haar.
p. cm. — (Princeton series in physics)
Includes bibliographical references and index.
ISBN 0-691-02141-4 (cloth : alk. paper)
1. Kramers, Hendrik Anthony, 1894–1952. 2. Quantum theory—History.
3. Physicists—Netherlands—Biography. I. Title. II. Series.
QC16.K69H33 1998
530.12—dc21   97-22356

The publisher would like to acknowledge the author of this volume for providing the camera-ready copy from which this book was printed

Princeton University Press books are printed on acid-free paper and meet the guidelines for permanence and durability of the Committee on Production Guidelines for Book Longevity of the Council on Library Resources

http://pup.princeton.edu

Printed in the United States of America

10 9 8 7 6 5 4 3 2 1

Contents

QC
16
.K69
H33
1998

Preface	vii
1. Introduction	3
2. The Old Quantum Theory	9
2.1. Applications of the Old Quantum Theory to Spectra	10
2.2. The BKS Paper	21
2.3. The Dispersion Formulæ	23
3. Quantum Mechanics	31
3.1. Kramers's Contributions to the General Theory of Non-Relativistic Quantum Mechanics	32
3.2. Electron Spin and the Relativistic Wave Equation	39
3.3. The Symbolic Method	45
3.4. Applications of Quantum Mechanics to Specific Problems	47
3.5. Kramers Degeneracy and Charge Conjugation	54
4. Quantum Electrodynamics	59
5. Statistical Mechanics, Solid-State Physics, and Low-Temperature Physics	67
5.1. Kramers's Contributions to the General Theory of Equilibrium Statistical Mechanics	68
5.2. Phase Transitions	72
5.3. Solid-State and Low-Temperature Physics	88
6. The Kramers Problem and Polymer Physics	93
6.1. The So-Called Kramers Problem	94
6.2. Flow of Polymers	102
7. Miscellaneous Topics	109
7.1. Kinetic Theory of Gases and Gas Dynamics	109
7.2. Other Problems	116
Reprinted Papers	
A. On the Scattering of Radiation by Atoms (with W. Heisenberg; *Zs. Phys.* **31**, 681 [1925]))	121

B. Some Remarks on Heisenberg's Quantum Mechanics
   (*Physica* **5**, 369 [1925])   145
C. Wave Mechanics and Half-Odd-Integral Quantisation
   (*Zs. Phys.* **39**, 828 [1926])   151
D. The Scattering of Light by Atoms (*Atti Congr.
   Intern. Fis., Como-Pavia-Roma*, II, 545 [1927])   163
E. General Theory of Paramagnetic Rotation in Crystals
   (*Amsterdam Proc.* **33**, 959 [1930])   173
F. Classical Relativistic Spin-Theory and Its Quantization
   (*Verh. Zeeman Jubil.* 403 [1935])   187
G. On the Eigenvalue Problem in a One-Dimensional Field
   of Force (*Physica* **2**, 483 [1935])   195
H. The Use of Charge-Conjugated Wavefunctions in the Hole
   Theory of the Electron (*Amsterdam Proc.* **40**, 814 [1937])   203
I. Brownian Motion in a Field of Force and the Diffusion
   Model of Chemical Reactions (*Physica* **7**, 284 [1940])   213
J. Statistics of the Two-Dimensional Ferromagnet
   (with G. H. Wannier; *Phys. Rev.* **60**, 252 [1941])   231
K. Fundamental Difficulties of a Theory of Particles
   (*Ned. Tijds. Natuurk.* **11**, 134 [1944])   249
L. The Behavior of Macromolecules in Inhomogeneous
   Flow (*J. Chem. Phys.* **14**, 415 [1946])   257

References   273

Publications of H. A. Kramers   277

Index   283

Preface

Hendrik Anthony Kramers was one of the giants of theoretical physics of the first half of the twentieth century. Dresden (1987) has written a biography of him that describes Kramers as a man of his time and, as far as his scientific work is concerned, concentrates on a few fields in which Kramers has made significant contributions. However, in my opinion, his book did not do justice to Kramers as a physicist who mastered all branches of physics of his time. The aim of the present book is to remedy this and to present an account of Kramers's scientific œuvre in its entirety. I have discussed in some detail all the papers contained in his *Collected Scientific Papers* (Casimir et al. 1956), and in order for the reader to get an idea of Kramers's style, reprinted twelve of his most important papers — in translation in those cases where they were originally in a different language.

I feel that this still leaves one more book to be written about Kramers, namely, an account of his philosophical and popular papers as well as of his activities as a diplomat in the service of the Netherlands; the latter have been briefly discussed both by Dresden in his book and recently by Goedkoop (1996), but they deserve a more detailed discussion.

In writing this book I have benefited from the advice and assistance of many people. My thanks go out to all of them, first of all to the North-Holland Publishing Company for their permission to reprint the papers from Kramers's *Collected Scientific Papers*, to Kramers's children for providing me with the photos that are reproduced here, and especially to his son, who has read most of the manuscript and made many extremely useful suggestions for improvements. I am also grateful to Professor H.B.G. Casimir, Dr. N. E. Frankel, and Professor J. Kistemaker for helping me to clarify some points of historical interest. My greatest debt is to a fellow-pupil of Kramers, Professor N. G. van Kampen; he helped me to choose the papers to be reprinted and has read through the whole of the manuscript, making numerous suggestions for additions or deletions which, I am sure, have greatly enhanced the readability of the text.

Petworth, October 1996   D. ter Haar

H. A. Kramers, circa 1940.

Master of Modern Physics

1 Introduction

The nature of physics has changed fundamentally during the second half of the twentieth century.[1] Up to the Second World War a physicist could know and understand all of physics and, if he were good enough, contribute to every field in that subject, and up to, say, 1960 it was possible for someone to be acquainted with practically all of physics. Nowadays, however, physicists are lucky if they have worked in more than one more or less specialised part of the subject, and it is becoming increasingly difficult for a specialist in one field to understand what a colleague working in another field is doing. The great physicists of the twentieth century, such as, for instance, Fermi or Landau,[2] to name but two, all made significant contributions to widely varying fields. As we shall show, another physicist who made significant contributions to most fields of physics is Hendrik Anthony Kramers, whose scientific work is the subject of the present book. One might wonder why he has not received the same recognition as some of his contemporaries such as Dirac, Fermi, or Pauli. This is partly due to the fact that he did not receive the Nobel prize. It is widely rumoured that, if he had not died in 1952 at the relatively young age of fifty-seven, he would have shared the Nobel prize for that year with Max Born — who himself had to wait until he received it belatedly in 1954, when he was seventy-two years old. It is also due to the fact that Kramers hid his light under a bushel. Dresden, in his thorough biography of Kramers (Dresden 1987), males it clear that he did not believe in self-advertisement. He also did not want to publish anything that was not of the highest standard, and was often too much aware of the fact that there were still flaws and lacunæ in some of the arguments. In his article (Kramers 1938b) written to commemorate Fokker's doctorate twenty-five years earlier, he quotes the

---

[1] An excellent description of the explosive growth of physics during the twentieth century can be found in Pais's account (Pais 1986) of the ideas about matter and force.

[2] Arguably the two greatest physicists of the twentieth century are Bohr and Einstein. We may remind the reader that in Einstein's annus mirabilis, 1905, he published his three seminal papers on three different topics: Brownian motion, light quanta, and special relativity, and that Bohr before he published his 1913 paper on quantum theory had already written a prize essay on the surface tension of liquids and published papers on that subject, as well as on the theory of metals and α-particle absorption.

remark by Fokker that too often in his thesis there is a *credo* instead of an *ecce*, and then states that quantum theory is a theory for which it still is true that all *ecces* are really *credos*—a strongly held opinion as is clear from a careful reading of his own textbook (Kramers 1933a, 1938e). This hesitancy to publish has undoubtedly contributed to his being relatively unknown to present-day physicists. It is hoped that the present book may do something to remedy this.

In my opinion, Kramers must be counted among the great theorists of the first half of the twentieth century. After all, Kramers is still mentioned whenever people write about the Kramers opacity formula, the Kramers dispersion theory, the Kramers-Heisenberg dispersion formulæ—which in all fairness should be called the Kramers dispersion formulæ—the Kramers-Kronig relations, the Kramers degeneracy, the WKB approximation method, the Kramers equation, or even the "Kramers problem" (Mel'nikov 1991). If this is not enough we may add the fact that it was Kramers who first introduced the thermodynamic limit and, as discussed by Dresden in his biography of Kramers, it was also Kramers who, independently of Dirac, derived the relativistic wave equation[3] for the electron (Dresden 1987, p. 64) and, independently of Debye, predicted the Compton effect (Dresden 1987, ch. 14).[4] In the case of the relativistic wave equation Dirac's paper just preceded a possible publication of Kramers's result, while in the case of the Compton effect Bohr's opposition to the idea of photons was so strong that Kramers was prevented from publishing his result. In fact, Dresden devotes a separate chapter, in which he gives a beautiful blow-by-blow account, to the Compton effect episode. From Dresden's biography it is also clear that not only was Kramers's interest in physics extremely broad, his interests in general were widespread: he was an accomplished linguist, cellist and pianist, and poet, and was interested not only in the interaction of physicists with society and the way scientific research is carried out, but also in subjects as varied as religion, literature, and history. This also meant that he did not single-mindedly pursue his own interests in the world of physics.[5]

One might wonder whether there is a place for another book about Kramers when Dresden (1987) has written what in many respects looks like the definitive biography. However, although Dresden discussed in great detail Kramers's development as a physicist and all personal aspects of his life, he was restricted by the requirement to limit the size of his book to manageable proportions, and thus he was able to discuss in detail only a limited number

---

[3] Kramers's derivation of this equation was not published at the time he produced it, but can be found in his contribution to the Zeeman Jubilee volume (Kramers 1935b) or in the second part of his textbook on quantum mechanics (Kramers 1938e).

[4] This derivation was never published.

[5] H.B.G. Casimir, who was a personal friend of Kramers, gives in his autobiography (Casimir 1983) a very good picture of Kramers, both as a physicist and as a person.

of Kramers's scientific papers, namely, those that he considered to be most important for the development of Kramers as a leading physicist of his time; this meant especially Kramers's work in the 1920s and his work on renormalisation. He followed up his biography by an article in *Physics Today* (Dresden 1988) in which he outlined Kramers's contributions to statistical mechanics. In the present book I am hoping to cover Kramers's scientific output more completely. Since this book will be exclusively devoted to Kramers's scientific work, it will be to reprint some of Kramers's papers, either in the original, or in English translation in those cases where the originals were in German, French, or Dutch. This will give the reader an opportunity to see for himself the particular flavour of a typical Kramers paper. I would have liked to include more papers than are to be found at the end of the book, but here also there are limitations of size. This is also the reason why of some of the papers have been printed here only in part.

Before considering Kramers's scientific work, I want to take issue with Dresden on two points. First, I feel that although he seems to have changed his attitude to Kramers slightly by the time he wrote the article on the statistical mechanics papers, in his book he seems to put too much emphasis on what he calls Kramers's "near misses,"[6] even to the extent of giving as a kind of dedication the following quotation from Whitehead (1974): "To come very near a true theory and to grasp its precise application are two very different things." I feel it is worthwhile to consider what seems in Dresden's opinion to be a "near miss." As far as I can judge, Dresden feels that a near miss is the failure to follow up a result in such a way that it is clear to everybody what has been achieved and how important the breakthrough is that has been obtained. I feel, however, that one can only talk about a "near miss" if the breakthrough itself is missed. There are some indications that Dresden has some sympathy for this opinion. In fact, he gives nine or ten instances of "near misses" but in some cases he himself queries whether they should truly be called this. I feel that among those cases there are only two in which one might speak of a near miss, namely, not developing matrix mechanics after having discovered the dispersion formulæ, and not finding the exact solution to the two-dimensional Ising model after having applied the transfer matrix method to this problem. As far as these two cases are concerned, I should like once again to quote van Kampen (1988): "Is it necessary to explain that if one has with great effort and anguish used the correspondence principle to construct a

---

[6] In this connection I would also like to quote van Kampen (1991): "Dresden emphasizes that in many subjects Kramers did all the groundwork but failed to make the decisive step. There is a grain of truth in it, but to consider that as a basic flaw of Kramers as a physicist seems to me unfair. Does one blame Lorentz for not taking the final step to relativity? Einstein for not going on to discover quantum mechanics? Columbus for stopping half way on his voyage to India? Rather I think that it is in the nature of things that those who laboriously lay the foundations for the new development no longer have the freshness of mind needed to initiate an entirely new approach."

dispersion formula, it is impossible to forget this background and that it needs a fresh mind to take the next step?"[7] and "Does one need an explanation for the fact that Kramers and Wannier after having invented the transfer matrix for the two-dimensional Ising model did not produce Onsager's tour de force?"[8] None of the other cases, in my opinion, deserves to be called a near miss. I do not want to discuss them in detail, but restrict myself to the one that Dresden considers the greatest miss: Kramers's failure to construct the formal renormalisation theory for which Tomonaga, Schwinger, and Feynman received their Nobel prize. This shows, in my opinion,[9] how much Dresden has misunderstood Kramers's attitude to results that were not fully justified. Kramers had a healthy doubt about the legitimacy of perturbation theory, and the idea that by using perturbation theory and, if necessary, equating ∞ with 0, one could renormalise would have been repellent to him.[10] Altogether, I feel that a careful consideration of Dresden's list of "near misses" shows that his emphasis on this aspect of Kramers's work was mistaken and almost completely based on his misunderstanding of Kramers's attitude to physics, and that he was misguided to make it one of the cornerstones of his otherwise admirable portrait of Kramers.[11]

More seriously, I feel that Dresden misreads the course of events in the summer of 1925.[12] Dresden seems to suggest (Dresden 1987, p. 52) that although Kramers knew all about Heisenberg's paper (Heisenberg 1925) in which he introduces matrix mechanics, since he visited Göttingen at the end of June and met Heisenberg, he did not tell Bohr about it. This seems completely at variance with all we know about Kramers. Dresden makes clear in his book and emphasises how extraordinarily unselfish and unself-seeking

---

[7] In this connection I should also like to quote Bohr (1952): "This work (Kramers's dispersion theory) actually proved a stepping stone for Heisenberg who ... accomplished a rational formulation of quantum mechanics." on which Opechowski (1988) comments: "In fact, one may argue that Kramers' work played a somewhat analogous role in Heisenberg's creation of matrix mechanics as de Broglie's work did in the creation of Schrödinger's wave mechanics."

[8] One should perhaps also mention that the Kramers-Wannier paper appeared in 1941 and Onsager's paper in 1944, and that the Netherlands were occupied during the period from 1941 to 1944, producing an atmosphere hardly conducive to sustained scientific research. Professor Dresden has kindly informed me that, according to Uhlenbeck, Kramers had already found the transfer matrix method in 1937 and had mentioned it at the Van der Waals Conference in 1937.

[9] An opinion apparently shared by van Kampen: see van Kampen 1988.

[10] See in this connection also Pais's account (Pais 1986, p. 449) which shows why Kramers did not pursue the line Dresden feels he should have pursued.

[11] I should also like to quote what Opechowski, another of Kramers's collaborators, wrote in his review of Dresden's book (Opechowski 1988): "I think that most of them are really not 'near misses.' What really mattered to Kramers in his research was to achieve an 'approbation intérieure.' This is a phrase he found in a letter of Fresnel to Thomas Young, and I heard him quote it more than once. But an 'approbation intérieure' does not necessarily imply a worldly success."

[12] The events of June 1925 are told in detail not only by Dresden but also by Pais (1991).

Kramers was. He totally lacked the ruthlessness that characterised Heisenberg and even, to a large extent, Bohr. I therefore wonder whether Heisenberg, in fact, told Kramers about the results of his paper. First, the paper was not finished until about ten days after Heisenberg and Kramers met. Second, it is interesting to note that although Heisenberg wrote to Bohr that he had submitted this paper he did not mention the contents (Dresden loc. cit.). Heisenberg, in contrast to Kramers, was very ambitious and would, certainly in his youth, use all means to promote his own interest;[13] he may have wanted to have some time to develop his ideas further by himself without competition from Kramers. Therefore he may well not have told Kramers the details of his paper, especially as at the time of meeting Kramers he had not yet completed the paper[14] and as, even when he completed the paper, he felt the need for "a more intensive mathematical investigation of the method which has been quite superficially employed here"[15] — and Kramers was the person who more than anybody else might have done this.[16]

In attempting to assess Kramers's contributions to physics we must mention that Kramers made important contributions to physics not only through his published papers but also through communicating to others some of his unpublished knowledge of the subject, and by making suggestions that he himself did not follow up for one reason or another. I may mention here three instances, two of which are mentioned by Dresden. The first is the importance of the thermodynamic limit in the theory of phase transitions (Dresden 1987, p. 321). The second is the importance of the analytical properties of the so-called S-matrix introduced by Heisenberg (Dresden 1987, § 17 IIC). The final instance is the explanation of the Third Law of Thermodynamics (Casimir 1963).

Finally, we must mention that in order properly to evaluate the importance of Kramers's work it is necessary to read his published work with great care. There are many instances in which he buried deep insights into various problems without explicitly emphasising the significance of what he wrote — not so much, I suggest, because he did not realise their importance but because he was averse to self-advertisement.

As was emphasised by Dresden, Kramers's personality was complicated, and he was greatly influenced by his Dutch upbringing. Although I do not agree with Dresden in all details of his analysis, as I have already pointed out, the reader should consult his book together with the present one in order fully to comprehend Kramers's importance as a physicist.

---

[13] See, for instance, the history of how he got his name on the Kramers-Heisenberg (1925) paper as recounted by Dresden (1987, p. 273).

[14] Heisenberg met Kramers on 21 June (Dresden loc. cit.), but on the 24th he wrote to Pauli that "everything is still unclear to me" (Pais 1991, p. 275).

[15] The translation is by Pais (Pais 1991, p. 276).

[16] In this connection it is also interesting to quote what Born writes about it (Born 1978, p. 216): "Heisenberg pursued some work of his own, keeping its idea and purpose somewhat dark and mysterious."

In the list of publications of H. A. Kramers on p.277 there are altogether 147 entries, 66 of which are reprinted in his *Collected Scientific Papers* (Casimir et al. 1956). Of the 81 entries that cannot be found in the Collected Papers three are his books (Kramers and Holst 1922, Kramers 1933a, 1938e). Then there are various versions of the same paper; for instance, the paper by Bohr, Kramers, and Slater which Kramers and Bohr considered to be so important that they submitted it both in German (Bohr, Kramers, and Slater 1924a) and in English (Bohr, Kramers, and Slater 1924b), while the work with de Haas and Wiersma on the adiabatic demagnetisation experiments appeared in five versions (de Haas, Wiersma, and Kramers 1933a, b, c, d, 1934). Finally, there are the many popular papers about physics for non-specialists, as well as papers describing the work of other physicists on the occasion of their being awarded prizes, their deaths, or their anniversaries.

I have split Kramers's work, rather arbitrarily, into six categories, and shall devote a separate chapter to each of those categories. Chapter 2 considers his papers — and his book with Holst — dealing with the old quantum theory, that is, quantum theory before Heisenberg and Schrödinger introduced matrix and wave mechanics which, to distinguish it from the old quantum theory, we shall call quantum mechanics, Chapter 3 is devoted to his papers — and his monumental textbook — on quantum mechanics. Chapter 4 deals with Kramers's papers on renormalisation, as well as the relevant sections of his quantum mechanics textbook. Chapter 5 is devoted to the papers on statistical mechanics and to papers dealing with solid-state and low-temperature physics. In Chapter 6 I discuss Kramers's classical paper on Brownian motion as well as his paper on polymer flow, while in the last chapter I consider the remaining papers, dealing with a variety of topics, ranging from general relativity to astrophysics. There is, quite naturally, a certain amount of overlap between the categories, so that my assignment of papers to the various chapters is a rather personal one.

2 The Old Quantum Theory

Dresden (1987) has told the story[1] of how Kramers in September 1916 started to work with Bohr just after the latter had returned from his second stay in England to take up a newly created Chair of Theoretical Physics. Kramers stayed in Copenhagen until 1926, when he returned to the Netherlands to take up a physics chair at the University of Utrecht. This decade almost exactly coincides with the period during which the old quantum theory was developed, reached its peak, and was finally replaced by matrix and wave mechanics.[2]

During these years Copenhagen was one of the main centres of theoretical physics — if not **the** main centre — where Kramers very soon after his arrival became Bohr's assistant and trusted second-in-command. As a result this is also the period when some of Kramers's best and most seminal work was done. Of his *Collected Scientific Papers* (Casimir et al. 1956), thirteen were written in Copenhagen, all but two dealing with topics in the old quantum theory. The two exceptions — one paper on general relativity (Kramers 1921) and one on chemical reactions (Christiansen and Kramers 1923) — will be discussed in Chapter 7. Apart from those papers, during his stay in Copenhagen Kramers wrote a book with Holst, explaining in semi-popular language the basic ideas of the old quantum theory, as well as several more or less popular accounts of this theory (see Kramers 1922a, b, c, d, e, 1923a, b, 1924c, d, 1925a, b).

The eleven papers dealing with topics from the old quantum theory that are reprinted in the Collected Papers fall roughly into four categories:

i. Papers dealing with applications of the theory as it existed when Kramers arrived in Copenhagen. These include his Leiden thesis (Kramers 1919), a paper on the theory of the helium atom (Kramers 1923c), and two papers on molecular spectra (Kramers 1923d, Kramers and Pauli 1923);
ii. The paper on X-ray absorption (Kramers 1923e) which extends the old quantum theory to deal with continuous spectra;
iii. The paper containing the ill-fated BKS proposal (Bohr, Kramers, and Slater 1924a, b); and

---
[1] See also Pais (1991), §9(c).
[2] See, for instance, the accounts by Dresden (1987) and Pais (1991).

iv. The papers on dispersion theory (Kramers 1924a, b, 1925c, Kramers and Heisenberg 1925) of which the last is reprinted at the end of the book as paper A.

In the present chapter we shall discuss these eleven papers — those in categories i and ii in §2.1, the BKS paper in §2.2, and those in category iv in §2.3 — as well as those (Kramers 1927b, 1929d) dealing with the so-called Kramers-Kronig relations, since Kramers had derived the main results during the period covered in the present chapter.[3]

2.1. Applications of the Old Quantum Theory to Spectra

When Kramers arrived in Copenhagen in 1916 the old quantum theory was at the stage where its validity was not yet accepted by all physicists and where its rules were still being developed. The basic ideas were there: atomic systems will be found in stationary states in which no emission of radiation will take place, even though the system according to classical electrodynamics should emit such radiation, and emission takes place when the system makes a transition from one stationary state to another at a frequency ν given by the difference of the energies in the two states divided by Planck's constant,

$$\nu = \frac{E_1 - E_2}{h}. \tag{2.01}$$

To find the stationary states one has to use quantisation rules and Sommerfeld (1915, 1916) and Wilson (1915) had introduced their quantisation rules for multiply periodic systems — which were a generalisation of the rule introduced by Bohr in his original paper (Bohr 1913b) —

$$J_k = \oint p_k \, dq_k = n_k h, \tag{2.02}$$

where the q_k and p_k are the generalised coordinates and momenta of the system and the n_k are integers, the so-called *quantum numbers*. The integration in Eq. (2.02) is over one period of the q_k. The J_k are *action variables*, which were shown by Burgers (1917) to be adiabatic invariants. This is an essential property of the J_k, since in order to find the stationary states we must have quantities that do not change when the system is subject to a very slow — adiabatic — change. It should be noted that in order for one to use the quantisation rules it is necessary to have a system for which the coordinates can be separated.[4]

---

[3] The first of those two papers is reprinted at the end of this book as paper D.
[4] For such systems the Hamilton-Jacobi function, or action, is additive — not necessarily the Hamiltonian as stated erroneously by Dresden (1987, p. 108).

During the first few years of Kramers's stay in Copenhagen Bohr had introduced the *correspondence principle*, according to which quantum mechanics should go over into classical theory in the limit of large quantum numbers.[5]

Finally, in that period Einstein (1917) published his paper on the absorption and the spontaneous and induced emission of radiation, in which he had introduced what are now known as the Einstein coefficients. The important feature of these coefficients is that they depend on both the initial and the final state of the atom in the transition producing the emission, whereas classically one expects dependence only on the initial state.

The old quantum theory had been able to predict the energy levels of a few atoms, especially hydrogen, but at that time it had not yet tackled the question of the polarisations and the intensities of the various spectral lines, although Bohr (1918) had suggested that one could obtain information about these properties by looking at the amplitudes of the harmonic vibrations in which the motion of the particles in the system may be resolved. Bohr suggested this topic as the subject of Kramers's thesis. In his thesis Kramers applied Bohr's ideas to the case of a hydrogen atom by itself and in an electric field. This paper established Kramers as one of the leading theoretical physicists of the time, and it is not difficult to see why, since it shows him as a master of both classical mechanics and of the new atomic theory of Bohr as well as a consummate mathematician. His thesis, at the time it was published, was right at the front of the developments of the old quantum theory.

The thesis consists of two parts, each of four sections. The first part is purely classical mechanics and a beautiful application of Hamilton-Jacobi theory.[6] In it Kramers first discusses the general theory of multiply periodic systems, which he called conditionally periodic systems, that is, systems for which the separation of variables method can be used, and he shows how one can for such systems express the coordinates which describe the classical orbits as Fourier series in the classical frequencies. Kramers then applies this theory successively to the hydrogen atom governed by relativistic mechanics, to the hydrogen atom in an electric field which is so large that the relativistic corrections can be considered to be a small perturbation, and to the hydrogen atom governed by relativistic mechanics in a very weak electric field, so that the effect of the field can be considered to be a small perturbation. In all these cases one is still dealing with systems where the coordinates, if chosen suitably, can be separated.

In the second part, quantum theory is introduced. Kramers first sketches what one can deduce about the polarisation and intensities of the spectral lines emitted by an atomic system if one knows its stationary states. He

---

[5] This principle was introduced in the first of Bohr's comprehensive survey papers (Bohr 1918, 1922) when he called it the analogy principle; the name "correspondence principle" was not used by him until 1920 (Bohr 1920).

[6] Details of the Hamilton-Jacobi theory and of classical perturbation theory can be found in many textbooks (see, for instance, ter Haar 1971, ch. 6).

then applies this in the later sections to the hydrogen atom for the three cases considered in the first part, and he concludes the thesis with a brief discussion of the effect of an external magnetic field on a hydrogen atom.

To appreciate the difficulties of this work one must read the thesis, and we can here only sketch the main ideas of the calculations. One first writes down the Hamiltonian of the system in terms of suitably chosen generalised coordinates q_i and momenta p_i and solves the canonical equations of motion. This is done by introducing the Hamilton-Jacobi function $S(q_i)$, which is a function of the q_i and of new generalised momenta α_k. The original coordinates were chosen in such a way that one could separate the variables, that is, write the Hamilton-Jacobi function as a sum of terms, each depending on one of the q_i only.

The Hamilton-Jacobi function is found as the solution of a partial differential equation,

$$H\left(\frac{\partial S}{\partial q_i}, q_i\right) = \alpha_1, \qquad (2.03)$$

the so-called Hamilton-Jacobi equation which is obtained by replacing the momenta p_i in the Hamiltonian by $\partial S/\partial q_i$, using part of the equations governing the canonical transformation from the p_i, q_i variables to new variables α_k, β_k (compare Eq. (2.05)). We see that the first of the α_k is the energy, while the other α_k are integration constants arising when Eq. (2.03) is solved.

We now take into account that we are dealing with a multiply periodic system, that is, a system which is periodic in each of the q_i. We introduce a new set of variables J_k—the action variables of Eq. (2.02)—through the equations

$$J_k = \oint p_k \, dq_k. \qquad (2.04)$$

We see that as the p_k are equal to $\partial S/\partial q_k$ they are functions of the q_k and the α_k and therefore the J_k are functions of the α_k and Eq. (2.04) is a straightforward coordinate transformation from the α_k to the J_k. If we now express the α_k in terms of the J_k, the Hamilton-Jacobi function becomes a function of the q_i and the J_k. One can use this new Hamilton-Jacobi function to perform a canonical transformation,

$$p_k = \frac{\partial S}{\partial q_k}, \qquad w_k = \frac{\partial S}{\partial J_k}, \qquad (2.05)$$

from the p_i, q_i to the action and angle variables J_k, w_k. From Eq. (2.03) and the fact that the action variables depend on the α_k only, it follows that the Hamiltonian will depend only on the J_k and not on the angle variables, the w_k. The solution of the canonical equations of motion corresponds therefore to all J_k being constants and the w_k being linear functions of the time:

$$J_k = \text{const}, \qquad w_k = \omega_k t + \delta_k, \qquad \omega_k = \frac{\partial H}{\partial J_k}, \qquad (2.06)$$

where the δ_k are constants. One can show (see, e.g., Kramers 1919, §1) that the solutions for the orbits are periodic in the three w_k, all three periods being equal to 1. Using Eq. (2.06) we can thus write the solution for the original coordinates q_k in the form

$$q_k = \sum_{\tau_1,\tau_2,\tau_3=1}^{\infty} C^{(k)}_{\tau_1,\tau_2,\tau_3} e^{2\pi i (\tau_1 \omega_1 + \tau_2 \omega_2 + \tau_3 \omega_3)t}. \qquad (2.07)$$

Equation (2.07) gives us the required expression for the coordinates as functions of the classical frequencies, ω_k, which, according to the last of Eqs. (2.06), are related to the energy E of the orbit through

$$\omega_k = \frac{\partial E}{\partial J_k}. \qquad (2.08)$$

The next step is to note that classically the intensity of the radiation corresponding to a frequency $\tau_1 \omega_1 + \tau_2 \omega_2 + \tau_3 \omega_3$ will be proportional to the square of the corresponding coefficient, C_{τ_1,τ_2,τ_3}.

So far the arguments have been purely classical and the important step is now to introduce quantum theory. There are several aspects of this. First, one selects the orbits that are allowed according to quantum theory by using the quantisation rules to put the J_k equal to $n_k h$ with integer values of the n_k. Second, one takes into account that, in contrast to the classical theory where the radiation takes place at the frequencies corresponding to a given orbit, in quantum theory the frequency of the radiation is determined by Bohr's condition (2.01), that is, it depends on both the initial and the final orbit. Let us consider a transition between an initial and a final state which are characterised, respectively, by the quantum numbers n'_k and n''_k. If the classical states lying "between" these two states have values of the J_k equal to $\{\lambda n''_k + (1-\lambda) n'_k\} h$ ($0 \leq \lambda \leq 1$), one can prove that the frequency ν of the radiation emitted during the transition is given by the equation

$$\nu = \frac{1}{h} \int_{\lambda=0}^{\lambda=1} \delta E = \frac{1}{h} \int_{\lambda=0}^{\lambda=1} [\omega_1 \delta J_1 + \omega_2 \delta J_2 + \omega_3 \delta J_3]$$

$$= \int_0^1 \{(n'_1 - n''_1)\omega_1 + (n'_2 - n''_2)\omega_2 + (n'_3 - n''_3)\omega_3\} \, d\lambda, \qquad (2.09)$$

where we have used Eq. (2.08) to write

$$\delta E = \omega_1 \delta J_1 + \omega_2 \delta J_2 + \omega_3 \delta J_3. \qquad (2.10)$$

In the region of stationary states with quantum numbers so large that for small differences $n'_k - n''_k$ the motions in the initial and the final state differ relatively little, the ω_k may considered to be constant when λ varies from 0 to 1 so that the frequency ν, given by Eq. (2.09), asymptotically approaches the frequency $(n'_1 - n''_1)\omega_1 + (n'_2 - n''_2)\omega_2 + (n'_3 - n''_3)\omega_3$. This frequency is present in the motion of the system, and the correspondence principle then leads to the conclusion that the polarisation and intensities of the spectral lines emitted in the region of large quantum numbers will asymptotically be equal to the polarisation and intensities of the corresponding radiation emitted according to classical electrodynamics. These properties will be determined by the relevant coefficients $C_{n''_1 - n'_1, n''_2 - n'_2, n''_3 - n'_3}$ with the polarisation being determined by the ratios of these coefficients in the series for the x, y, and z components of the motion, and the intensities being proportional to the squares of the coefficients. For instance, if for a given transition there is no term in the series for the x and y components, while there is such a term in the z component, we can conclude that the radiation will be plane polarised.

The question now arises of what happens when the n_k are not large. One would expect that there will still be a close connection between the relevant coefficients and the a priori probabilities for transitions between the states considered, that is, the Einstein coefficients governing the intensities of the spectral lines. As far as the polarisation is concerned, this connection will be similar to that in the asymptotic case and the absence, or presence, of the relevant terms in the series for the x, y, and z components will tell us whether the radiation is, say, plane polarised. As to the intensities, the situation is much more complicated since one now has to take into account that E, and hence the ω_k, will depend on λ. This means that one has to take somehow an average of the coefficients over λ, and the choice of averaging process will be to a large extent arbitrary. However, as Kramers pointed out, there is one kind of problem where one may hope to get reasonable results. This is the case where one compares the intensities of lines belonging to a multiplet.

In the case of the non-relativistic hydrogen atom—or hydrogen-like atoms like the He ion— the stationary states are, in the absence of external fields, degenerate—that is, instead of three independent frequencies there is only a single one, and the stationary states depend only on a single quantum number, n. However, if there is an external electric field, there will be three frequencies and, correspondingly, three quantum numbers n_1, n_2, and n_3. As a result the single spectral line corresponding to a transition $n' \to n''$ will now be split into several lines, corresponding to different possibilities $n'_1, n'_2, n'_3 \to n''_1, n''_2, n''_3$

showing Stark splitting. It seems reasonable to assume that the averaging over λ for the different lines of these multiplets will not significantly differ from that for the single line and hence will be practically the same for all lines in a given multiplet. Kramers concludes: "We are therefore led to expect that *it will be possible to form an idea of the relative intensities with which the different components of the Stark effect will appear, by comparing the intensity of each component with the values of the squares of the amplitudes of the corresponding harmonic vibrations in the initial state and in the final state and in the mechanically possible states lying 'between' these states.*"[7]

Using expressions obtained earlier in his thesis Kramers calculated the relative intensities of the Stark effect components and summarised his results as follows: "On the whole it will be seen that it is possible on Bohr's theory to account in a convincing way for the intensities of the Stark effect components." Kramers also applied the same procedure to the hyperfine components of hydrogen and ionised helium, and again found good agreement with experimental data. He emphasised, however, the incomplete and preliminary nature of these calculations. A year later he considered (Kramers 1920) the general case of a hydrogen atom in an electric field — when it is not always possible to find a system of coordinates where the variables can be separated — thus completing the discussion of the Stark effect of a hydrogen atom, although many of the details that were promised to be given in a paper with O. Klein were never published.

The old quantum theory clearly was very successful when dealing with one-electron systems. The question then arose whether it could also deal with systems with two or more electrons. Bohr was able qualitatively to account for the shell structure of atoms, which leads to the arrangement of the various elements in the periodic system. How this is done is described by Kramers and Holst (1922) in their semi-popular book on the Bohr theory. However, these considerations were purely qualitative and a proof of Bohr's remarkable intuitive feel for the physics of atomic systems. As soon as attempts were made for a quantitative theory, difficulties arose. The simplest many-electron system is the helium atom. Many physicists tried to give a quantitative theory of the helium spectrum, but the great difficulty was to find the proper system of coordinates that would allow a separation of variables and hence an application of the quantisation rules. A most ingenious attempt was the one made by Kramers (1923c). He pointed out that all attempts using models in which the two electrons moved in the same plane had been unable to describe successfully the ground state of the helium atom, so that it was necessary to consider a model with the two electrons moving in planes at an angle to one another. The angle between the two planes is

---

[7] The italics are Kramers's.

chosen such that the total angular momentum of the atom is equal to $\hbar$.[8] The phase difference between the two electrons, both moving in an s-orbit, is chosen so as to minimise the average energy of the interaction between the two electrons. Once that is done, the model is completely determined, and by an ingenious splitting of the total Hamiltonian into one part which admits an exact solution so that the quantisation rules can be applied, and an additional part which is small so that it can be treated as a perturbation, Kramers succeeds in solving the quantum theoretical problem. However, it turns out that not only is the binding energy of the helium atom found in this way out by 3.8 V out of the exact value of 24.6 V, but the ground state configuration found in this way is unstable. Rather than blaming the quantum theory for this failure, Kramers concludes that classical mechanics can no longer be trusted when one is considering many-electron systems.

The next problem tackled by Kramers, using the old quantum theory, was that of molecular spectra. In the first of two papers, the second of which was written in collaboration with Pauli (Kramers 1923d; Kramers and Pauli 1923), he considers a molecular model consisting of a rigid body which contains in it a symmetric top that is free to rotate around its axis, which is fixed to the body. He considers the quantisation of the force-free motion of this model, which represents a molecule for which the electronic angular momentum is neither zero nor aligned with the molecular axis as in earlier models, which were all restricted to diatomic molecules. This model has four degrees of freedom: the three rotational degrees of freedom of the rigid body and the rotational degree of freedom of the top. It is a multiply periodic system so that the quantisation rules can be applied.

Since Kramers considers a rigid body motion, the vibrational degrees of freedom are neglected, and one would expect just to obtain the rotational bands of the molecular spectra. In the second paper, Kramers and Pauli discuss in some detail the band spectra of diatomic molecules, especially those of the halogenhydrides and oxygen.

Kramers's work on the Stark effect in hydrogen, the helium atom, and molecular spectra is of interest because it shows his mastery of classical mechanics, but the results had no lasting value since these problems, when solved by quantum mechanics, obtained different solutions. In contrast, his work (Kramers 1923e) on the absorption of X-rays and the continuous X-ray spectrum—the last of his papers applying the old quantum theory to spectral problems—did produce results some of which are still relevant today.[9] The paper was an an instant success when it appeared, not least because it was able to explain many experimental data. In that paper Kramers again combined his knowledge of classical mechanics and his mastery of mathemati-

---

[8] It is interesting to note that Kramers uses a new symbol, H, for Planck's constant divided by 2π.

[9] Unfortunately, lack of space prevented me from including this paper among the ones reprinted at the end of the book; it was the next one on my list.

cal techniques with Bohr's correspondence principle and Einstein's treatment of spontaneous and stimulated emission and absorption of radiation.

Apart from a term corresponding to the scattering of the X-rays, which is in general small and can be satisfactorily explained by Thomson scattering, and which will not concern us here, the absorption cross-section, σ_{abs}, of X-rays corresponds to the absorption of a light quantum combined with the ejection of an electron from the atom — thus producing an ion and a free electron. Experimentally it is found that it can be written in the following form:

$$\sigma_{\text{abs}} \cong CZ^4\lambda^3, \tag{2.11}$$

where λ is the wavelength of the X-rays and Z the atomic number of the material absorbing the X-rays. The experimental data used by Kramers showed that the coefficient C was equal to about 0.02 for wavelengths on the short-wavelength side of the K absorption limit and about 5 to 7 times smaller on the long-wavelength side of that limit.

The continuous X-ray spectrum is, apart from a small correction term, given by the expression

$$I_\nu = DZ(\nu_0 - \nu), \tag{2.12}$$

where $I_\nu \, d\nu$ is the intensity of the spectrum with frequencies lying between ν and $\nu + d\nu$, D is a constant, and ν_0 is the Duane-Hunt limit corresponding to the relation

$$h\nu_0 = eV, \tag{2.13}$$

where V is the applied voltage and e the elementary charge.

In the old quantum theory there was no theory of electromagnetic radiation, so that Kramers had to use a very ingenious method to find expressions for σ_{abs} and I_ν. He combined Einstein's ideas about the relations between induced emission and absorption with the correspondence principle. His approach was the following. We are interested in the cross-section for the following reaction:

$$\text{radiation} + \text{atom} \;\to\; \text{ion} + \text{electron}. \tag{2.14}$$

However, in the old quantum theory we know how to deal with orbits, which means that we should possibly be able to find the cross-section for the reaction

$$\text{ion} + \text{electron} \;\to\; \text{radiation} + \text{atom}. \tag{2.15}$$

The cross-section for reaction (2.14) is σ_{abs}, which means that the probability, $P_{\text{rad}+\text{at}\to\text{ion}+\text{el}}$, for that reaction to occur per unit time in a radiation field $\varrho(\nu)$ with the electron having a velocity between v and $v+dv$ will be equal to

$$P_{\text{rad}+\text{at}\to\text{ion}+\text{el}} = \sigma_{\text{abs}} \frac{\varrho(\nu)c}{h\nu} \frac{d\nu}{dv} dv, \tag{2.16}$$

where c is the velocity of light and where ν and v are related to one another through the equation

$$h\nu = W + \tfrac{1}{2}mv^2, \tag{2.17}$$

with W the energy needed to remove the electron from the atom while m is the electron mass.

On the other hand, the probability, $P_{\text{ion}+\text{el}\to\text{rad}+\text{at}}$, for the reaction (2.15) to occur per unit time if there is a single atom in a volume V will be given by the equation

$$P_{\text{ion}+\text{el}\to\text{rad}+\text{at}} = \frac{\beta v}{V}, \tag{2.18}$$

where β is the cross-section for this reaction. The point of Kramers's calculations now is that whereas it is impossible to evaluate σ_{abs} directly using the old quantum theory, it is possible to evaluate β and by using the same arguments that Einstein (1917) used in his paper which introduced the Einstein coefficients for stimulated and spontaneous emission, one can relate the two cross-sections, thus obtaining σ_{abs}, once β has been found. The argument is that in thermal equilibrium the number of reactions (2.14) and (2.15) must be the same, or

$$P_{\text{at}} P_{\text{rad}+\text{at}\to\text{ion}+\text{el}} = P_{\text{ion}} P_{\text{ion}+\text{el}\to\text{rad}+\text{at}}, \tag{2.19}$$

where P_{at} and P_{ion} are the relative probabilities of finding the atom or the ion with an electron with a velocity v between v and $v+dv$ present. We only need their ratio, which in thermal equilibrium at a temperature T is given by the equation

$$\frac{P_{\text{at}}}{P_{\text{ion}}} = \frac{gh^3 e^{h\nu/k_B T}}{4\pi V m^3 v^2 dv}, \tag{2.20}$$

where g is the a priori weight of the state of the atom. Of course, in Eq. (2.16) we must substitute for $\varrho(\nu)$ the thermal equilibrium expression for the radiation density:

$$\varrho(\nu) = \frac{8\pi h\nu^3}{c^3} e^{-h\nu/k_B T}, \tag{2.21}$$

where $k_{\rm B}$ is Boltzmann's constant and where we have assumed for simplicity that $h\nu \gg k_{\rm B}T$.[10]

From Eqs. (2.16) to (2.21) we finally find

$$\sigma_{\rm abs} = \frac{m^2 v^2 c^2}{2gh^2\nu^2}\beta, \qquad (2.22)$$

where ν and v are related through Eq. (2.17), which we have also used to find $d\nu/dv$.

To calculate β Kramers uses as his model the interaction between a bare nucleus of charge Ze and an electron. The classical motion of the electron is along a Kepler orbit, and Kramers expands this orbit in a Fourier series, much as he did in his thesis for the orbits of the hydrogen atom in an electric field. He then calculates the spectrum emitted classically. As far as the application of the quantum theory to this problem is concerned, "we will assume that a free electron which collides with a positive nucleus may perform a transition to a stationary state of motion of less energy, accompanied by the emission of a quantum $h\nu$ of monochromatic radiation. In the latter state the electron may either still be free ... or the collision will result in the binding of the electron in one of the discrete series of ... stationary states known from Bohr's theory. ... The statistical result of a large number of collisions with free electrons of the same velocity will therefore be the emission of a continuous spectrum extending from very small frequencies to a limiting frequency ν_0 ... and of a spectrum of discrete lines."[11]

The next step is the use of the correspondence principle, assuming that "we should expect that every possible quantum transition corresponds to a certain frequency present in the motion of the electron, and that the probability for the occurrence of a transition process is closely connected with the amount of energy in the radiation which on the classical theory is correlated with the corresponding frequency." It is straightforward to find the corresponding frequency in the case of the continuous spectrum, but in the case of the line spectrum, Kramers suggests that one should integrate the classical intensity over a frequency range around the frequency of the actual quantum transition.[12] Classically the energy emitted in a frequency range $d\nu$ will be proportional to Z^2 and inversely proportional to ν, and the frequency range

---

[10]If this last condition is not satisfied the argument becomes slightly more complicated, but the final result, Eq. (2.22), remains unaltered as one can see from Einstein's paper.

[11]The frequency ν_0 corresponds, of course, to the Duane-Hunt limit defined by Eq. (2.13).

[12]Kramers, in fact, uses for the range corresponding to the transition of the electron to the nth Bohr orbit the frequencies between those which would correspond to bound states with principal quantum numbers $n - \frac{1}{2}$ and $n + \frac{1}{2}$.

will be proportional to the relevant Rydberg constant, that is, proportional to Z^2. We thus find from Eq. (2.22) that, in agreement with the experimental data (2.11), the absorption cross-section will be proportional to $Z^4\lambda^3$. It turns out that the constant in the expression for σ_{abs}, obtained after averaging over the impact parameter, also is in reasonable agreement with the experimental value.

If we want to find the continuous spectrum we must consider free-free transitions. In this case the frequency in the spectrum will be the same as that in the expansion of the classical radiation. To find the spectral density I_ν we must take into account that the continuous spectrum is produced by a thick target, whereas so far we have considered only the scattering by a single atom. The first step is to average the expression for the intensity produced by a single atom, i_ν, over all impact parameters. The result is the following:

$$i_\nu = \frac{KZ^2}{v^2}, \qquad (2.23)$$

where K is a constant. The spectral density is now obtained by integrating expression (2.23) over the possible velocities of the electron in the target:

$$I_\nu = \int_{v_0}^{v_\nu} i_\nu \frac{dx}{dv} \, dv, \qquad (2.24)$$

where v_0 is the original velocity, related to the Duane-Hunt limit by $h\nu_0 = \tfrac{1}{2}mv_0^2$, and v_ν is the velocity for which we have $h\nu = \tfrac{1}{2}mv_\nu^2$. Kramers uses for the rate at which the electron loses energy Bohr's expression (Bohr 1913a, 1915)

$$\frac{dv}{dx} = -\frac{K'Z}{v^3}, \qquad (2.25)$$

where K' is a constant. Using that expression it follows from Eq. (2.24) that

$$I_\nu = K''Z\left(v_0^2 - v_\nu^2\right) = DZ(\nu_0 - \nu), \qquad (2.26)$$

in agreement with the experimental expression (2.12), including the numerical value of the constant D. Kramers also explains why some of the approximations made in his calculations are responsible for the non-appearance of the small correction mentioned in connection with Eq. (2.12).

2.2. The BKS Paper

In the preceding subsection we mentioned that there was no theory of electromagnetic radiation in the old quantum theory, while in the previous chapter we mentioned Bohr's opposition to the idea of photons.[13] Pais (1995) suggests that Bohr's opposition to the photon concept was due to the fact that it could not be made to fit in with the correspondence principle. This opposition to the photon, to which Kramers had been converted by Bohr when the latter talked him out of his prediction of the Compton effect, can be seen as the result of the fact that the idea of complementarity — which after all is not exclusively a quantum effect, cropping up whenever one talks about wave trains — had not yet entered the subconscious of the theoretical physicists so that the discontinuous nature of the quantum transitions seemed to be in contradiction to the continuous nature of electromagnetic radiation. This led several physicists,[14] among them Bohr (1923), to consider the possibility that energy (and momentum) conservation is not true for every individual separate atomic process, but is only true statistically, averaged over a large number of such processes. It should be emphasised that before 1925 there was no experimental evidence for energy and momentum conservation in individual atomic processes.

This was the position toward the end of 1923, when Slater (1924) suggested that atoms in the kth excited stationary state would be accompanied by a virtual electromagnetic field which formally could be seen as the field produced by harmonic oscillators with frequencies ν_{kl} ($E_k > E_l$) corresponding to the Bohr frequencies (2.01). Slater proposed that this virtual field should govern the emission and absorption processes of the atoms in such a way that the energy density of the field would be proportional to the probability of finding a "photon" and that the probability for the emission of a "photon" is determined by the Poynting vector of the field. We must mention that Slater's "photons" are not the usual light quanta; the virtual field guides them, and their path is not necessarily rectilinear and their speed not necessarily the speed of light. The introduction of the oscillators is similar to the one by Ladenburg and Kramers in their discussion of dispersion phenomena (see § 2.3).

When Slater arrived in Copenhagen, just before Christmas 1923, and when he told Bohr about his ideas, Bohr became extremely enthousiastic about the concept of the virtual field, but he did not want to have anything to do with photons. A hectic few weeks followed and the resulting paper was finished within one month of Slater's arrival in Copenhagen, and submitted to the *Philosophical Magazine*. It was deemed so important that it was translated into German and within another few weeks submitted to the

---

[13] Excellent accounts of the nature, the importance, and the consequences of the BKS paper have been given by Pais (1982, 1986, 1991) and Dresden (1987). Dresden, in particular, goes into much more detail than we do here.

[14] See Pais's account (Pais 1982, Ch. 22) of the history of the BKS paper.

Zeitschrift für Physik. This speed is even the more remarkable if one bears in mind that Bohr usually took months, if not years, to put the final touches to his papers.[15] This paper is the famous BKS paper which — as Pais (1991) points out — contains not so much a theory as a proposal for a theory for the interaction between matter and radiation.

The basic features of the paper are the following. Following Slater's idea it is assumed that each atom in a stationary state, say the kth state, is accompanied by a virtual classical field which is formally equivalent to the electromagnetic field produced by a set of classical harmonic oscillators with frequencies ν_{kl}, where now — in contrast to Slater's field — E_l can correspond to any of the stationary states of the atom. This field carries neither energy nor momentum. The total virtual field at the position of an atom will be the superposition of its own virtual field and the virtual fields of all other atoms in the system. The intensity of the klth component of the virtual field will be proportional to the probability that the atom will make the $k \rightarrow l$ transition, which may be accompanied by either the emission or the absorption of electromagnetic (real) radiation at the frequency $|\nu_{kl}|$. The transitions of the various atoms are independent, so that there is no conservation of energy and also there is a loss of causality. However, after the transition the virtual field of the atom concerned will have changed, and with it the probabilities for transitions of the other atoms. The next feature of the BKS proposal is that the virtual field of an atom will as the result of the virtual fields of other atoms become the source of secondary virtual radiation — following Huygens's principle in the form of spherical waves — and that if a real transition has taken place, the new virtual field will be such that, if the first transition had increased the total energy, transitions which will decrease the total energy will become more probable. This can be done by fixing the phase relations between the original and the secondary fields and it will lead to a statistical energy conservation. We note that the use of spherical waves, in accordance with Huygens's principle, will also lead to non-conservation of momentum.

As the Compton effect — the change in the wavelength of X-rays when they were scattered by electrons — which was discovered in 1923 (Compton 1923) had been explained both by Compton and by Debye (1923) as being the scattering of a photon by an electron, thus invoking both the photon concept and energy and momentum conservation it was necessary to show that the BKS proposal was compatible with this effect. In the BKS paper it was argued that, indeed, the experiments as they stood at the end of 1923 could be interpreted as showing the effect to be a statistical one.

The BKS paper marked a very important stage in the development of quantum mechanics. The world of theoretical physics was divided: some theorists, such as Born and Schrödinger, believed it to be correct, while others,

---

[15] His famous paper on the scattering and stopping of particles in matter was started in 1940, submitted to be printed in 1946, but not released to be published in its final form until 1948.

such as Einstein and Pauli, felt it to be wrong. It stimulated experimental efforts to prove that neither the lack of causality nor the non-conservation of energy and momentum was in accordance with experimental data. These efforts were successful. In the BKS explanation of the Compton effect, the recoil electron — produced by a discontinuous quantum process — and the secondary radiation — a continuous phenomenon — were uncorrelated, whereas in the photon version of the Compton effect they are produced simultaneously, as demanded by causality. Bothe and Geiger (1925) showed that the two events were, indeed, simultaneous — within the experimental accuracy. At about the same time Compton and Simon (1925) used cloud chamber experiments to study individual scattering processes, and they showed that in those the conservation laws were satisfied. These experiments meant the death of the BKS proposal.

2.3. The Dispersion Formulæ

Looking at Kramers's papers from his Copenhagen period from the point of view of how important they were for the development of quantum mechanics one must say that by far the most important papers are the ones he wrote in the period between the publication of the BKS proposal and his departure to Utrecht: his papers on the dispersion formulæ. These consist of four papers: two short notes in Nature (Kramers 1924a, b), the Kramers-Heisenberg paper (Kramers and Heisenberg 1925),[16] and the paper announcing the Kramers-Kronig relations (Kramers 1927b), the results of which date from 1925. These papers, the last two of which are reprinted as papers A and D at the end of the book, are the subject of the present subsection.

Although the first paper on the dispersion formulæ appeared after the BKS paper, from Dresden's account[17] of Kramers's dispersion theory it is clear that the main ideas had been developed by Kramers before Slater arrived in Copenhagen;[18] in fact, although Kramers in his papers refers to the

---

[16] From the account given by Dresden about the history of this paper (Dresden 1987, especially pp. 150, 224, 273, and 274) it is clear that practically the whole of that paper is due to Kramers and that, but for Bohr's intervention, Heisenberg's contribution to the paper would have been in the form of a footnote of the kind "I am grateful to Dr Heisenberg for pointing this out to me," and an acknowledgement at the end of the paper. Dresden (1987, p. 252) makes the interesting suggestion that if this paper had been published under Kramers's name alone and then been followed by Heisenberg's paper introducing matrix mechanics, the credit for the discovery of matrix mechanics (and the 1932 Nobel prize?) might well have been shared by Kramers and Heisenberg.

[17] See Dresden 1987, § 13II, where one can also find a more detailed description of Kramers's three papers on this topic than we are giving here.

[18] Dresden indicates that Kramers published his first short note in order that Slater could refer to a published paper when using the results; the second short note was published in order to reply to a specific point raised by Breit, and the paper containing all details of the theory became the Kramers-Heisenberg paper after Bohr's intervention.

BKS paper, this is paying lip service to this paper to which, as we saw, Kramers was very much attached, and the ideas stand completely independent of the BKS philosophy.

We have mentioned several times the fact that electromagnetic radiation did not fit naturally into the framework of the old quantum theory. The task Kramers set himself was to fit dispersion phenomena into this framework, that is, to use the correspondence principle to derive expressions describing the reaction of material media to electromagnetic radiation. In fact, he successfully derived such formulæ which, moreover, contained only observable quantities and no longer contained attributes specific for the multiply periodic systems that he used as his model for the atoms. In his first note he mentions that one can compare the reaction of the system to the incoming radiation to that of a set of virtual harmonic oscillators (compare Eq. (2.29)), but emphasises in his second note that these oscillators are introduced purely for illustrative purposes.

In his first paper he mentions that "it is well known that a consistent description of the phenomena of dispersion, reflection, and scattering of electromagnetic waves by material media can be given on the fundamental assumption that an atom, when exposed to radiation, becomes a source of secondary spherical wavelets, which are coherent with the incident waves." Classically, one proceeds as follows. Consider an atom consisting of a single electron of charge $-e$ and mass m bound isotropically to an equilibrium position which is irradiated by a wave with an electric field vector[19]

$$E(t) = E \cos 2\pi\nu t. \qquad (2.27)$$

The secondary wave will be due to a dipole moment P given by the expression

$$P(t) = \frac{e^2 E}{m} \frac{\cos(2\pi\nu t - \phi)}{4\pi^2 (\omega_1^2 - \nu^2)}, \qquad (2.28)$$

with ω_1 the eigenfrequency of the electron. Equation (2.28) holds provided ν is not too close to ω_1, and in that case the phase difference ϕ is very small. If, instead of having a single eigenfrequency, the atom has a number of absorption lines at the frequencies $\omega_1, \omega_2, \ldots$, the experimental data were found by Ladenburg to be well represented by a formula of the following shape (Sellmeier's formula):

$$P = E \frac{e^2}{4\pi^2 m} \sum_i \frac{f_i}{\omega_i^2 - \nu^2}, \qquad (2.29)$$

where P is the coefficient of $\cos(2\pi\nu t - \phi)$ in Eq. (2.28) and where the f_i are constants.

---

[19] For the sake of simplicity we consider here a one-dimensional case.

To derive a formula similar to Eq. (2.29) Kramers proceeded very much as he had done so many times before. He considered a multiply periodic system so that the dipole moment $\mathfrak{P}_0$ in the absence of radiation is described by an equation like Eq. (2.07):

$$\mathfrak{P}_0(t) = \sum_{\tau_1 \ldots \tau_s} \mathfrak{C}_{\tau_1 \ldots \tau_s} e^{2\pi i(\tau_1 \omega_1 + \ldots + \tau_s \omega_s)t}, \tag{2.30}$$

where the summation is over all positive and negative integral values of the τ_i and where the eigenfrequencies ω_i of the system are given by Eqs. (2.06) or (2.08).[20] If the system is now subject to radiation described by

$$\mathfrak{E}(t) = \text{Re}\{\mathfrak{E}\, e^{2\pi i\nu t}\}, \tag{2.31}$$

where we assume ν to be positive, one must add an extra term to the Hamiltonian of the system. In order to find the new dipole moment to first order in the radiation field one applies perturbation theory;[21] this leads to the following expression for the extra dipole moment $\mathfrak{P}$ which will give rise to the secondary radiation:

$$\mathfrak{P} = \text{Re}\left\{\sum_{\tau_1 \ldots \tau_s} \sum_{\tau'_1 \ldots \tau'_s} \left[\frac{\partial \mathfrak{C}}{\partial J'} e^{2\pi i\omega t} \frac{(\mathfrak{E}\cdot\mathfrak{C}')}{\omega' + \nu} e^{2\pi i(\omega'+\nu)t}\right.\right.$$
$$\left.\left. - \mathfrak{C} e^{2\pi i\omega t} \frac{\partial}{\partial J}\left(\frac{(\mathfrak{E}\cdot\mathfrak{C}')}{\omega' + \nu}\right) e^{2\pi i(\omega'+\nu)t}\right]\right\}, \tag{2.32}$$

where we have used the following abbreviations:

$$\omega \equiv \tau_1 \omega_1 + \ldots + \tau_s \omega_s, \qquad \omega' \equiv \tau_1 \omega'_1 + \ldots + \tau_s \omega'_s, \tag{2.33}$$

and

$$\frac{\partial}{\partial J} \equiv \tau_1 \frac{\partial}{\partial J_1} + \ldots + \tau_s \frac{\partial}{\partial J_s}, \qquad \frac{\partial}{\partial J'} \equiv \tau_1 \frac{\partial}{\partial J'_1} + \ldots + \tau_s \frac{\partial}{\partial J'_s}. \tag{2.34}$$

In the expression for $\mathfrak{P}$ there are many terms; some of them have the same frequency as the incoming radiation while the others have frequencies that differ from ν by amounts $\pm|\omega|\pm|\omega'|$, where ω and ω' are frequencies of the unperturbed motion. Note that in the case where the signs in front of ω and ω' are different, the difference corresponds to the frequency of a Bohr transition, that is, a frequency from the spectrum of the unperturbed system. The radiation due to the first

---

[20] One should note that our notation differs slightly from the one used by Kramers in paper A.
[21] For details of canonical perturbation theory see, for instance, ter Haar 1971, §7.2.

class of terms is coherent[22] and the remaining part is incoherent. The terms corresponding to the coherent radiation are those for which $\tau_i = -\tau'_i$ for all i. In what follows we shall concentrate on those terms. If we drop the incoherent part, we find that the coherent secondary radiation comes from the (induced) dipole moment

$$\mathfrak{P} = \mathrm{Re}\left\{\sum_{\tau_i}{}' \frac{\partial}{\partial J}\left[\frac{\mathfrak{C}(\mathfrak{E}\cdot\mathfrak{C}^*)}{\omega-\nu} + \frac{\mathfrak{C}^*(\mathfrak{E}\cdot\mathfrak{C})}{\omega-\nu}\right]e^{2\pi i\nu t}\right\}, \qquad (2.35)$$

where the prime on the summation sign indicates that the summation is over all combinations of the τ_i such that ω is positive and where $\mathfrak{C}^*$ is the complex conjugate of $\mathfrak{C}$.

So far we have only used classical arguments. Kramers now introduces quantum theory by invoking, as so often before, the correspondence principle and first looking at the case of large quantum numbers. The quantum numbers are introduced through the quantisation rules (2.02). Let us compare the expression for the classical frequency ω, obtained using Eqs. (2.33) and (2.06),

$$\omega = \left(\tau_1\frac{\partial}{\partial J_1} + \cdots + \tau_s\frac{\partial}{\partial J_s}\right)H = \frac{\partial H}{\partial J}, \qquad (2.36)$$

with that for the Bohr transition frequency (2.01), which in the limit of large quantum numbers we can approximately write in the form[23]

$$\nu_{\mathrm{qu}} = \frac{\Delta H}{h} = \left(\frac{\Delta J_1}{h}\frac{\partial}{\partial J_1} + \cdots + \frac{\Delta J_s}{h}\frac{\partial}{\partial J_s}\right)H. \qquad (2.37)$$

We see that the correspondence principle suggests that in the limit of large quantum numbers one must replace differentials by differences.[24] This is the first step to get from Eq. (2.35) to a formula like Eq. (2.29). The next step is to replace the classical amplitudes $\mathfrak{C}$ by the amplitudes of the harmonic components of the radiation, $\mathfrak{A}$, and relate those to the Einstein transition probabilities A for absorption or stimulated emission. Using the classical expression for the emitted radiation the latter relation is

---

[22] We see from Eq. (2.35) that the coefficients $\mathfrak{C}$ from Eq. (2.30) occur in pairs: one $\mathfrak{C}$ and one $\mathfrak{C}^*$ so that $\mathfrak{P}$ and $\mathfrak{E}$ are in phase.
[23] Compare Eq. (2.10).
[24] This replacement of differentials by differences is one of the most essential steps in Kramers's theory and he devoted a lecture to the Scandinavian Mathematical Congress (Kramers 1925c) to it.

$$h\nu_{qu} A = \frac{(2\pi\nu_{qu})^4}{3c^3}(\mathfrak{A}\cdot\mathfrak{A}^*). \tag{2.38}$$

The first step leads to the following expression for $\mathfrak{P}_{qu}$:

$$\mathfrak{P}_{qu} = \mathrm{Re}\left\{\left[\sum_a \frac{1}{4h}\left(\frac{\mathfrak{A}_a(\mathfrak{E}\cdot\mathfrak{A}_a^*)}{\nu_a - \nu} + \frac{\mathfrak{A}_a^*(\mathfrak{E}\cdot\mathfrak{A}_a)}{\nu_a + \nu}\right)\right.\right.$$
$$\left.\left.- \sum_e \frac{1}{4h}\left(\frac{\mathfrak{A}_e(\mathfrak{E}\cdot\mathfrak{A}_e^*)}{\nu_e - \nu} + \frac{\mathfrak{A}_e^*(\mathfrak{E}\cdot\mathfrak{A}_e)}{\nu_e + \nu}\right)\right]e^{2\pi i \nu t}\right\}, \tag{2.39}$$

where the first sum is over all frequencies ν_a corresponding to possible absorptions by the system and the second over all frequencies ν_e corresponding to possible spontaneous emissions.

If we now assume that the vectors $\mathfrak{E}$, $\mathfrak{A}_a$, and $\mathfrak{A}_e$ are all real and parallel to one another, introduce the lifetime τ_ν of an electron oscillating with a frequency ν,

$$\tau_\nu = \frac{3mc^3}{8\pi^2 e^2 \nu^2}, \tag{2.40}$$

and define the oscillator strength f of a transition by the relation

$$f = A\tau_\nu, \tag{2.41}$$

it follows from Eq. (2.39) that the induced dipole moment is equal to

$$\mathfrak{P}_{qu} = \mathfrak{E}\frac{e^2}{4\pi^2 m}\left[\sum_a \frac{f_a}{\nu_a^2 - \nu^2} - \sum_e \frac{f_e}{\nu_e^2 - \nu^2}\right]\cos 2\pi\nu t. \tag{2.42}$$

In the case where the system is in its ground state, so that the second sum is absent, Eq. (2.42) is the same as Eq. (2.29).

If the three vectors are not parallel the final expression cannot be simplified, but the fact remains that, as Kramers emphasised in his second note, the formula for the induced dipole moment "contains only such quantities as allow of a direct physical interpretation on the basis of the fundamental postulates of the quantum theory of spectra and atomic constitution, and exhibits no further reminiscence of the mathematical theory of multiple periodic systems."

We shall not discuss the incoherent part of the induced dipole, but merely mention that it explains the Smekal-Raman effect.

The final paper we want to discuss in the present chapter is the one introducing the so-called Kramers-Kronig relations, that is, the equations that allow one to find the imaginary (real) part of a susceptibility, permeability, or permittivity as function of the frequency once one knows the real (imaginary) part. This paper, reprinted at the end of this book as paper D, was published in 1927, but Kramers mentions that he derived the relations in 1925 when he presented them to the Danish Academy.[25] From paper D it is clear that he considers this work to be a natural application of his work on the dispersion theory.

In an isotropic medium the dielectric permittivity ε will be a complex quantity. If we write

$$\sqrt{\varepsilon} = n(1 - i\varkappa), \tag{2.43}$$

with n and $\varkappa$ real quantities, the propagation of an electromagnetic wave with frequency ω will be characterised by a factor

$$e^{i\omega(t - xn/c)} e^{-\varkappa x n/c}, \tag{2.44}$$

from which we see that the velocity of the waves is determined by the refractive index n while the absorption depends on $n\varkappa$.

To evaluate ε we proceed as before, assuming for the sake of simplicity that the components E_k of the electric field are of the form

$$E_k = \mathrm{Re}\{A_k e^{i\omega t}\}, \tag{2.45}$$

and that the components of the resulting induced atomic dipole moment are

$$P_k = \mathrm{Re}\{B_k e^{i\omega t}\}. \tag{2.46}$$

For not too strong fields the vector **B** will be a linear function of the vector **A**. In the case of an isotropic medium their relation will be

$$B_k = \zeta A_k, \tag{2.47}$$

and electromagnetic theory then tells us that for a dilute medium for which the dielectric permittivity ε is close to unity, ε will be given by the equation

$$\varepsilon = 1 + 4\pi N \zeta, \tag{2.48}$$

where N is the number of atoms per unit volume. We have already seen that we must expect ζ to be a complex quantity:

$$\zeta = \xi + i\eta. \tag{2.49}$$

---

[25] *Nature* **117**, 775 (1926) reports a talk on the subject by Kramers given at the Royal Danish Academy of Science and Letters on November 27, 1925. Kramers also presented his work at a meeting in Zürich in July 1929 and an excerpt of his talk was published (Kramers 1929d); it should be noted that the text of this excerpt was written not by Kramers himself, but by F. Bloch.

In the case where we can represent the atom by a (virtual) harmonic oscillator of frequency[26] ω_1 we have the following generalisation of Eq. (2.28):

$$\zeta = \frac{e^2}{m} \frac{1}{\omega_1^2 - \omega^2 + i\delta}, \tag{2.50}$$

where the term $i\delta$ in the denominator takes into account the radiation damping (compare Eq. (2.40)):

$$\delta = \frac{2e^2\omega^3}{3mc^3}. \tag{2.51}$$

In the case where $\delta/\omega \ll |\omega_1 - \omega|$ we retrieve Eq. (2.28). If there are several eigenfrequencies in the system we have a generalisation of Eq. (2.29).

We now relate the imaginary part of ζ to the atomic absorption coefficient $\alpha(\omega)$. We see from Eq. (2.44) that

$$\alpha(\omega) = -\frac{4\pi\omega}{c}\eta. \tag{2.52}$$

By arguments similar to the ones used when discussing his dispersion theory[27] Kramers finds that the oscillator strengths f_i are given by the following equation:

$$f_i = \frac{mc}{2\pi^2 e^2} \int \alpha(\omega)\,d\omega, \tag{2.53}$$

where the integration is over a small frequency range near ω_i. Using this relation one then finds from Eq. (2.29) that

$$\xi(\omega) = \frac{c}{2\pi^2} \mathcal{P} \int_0^\infty \frac{\alpha(\omega')\,d\omega'}{\omega'^2 - \omega^2}, \tag{2.54}$$

where $\mathcal{P}$ indicates that one must take the principal value of the integral.

If one now uses the relation between α and η and defines ξ and η, which so far have been defined only for positive values of ω, also for negative values of ω in such a way that ξ is an even and η an odd function of the frequency, Eq. (2.54) becomes

$$\xi(\omega) = \frac{\mathcal{P}}{\pi} \int_{-\infty}^{+\infty} \frac{\eta(\omega')\,d\omega'}{\omega - \omega'}. \tag{2.55}$$

---

[26] Note that the ω_1 introduced here differs from the one in Eq. (2.29) by a factor 2π.

[27] See paper D for details.

This tells us that we can find the real part of the polarisation coefficient, once we know its imaginary part. The inverse formula,

$$\eta(\omega) = -\frac{\mathcal{P}}{\pi} \int_{-\infty}^{+\infty} \frac{\xi(\omega')\,d\omega'}{\omega - \omega'}, \qquad (2.56)$$

determines the imaginary part, once the real part is known. Equations (2.55) and (2.56) together are the Kramers-Kronig relations, which were derived independently by Kronig (1926). These relations have had many applications, for instance, as criteria for the internal consistency of measured values of absorptive and dispersive components of susceptibilities.[28]

---

[28] The derivation sketched here is not the most straightforward one. The easiest way to derive them is indicated by Kramers in his contribution to the Zürich meeting; it uses causality and is sketched in somewhat more detail by Pais (1986, p. 499).

3 Quantum Mechanics

In the previous chapter we discussed Kramers's contributions to the old quantum theory. We saw how one of the guidelines for his work was the correspondence principle and the connection between classical mechanics and quantum theory, and also saw how much he was intrigued by the interaction between matter and radiation. These two strands are also found in his work on quantum mechanics. In the present chapter we shall consider both his papers dealing with some fundamental aspect of quantum mechanics and those in which he applies quantum mechanics to specific problems, postponing a discussion of his work on the interaction between matter and radiation to the next chapter.

A large part of Kramers's contribution to quantum mechanics can be found in his monumental monograph/textbook on the subject, the first part of which deals with the foundations of (non-relativistic) quantum theory and the second part of which discusses the quantum theory of the electron and of radiation.[1] Many important innovations were not published separately but can be found there, to some extent hidden. Typically, Kramers nowhere mentions, let alone emphasises, where his approach is novel; a typical example is his normalisation of the wavefunctions corresponding to the continuous spectrum (QM § 23).

One might ask why this book has not been more widely used. This is partly due to the fact that it was published in German—and just before the war—and by the time an English translation appeared in 1957 there were a large number of competing textbooks available. Another reason is the uncompromising style so completely different from that found in present-day textbooks. Kramers's book emphasises the difficulties, many of which still exist in present-day quantum mechanics, and he expects of his reader that he work out the details for himself—no spoon-feeding!

Apart from one short paper in Dutch, all the papers we shall discuss in this chapter can be found in his *Collected Scientific Papers* (Casimir et al. 1956). We shall consider Kramers's publications on quantum mechanics and its applications under five separate headings.

---

[1] In what follows we shall often refer to specific passages in this book and we shall indicate these passages by the page number in the English translation (Kramers 1957) to be cited as QM.

i. The largest part of this chapter will deal in §3.1 with his contributions to the general theory of non-relativistic quantum theory. To a great extent we shall follow Kramers himself in the way he discusses this theory in the first part of his book, since this allows us at the same time to highlight some of the particularly original approaches that can be found there but hardly anywhere else.
ii. Spin properties and the relativistic treatment of the electron will be discussed in §3.2.
iii. The next section, §3.3, is devoted to the so-called symbolic method.
iv. In §3.4 we shall deal with Kramers's application of quantum mechanics to spectral problems.
v. Finally, in §3.5 we discuss two problems in which symmetry plays a role, namely, charge conjugation and the Kramers degeneracy.

There is one paper, with Belinfante and Lubański (Kramers, Belinfante, and Lubański 1941), which I feel belongs to the present chapter but which does not fit into any of the five subsections. It discusses free particles with mass and with an arbitrary spin, that is, a spin that is not necessarily 0 or $\frac{1}{2}$. These particles are described by a wave equation that is a generalisation of the Dirac equation.

Of the papers discussed in the present chapter, we reprint at the end of the book as paper B a short paper — originally published in Dutch and not included in the Collected Papers volume — which appeared very soon after Heisenberg had introduced matrix mechanics, and which describes some aspects of Heisenberg's work; as paper C the paper about the semi-classical approximation, also called the WKB approximation; as paper E the paper discussing the Kramers degeneracy; as paper F a paper on the classical theory of the electron spin; as paper G a paper on the eigenvalue spectrum in a one-dimensional periodic potential;[2] and as paper H the paper introducing charge conjugation.

3.1. Kramers's Contributions to the General Theory of Non-Relativistic Quantum Mechanics

The last paper Kramers wrote in Copenhagen is a semi-popular paper in Dutch (Kramers 1925d, reprinted as paper B) in which he discusses the close connection between the matrix mechanics developed by Heisenberg, Born, and Jordan and the Kramers-Heisenberg theory.[3] One of the conclusions he reaches is that Born and Jordan's quantum conditions mean that an electron

---
[2] I have included this paper at the suggestion of Professor H.B.G. Casimir. Although it does not represent any important new results, it is an elegant derivation of the band structure of the energy spectrum in a periodic potential, and it was one of Kramers's own favourites.
[3] From this paper one can also get an interesting glimpse of Kramers's feelings about the fact that Heisenberg's matrix mechanics is clearly based upon Kramers's own dispersion formulæ.

that is bound in an atom behaves under the action of a high-frequency alternating electric field like a bound electron in the classical theory. He also points out that if one applies the methods used in his dispersion theory — especially the replacement of differentials by differences — to the classical Poisson brackets for the canonical variables, one is led to the quantum-mechanical commutation relations for these variables.[4]

The basic equation in wave mechanics is the Schrödinger equation. If we are considering the case of a single particle of mass m moving in a force field that can be derived from a potential energy $U(\mathbf{r})$ which we assume to be independent of the time, its time-independent form is

$$\frac{\hbar^2}{2m}\nabla^2\varphi(\mathbf{r}) + (E - U(\mathbf{r}))\varphi(\mathbf{r}) = 0, \tag{3.001}$$

where the function $\varphi(\mathbf{r})$ is the wavefunction. In Eq. (3.001) we are considering the case of a stationary state with energy E.[5]

Let us for the moment consider the one-dimensional Schrödinger equation:

$$\varphi'' + Cy(x)\varphi = 0, \tag{3.002}$$

where the prime indicates differentiation with respect to x and where

$$C = \frac{2m}{\hbar^2}, \qquad y(x) = E - U(x). \tag{3.003}$$

In what follows we shall first consider the case where $U(x)$ has a single minimum, say at $x = 0$, for which we assume the value $U(0) = 0$ while U tends to $+\infty$ both as $x \to -\infty$ and as $x \to \infty$. For the sake of simplicity we shall also assume that $U' \leqslant 0$ for negative x and $U' \geqslant 0$ for positive x. In that case there are for any positive value of E two points (the classical turning points) x_1 and x_2 ($> x_1$) where $E = U(x)$. It is well known that for such a potential the eigenvalue spectrum of the Schrödinger equation is discrete and that all eigenvalues are non-degenerate.

The extremely original way in which Kramers shows the discreteness of the eigenvalue spectrum and at the same time indicates the nature of the eigenfunctions (QM §17) is worth sketching — it also leads us naturally to a discussion of Kramers's treatment of the quasi-classical approximation. We first of all notice that Eq. (3.002) is a

---

[4] Kramers mentions that this point had independently been made by Pauli. Dirac is always credited with this observation; in fact, the paper (Dirac 1926) in which he states the relation between Poisson brackets and commutators was submitted for publication in November 1925, at the same time as Kramers's paper.

[5] A point made by Kramers (QM, p. 41) — and in hardly any other textbook — is that one can see from the fact that the Schrödinger equation can be formulated only if the forces acting on the particle can be derived from a potential energy how fundamental is the role played in quantum mechanics by the conservation of energy.

homogeneous linear equation so that if φ is a complex function, both its real and its imaginary part must satisfy the equation so that we can without loss of generality assume φ to be real. We next notice that if we give the values of φ and of φ' at one point, the function will be completely determined by Eq. (3.002). We finally notice that for $x < x_1$ and for $x > x_2$ where $E < U$, the wavefunction will behave as an exponential, whereas for $x_1 < x_2$, where $E > U$, it will behave oscillatorily like a sine function. Moreover, in order that a function φ can represent a physical situation it must be normalisable, that is, we must have

$$\int_{-\infty}^{+\infty} |\varphi|^2 \, dx < \infty. \tag{3.004}$$

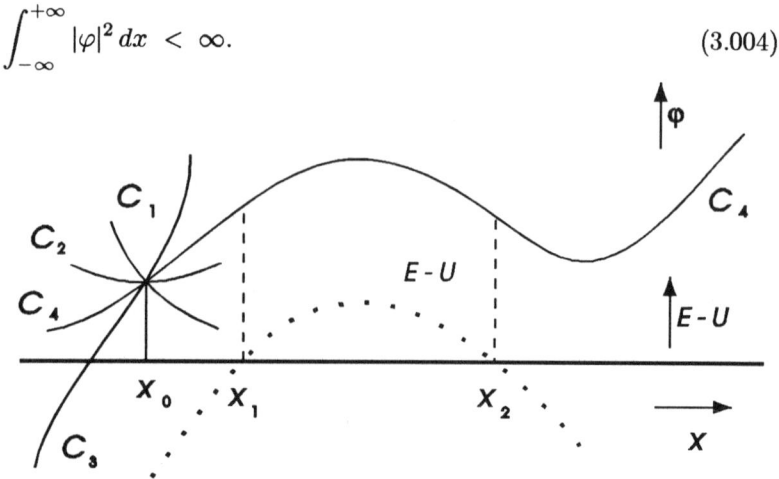

Fig. 3.1.

Consider now a particular positive value of E which is close to 0, and let x_0 be a point to the left of the two turning points. If we give the values, φ_0 and φ_0', of φ and φ' at x_0, the behaviour of the wavefunction will be completely determined by φ_0' (see Fig. 3.1): if φ_0' is negative or positive, but not very large (curves C_1 and C_2 in Fig. 3.1), the wavefunction will tend to $+\infty$ as $x \to -\infty$ and if φ_0' is positive and large (curve C_3 in Fig. 3.1), it will tend to $-\infty$. There is exactly one value of φ_0' for which the wavefunction will tend exponentially to zero as $x \to -\infty$ (curve C_4 in Fig. 3.1). Looking then at increasing values of x we see that at x_1 the wavefunction will cease to behave exponentially and become oscillatory until at x_2 it will again behave exponentially. If the interval $x_1 < x_2$ is not very large, as is assumed in Fig. 3.1, the value of φ' at x_2 will be insufficient to let the wavefunction tend to zero also as $x \to +\infty$. This means that for this particular value of E we can not find a normalisable wavefunction, that is, this value of E is not an eigenvalue of Eq. (3.002). If we now consider a larger value of E, the interval $x_1' < x < x_2'$ between the two turning points will be larger (see Fig. 3.2) and it may happen that the value of φ' at the right-hand turning point, x_2', will be too large, so that the new solution — which has been adjusted to vanish as $x \to -\infty$ — will tend to $-\infty$ as $x \to +\infty$ (curve C_4' in Fig. 3.2). This indicates that there will be an intermediate value of E such that the corresponding solution (curve C_4'' in Fig. 3.2) will tend to zero both as $x \to -\infty$ and as $x \to +\infty$. This solution will be the wavefunction corresponding to the lowest eigenvalue. Increasing the value E further, the interval between the turning points will become sufficiently large for the solution in that interval to cross the axis once; the corresponding solution, if it is such that it vanishes again both at $-\infty$ and at $+\infty$, is the wavefunction of the next lowest eigenvalue. Proceeding in that way, the next eigenfunction will have two zeroes between the turning points, and so on.

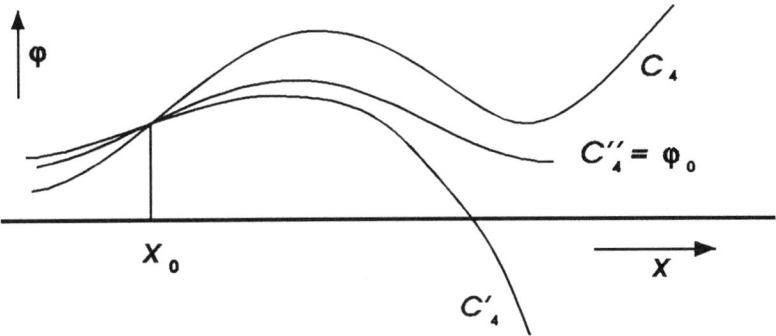

Fig. 3.2.

We shall now consider the semi-classical or quasi-classical approximation to the solution of Eq. (3.002). This approximation is also known as the WKB approximation after the physicists who independently introduced it into quantum mechanics: Wentzel (1926), Kramers (1926c; reprinted at the end of the book as paper C), and Brillouin (1926).[6] The way the wavefunction in the quasi-classical approximation is derived in most textbooks — and was derived by Wentzel and Brillouin — is by considering an approximation of the wavefunction — strictly speaking of the logarithm of the wavefunction — in the form

$$\varphi = e^{iS/\hbar}, \qquad (3.005)$$

---

[6] In papers by English authors it is occasionally referred to as the JWKB approximation, since Jeffreys (1924) considered a similar approximation, although not in connection with the Schrödinger equation.

with

$$S = S_0 + \hbar S_1 + \hbar^2 S_2 + \cdots. \tag{3.006}$$

From Eq. (3.006) it is clear that this approximation can be thought of as an expansion in powers of the quantum of action. In fact, the zeroth approximation leads to the Hamilton-Jacobi equation for S_0 from which we see the physical meaning of S_0: it is the classical action. Of course, the quantum of action is not a dimensionless quantity and, in fact, the expansion is in terms of

$$\frac{\hbar}{\sqrt{2mE}(x_2 - x_1)}. \tag{3.007}$$

The second approximation leads to the wavefunction that is usually associated with the WKB approximation, and we can now understand why this approximation is called the semi-classical approximation.

Kramers approaches the problem differently by considering in the spirit of the correspondence principle the case of large quantum numbers, that is, the case when the wavefunction has a large number of zeroes between the two classical turning points where it behaves oscillatorily. In that region a good approximation to the wavefunction will be a cosine function with practically constant wavelength and an amplitude that is proportional to the square root of the wavelength. One can see from Eq. (3.002) that in the turning points the instantaneous wavelength vanishes so that this solution will become infinite. Kramers therefore carefully considered the solution near those points, where the function $y(x)$ can be approximated by a linear function of x.[7] By requiring that the solution near the turning points turns into the solution already found between the turning points, and turns into an exponentially vanishing function at infinity, Kramers finds the (approximate) eigenfunctions of the Schrödinger equation and at the same time finds that the corresponding eigenvalues are given by the Bohr-Sommerfeld quantisation rule

$$2\int_{x_1}^{x_2} p(x)\,dx = (n + \tfrac{1}{2})h, \qquad p(x) = \sqrt{2my(x)}. \tag{3.008}$$

Let us assume for the wavefunction φ the form

$$\varphi(x) = g(x)\cos f(x), \tag{3.009}$$

where $g(x)$ and $f(x)$ are smooth functions of x. Assuming that the function $y(x)$ changes little over a wavelength $2\pi/\sqrt{Cy}$, we find that

$$f\left(x + \frac{2\pi}{\sqrt{Cy}}\right) - f(x) = 2\pi, \tag{3.010}$$

---

[7] This was also done by Jeffreys (1924), who expresses the solution in that region in terms of Airy functions.

or
$$\frac{2\pi}{\sqrt{Cy}} f'(x) = 2\pi, \tag{3.011}$$

whence we get

$$f(x) = \int^x \sqrt{Cy}\, dx. \tag{3.012}$$

To find the amplitude $g(x)$ Kramers considers the function $y(x)$ in the neighbourhood of a point x_0 and writes Eq. (3.002) in the approximate form

$$\varphi'' + C\{y(x_0) + y'(x_0)(x-x_0)\}\varphi = 0. \tag{3.013}$$

In the spirit of the approach used here, $y'(x_0)$ can be treated as a small quantity and to first order in $y'(x_0)$ one finds the solution

$$\varphi = \cos\left[\sqrt{Cy}(x-x_0)\right] - \frac{y'}{4y}\Big\{(x-x_0)\cos\left[\sqrt{Cy}(x-x_0)\right]$$
$$+ \sqrt{Cy}(x-x_0)^2 \sin\left[\sqrt{Cy}(x-x_0)\right]\Big\}, \tag{3.014}$$

where y and y' are the values of these functions at x_0. This means that to first order over an interval of the order of a wavelength the amplitude of the oscillating function is of the form $1 - y'(x-x_0)/4y$ so that $g(x)$ must satisfy the equation

$$\frac{g'}{g} = -\frac{y'}{4y}, \tag{3.015}$$

whence we find for φ:

$$\varphi = \frac{1}{\sqrt[4]{y}} \cos \int^x \sqrt{Cy}\, dx, \tag{3.016}$$

which is the WKB expression for the wavefunction in the classically accessible region.

The next question Kramers addressed was the behaviour of the wavefunctions in and near the classical turning points. In the turning points y vanishes and the solution (3.016) becomes infinite. In the immediate vicinity of a turning point we can approximate the function y by a linear function of x. Consider first the left-hand turning point x_1; in its neighbourhood Eq. (3.002) becomes

$$\varphi'' + C\alpha\xi\varphi = 0, \tag{3.017}$$

where α is the value of y' at the turning point and $\xi = x - x_1$. The solution of Eq. (3.017) can be written in the form of Airy functions, but Kramers proceeds differently in order to find the solution that goes over into expression (3.016) in the classically accessible region and into the exponentially vanishing solution at infinity. To do this he writes the solution of Eq. (3.017) in the form of a complex integral:

$$\varphi = K \int_{\mathcal{C}} e^{\sqrt{C\alpha}\xi t + \frac{1}{3}t^3} \, dt, \tag{3.018}$$

where K is a constant and where the contour $\mathcal{C}$ in the complex t plane is along a curve such as W_1 in Fig. 3.3, along which t asymptotically has the arguments $-\pi/3$ and $+\pi/3$. Of interest is the asymptotic behaviour of φ as ξ is large and negative or large and positive. The corresponding expressions are found by choosing a suitable contour and using the method of steepest descent. If ξ is large and negative one chooses the hyperbole W_1, the saddle point is the point A for which $t = \sqrt{(C\alpha)^{1/3}|\xi|}$, and the result is

$$\varphi \propto |\xi|^{-1/4} e^{-\frac{2}{3}\sqrt{C\alpha}|\xi|^{3/2}}, \tag{3.019}$$

with the correct asymptotic behaviour as $\xi \to -\infty$.

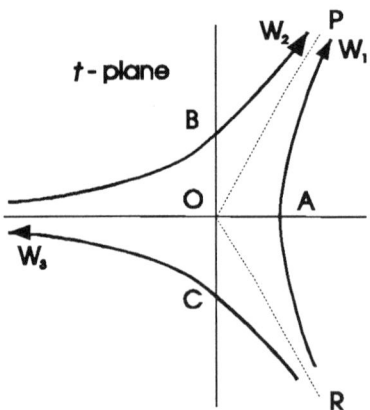

Fig. 3.3.

If ξ is large and positive one chooses the hyperboles W_2 and W_3, there are two saddle points in the point B and C for which $t = \pm i\sqrt{(C\alpha)^{1/3}\xi}$, and the result is

$$\varphi \propto \xi^{-1/4} \cos\left[\frac{2}{3}\sqrt{C\alpha}\xi^{3/2} - \frac{\pi}{4}\right], \tag{3.020}$$

which is the same as expression (3.016)—taking into account that $y = \alpha\xi$—which we now can write in a form where the lower limit of the integral is determined:

$$\varphi = \frac{1}{\sqrt[4]{y}} \cos\left[\int_{x_1}^{x} \sqrt{Cy}\, dx - \frac{\pi}{4}\right]; \qquad (3.021)$$

to fix the ideas we have assumed that $\varphi(x_1)$ is positive.

Repeating the procedure for the other turning point, x_2, we find that we can write

$$\varphi = \frac{1}{\sqrt[4]{y}} \cos\left[\int_{x_2}^{x} \sqrt{Cy}\, dx - \frac{\pi}{4} \pm \frac{\pi}{2}\right], \qquad (3.022)$$

where the choice between the plus or minus sign depends on whether $\varphi(x_2)$ is positive or negative.

Requiring that Eqs. (3.021) and (3.022) give us the same function φ then leads directly to Eq. (3.008).[8]

Another general problem connected with the Schrödinger equation (3.002) considered by Kramers is the one where the potential $U(x)$ is periodic. The paper devoted to this problem (Kramers 1935f), which was one of Kramers's own favourites, is reprinted as paper G. Kramers gives a simple, straightforward proof for the fact that for any periodic potential the energy spectrum consists of an infinite series of bands: a range of allowed energies is separated by a range of forbidden energies from the next range of allowed energies.[9]

Finally we mention a short note (Kramers 1948) in the Courant Anniversary Volume in which Kramers compares different ways to write the perturbation theory series.

3.2. Electron Spin and the Relativistic Wave Equation

We mentioned in Chapter 1 that Kramers had, independently of Dirac, derived the relativistic wave equation for the electron. In this subsection we shall sketch this derivation, which is of interest as it illustrates once again how deeply rooted a lot of Kramers's work was in classical physics. Kramers's derivation of the relativistic wave equation was not published at the time he first produced it but can be found in his paper in the Zeeman Jubilee volume (Kramers 1935b) — reprinted as paper F at the end of the book — which is the follow-up of a short paper (Kramers 1934d) entitled "On the Classical Theory of the Spinning Electron," as well as in the second part of his

---

[8] It is interesting to note that the matching conditions for the wavefunction at both sides of a turning point depend on whether one is dealing with a truly one-dimensional problem or with the problem of the radial part, say, of a three-dimensional problem; this has been discussed recently by Popov, Karnakov, and Mur (1996).

[9] The eigenfunctions cannot be normalised but must satisfy the requirement that they remain finite for all values of x, as is the usual requirement for the eigenfunctions of a continuous spectrum.

textbook.[10] As we shall see again in the next chapter, Kramers strongly felt that one should not embark upon a quantum theory of a phenomenon before the corresponding classical theory was satisfactory — in stark contrast to the modern idea that classical physics is merely the zeroth approximation of quantum theory.

In the first paper Kramers stresses that the concept of a rigid body is irreconcilable with the theory of relativity, so that if one wants to construct a classical theory of the electron spin one must formulate it independently of any detailed model of the electron. He then shows how this is possible by requiring that the equations of motion of the electron angular momentum — the electron spin — are Lorentz invariant. If this is done these equations lead both to the gyromagnetic ratio — strictly speaking the magnetogyric ratio — e/mc and to the so-called Thomas factor (Thomas 1926) in the equation of motion of the electron spin.

Non-relativistically the equation of motion of the angular momentum **A** of an electron at rest in a magnetic field **H** is

$$\frac{d}{dt}\mathbf{A} = \alpha[\mathbf{A} \wedge \mathbf{H}], \qquad (3.023)$$

where α is the ratio of the magnetic moment to the angular momentum — the magnetogyric ratio, usually incorrectly referred to as the gyromagnetic ratio. This equation is not relativistically invariant. In order to generalise it to a Lorentz invariant equation we assume that the components A_1, A_2, and A_3 of **A** are, respectively, the 23, the 31, and the 12 components of a skew symmetric second-rank tensor — a six-vector — the 14, 24, and 34 components of which are zero when the electron is at rest. These last three components define a vector that we denote by **B**. We now assume that the components of the six-vector

$$\mathbb{S} \equiv \mathbf{A} + i\mathbf{B} \qquad (3.024)$$

transform under a Lorentz transformation in the same way as the components of the six-vector $\mathbb{F}$ made up of the magnetic and the electric fields,

$$\mathbb{F} \equiv \mathbf{H} + i\mathbf{E}. \qquad (3.025)$$

The relativistic generalisation of Eq. (3.023) is then

$$\frac{d}{d\tau}\mathbb{S} = \alpha[\mathbb{S} \wedge \mathbb{F}], \qquad (3.026)$$

where $d\tau$ is an element of eigentime.

---

[10]Although the general ideas are the same in Kramers's treatment in his textbook as in the two earlier papers, there are some differences. Here we shall follow the slightly less sophisticated and more simple-minded approach of the two earlier papers.

Just as $\gamma\{\mathbf{E} + [\mathbf{v} \wedge \mathbf{H}]/c\}$, $i\gamma(\mathbf{E} \cdot \mathbf{v})$ is a Minkowski four-vector, where γ is the Lorentz factor,

$$\gamma = \frac{1}{\sqrt{(1 - v^2/c^2)}}, \qquad (3.027)$$

the same will be true of $\gamma\{\mathbf{B}+[\mathbf{v}\wedge\mathbf{A}]/c\}$, $i\gamma(\mathbf{B}\cdot\mathbf{v})$. Since, by definition, $\mathbf{B}$ is equal to 0 for $\mathbf{v} = 0$, the last Minkowski four-vector must vanish identically so that $\mathbf{B}$ is determined by $\mathbf{A}$ as follows:

$$\mathbf{B} = \frac{[\mathbf{A} \wedge \mathbf{v}]}{c}. \qquad (3.028)$$

Using Eq. (3.028) and the real part of Eq. (3.026), we find that in the rest frame of the electron we have

$$\frac{d\mathbf{B}}{d\tau} = \frac{1}{c}\left[\mathbf{A} \wedge \frac{d\mathbf{v}}{d\tau}\right]. \qquad (3.029)$$

On the other hand, from the imaginary part of Eq. (3.026) it follows that in the same frame we have

$$\frac{d\mathbf{B}}{d\tau} = \alpha[\mathbf{A} \wedge \mathbf{E}], \qquad (3.030)$$

and combining this with the equation of motion[11]

$$m\frac{d\mathbf{v}}{d\tau} = e\mathbf{E}, \qquad (3.031)$$

we find for the magnetogyric moment of the electron:

$$\alpha = \frac{e}{mc}. \qquad (3.032)$$

Substituting (3.028) and (3.032) into (3.025) we get

$$\frac{d\mathbf{A}}{d\tau} = \frac{e}{mc}\left\{[\mathbf{A} \wedge \mathbf{H}] - \frac{1}{c}[[\mathbf{A} \wedge \mathbf{v}] \wedge \mathbf{E}]\right\}$$

$$= \frac{e}{mc}\left\{[\mathbf{A} \wedge \mathbf{H}] + \frac{1}{2c}[\mathbf{A} \wedge [\mathbf{E} \wedge \mathbf{v}]] + \mathbf{T}\right\}, \qquad (3.033)$$

where, in the non-relativistic approximation where we can use Eq. (3.031), we have

$$\mathbf{T} = \frac{e}{mc^2}\left\{\mathbf{A}(\mathbf{v} \cdot \mathbf{E}) - \tfrac{1}{2}\mathbf{E}(\mathbf{v} \cdot \mathbf{A}) - \tfrac{1}{2}\mathbf{v}(\mathbf{E} \cdot \cdot \mathbf{A})\right\}$$

$$= \frac{1}{2c^2}\frac{d}{d\tau}[\mathbf{v} \wedge [\mathbf{A} \wedge \mathbf{v}]], \qquad (3.034)$$

---

[11] This equation holds in the approximation where, in agreement with Bohr's statement that it is impossible to detect the electron spin in external fields, the effect of the magnetic moment on the electron orbit is neglected.

which proves that the secular effects of **T** on the direction of the spin vanishes provided the electron velocity remains bounded.

If we restrict ourselves to secular changes in the spin direction, **T** can be neglected in Eq. (3.033) and that equation becomes

$$\frac{d\mathbf{A}}{d\tau} = \frac{e}{mc}\left\{[\mathbf{A}\wedge\mathbf{H}] + \frac{1}{2c}[\mathbf{A}\wedge[\mathbf{E}\wedge\mathbf{v}]]\right\}, \quad (3.035)$$

where the factor $\frac{1}{2}$ is the well-known Thomas factor.

In order to construct a quantum theory of the electron spin, Kramers constructs a Hamiltonian from which the equation of motion (3.026) follows as the canonical Hamiltonian equations. In this Hamiltonian he introduces explicitly the Pauli spin matrices,

$$\sigma_1 = \begin{Vmatrix} 0 & 1 \\ 1 & 0 \end{Vmatrix}, \quad \sigma_2 = \begin{Vmatrix} 0 & -i \\ i & 0 \end{Vmatrix}, \quad \sigma_3 = \begin{Vmatrix} 1 & 0 \\ 0 & -1 \end{Vmatrix}, \quad (3.036)$$

thus assuming a priori that the electron spin corresponds to a quantum number $\frac{1}{2}$. It then follows that the relativistic wave equation of an electron in an electromagnetic field characterised by a vector potential **A** and a scalar potential Φ reads as follows:[12]

$$[\varepsilon - \varrho_3(\boldsymbol{\pi}\cdot\boldsymbol{\sigma}) - \varrho_1 mc]\Psi = 0, \quad (3.037)$$

where ε and $\boldsymbol{\pi}$ are given by the equations

$$\varepsilon = \frac{E - e\Phi}{c}, \quad \boldsymbol{\pi} = \mathbf{p} - \frac{e\mathbf{A}}{c}, \quad (3.038)$$

with E the energy and **p** the momentum of the electron. In Eq. (3.037) Ψ is a four-component wavefunction, ϕ_{rs} ($r, s = 1, 2$),

$$\Psi = \begin{pmatrix} \phi_{11} \\ \phi_{12} \\ \phi_{21} \\ \phi_{22} \end{pmatrix}, \quad (3.039)$$

$\boldsymbol{\sigma}$ is a vector with as components the Pauli matrices σ_1, σ_2, and σ_3, which act on the second index of the ϕ_{rs}, and ϱ_1 and ϱ_3 are 2×2 matrices which have the same form as σ_1 and σ_3 and which act upon the first index of the ϕ_{rs}. Equation (3.037) has a form very similar to the Dirac equations, which are usually written in the form

$$[\varepsilon - \varrho_1(\boldsymbol{\pi}\cdot\boldsymbol{\sigma}) - \varrho_3 mc]\Psi = 0. \quad (3.040)$$

In fact, one can obtain Eq. (3.040) from Eq. (3.037) through a straightforward transformation.

---

[12] The vector potential **A** should not be confused with the angular momentum vector of Eq. (3.023).

We shall follow Kramers's discussion in paper F.[13] To find the Hamiltonian $\mathcal{H}_{\text{spin}}$ from which Eq. (3.026) follows as the canonical equation of motion we write, by analogy with the Hamiltonian of a dipole in a field,

$$\mathcal{H}_{\text{spin}} = -\alpha (\mathbb{S} \cdot \mathbb{F}). \tag{3.041}$$

We choose as canonical variables p and q defined as follows:

$$p = S_1, \qquad q = \arctan \frac{S_2}{S_3}. \tag{3.042}$$

The Poisson brackets of the components of $\mathbb{S}$ are

$$\left.\begin{aligned}\{S_1, S_2\} &\equiv \left[\frac{\partial S_1}{\partial p}\frac{\partial S_2}{\partial q} - \frac{\partial S_1}{\partial q}\frac{\partial S_2}{\partial p}\right] = -S_3, \\ \{S_2, S_3\} &= -S_1, \qquad \{S_3, S_1\} = -S_2,\end{aligned}\right\} \tag{3.043}$$

and the equations of motion

$$\frac{dS_k}{d\tau} = -\sum_l \{S_k, S_l\} \frac{\partial \mathcal{H}_{\text{spin}}}{\partial S_l}, \tag{3.044}$$

which after substitution of Eqs. (3.041) and (3.043) lead to Eq. (3.026).

We now have to add the spin Hamiltonian (3.038) to the Hamiltonian governing the motion of a spinless electron. The Hamiltonian equation in the latter case is

$$\mathcal{H}_0 \equiv \frac{1}{2m}\left[-\varepsilon^2 + (\boldsymbol{\pi} \cdot \boldsymbol{\pi})\right] = -\tfrac{1}{2}mc^2, \tag{3.045}$$

where ε and $\boldsymbol{\pi}$ are given by Eq. (3.038).

The total Hamiltonian equation will now be

$$\overline{\mathcal{H}} \equiv \frac{1}{2m}\left[-\varepsilon^2 + (\boldsymbol{\pi} \cdot \boldsymbol{\pi}) + \frac{e}{mc}(\mathbb{S} \cdot \mathbb{F})\right] = -\tfrac{1}{2}mc^2, \tag{3.046}$$

or, multiplying by $-2m$,

$$-2m\overline{\mathcal{H}} \equiv \mathcal{H} \equiv \varepsilon^2 - (\boldsymbol{\pi} \cdot \boldsymbol{\pi}) - \frac{2e}{c}(\mathbb{S} \cdot \mathbb{F}) = m^2c^2. \tag{3.047}$$

To quantise the electron motion governed by the Hamiltonian equation (3.047) we consider $\mathcal{H}$ as an operator by putting[14]

---

[13] We note that in this paper Kramers uses e for the charge of the electron whereas in his earlier paper e was the elementary charge.

[14] If we wish to obtain the time-dependent wave equation we also must in ε put $F - i\hbar\partial/\partial t$.

$$\mathbf{p} = -i\hbar\nabla, \qquad \mathbf{S} = \tfrac{1}{2}\hbar\boldsymbol{\sigma}, \tag{3.048}$$

where $\boldsymbol{\sigma}$ is again the vector that has the Pauli matrices as components. The wave equation will thus be

$$\mathcal{H}\psi = m^2c^2\psi, \tag{3.049}$$

where ψ is a wavefunction that depends both on $\mathbf{r}$ and t and on a spin variable, so that for a spin-$\tfrac{1}{2}$ particle it must be a two-component wavefunction — a so-called spinor,

$$\psi = \begin{pmatrix} \psi_+ \\ \psi_- \end{pmatrix}. \tag{3.050}$$

We now note that $\mathcal{H}$ can be factorised as follows:

$$\mathcal{H} = \left\{\varepsilon - \frac{2}{\hbar}(\boldsymbol{\pi}\cdot\mathbf{S})\right\}\left\{\varepsilon + \frac{2}{\hbar}(\boldsymbol{\pi}\cdot\mathbf{S})\right\}. \tag{3.051}$$

If we introduce another two-component wavefunction χ through the equation

$$\left\{\varepsilon + \frac{2}{\hbar}(\boldsymbol{\pi}\cdot\mathbf{S})\right\}\psi = mc\chi, \tag{3.052}$$

it follows from Eqs. (3.049), (3.051), and (3.052) that

$$\left\{\varepsilon - \frac{2}{\hbar}(\boldsymbol{\pi}\cdot\mathbf{S})\right\}\chi = mc\psi, \tag{3.053}$$

and we have recovered Eq. (3.037) with

$$\phi_{11} = \chi_+, \quad \phi_{12} = \chi_-, \quad \phi_{21} = \psi_+, \quad \phi_{22} = \psi_-. \tag{3.054}$$

The advantage of Kramers's derivation over the one by Dirac is that the Lorentz invariance of the equations follows directly, since the whole argument has been Lorentz invariant whereas Dirac has to prove that his equations satisfied this property.

In concluding this subsection we want to draw attention to Kramers's extremely elegant method of normalising the wavefunctions of the discrete energy spectrum of the hydrogen atom when the Dirac equation is used to describe it. He shows (QM, p. 306) that the normalisation integral can be related to the asymptotic behaviour of the eigenfunctions and their derivatives with respect to the energy parameter. We refer to QM for details.

3.3. The Symbolic Method

Kramers himself was very fond of the so-called symbolic method. In his lectures at the California Institute of Technology in 1947 he described this as "methods which do not seem to be very much used in the current literature or treated in the usual textbooks; methods which make many problems much easier. ... The technique ... reduces the labor of the calculation so that the formalism is usually no more complicated than the final result; however, with this technique one must think harder although he calculates less" (Kramers 1947b). As Casimir (1952) puts it in his address at the Kramers Memorial Meeting of the Dutch Physical Society: "It must be admitted that this formalism has only been used by a limited number of physicists, ... but it is certain that in the hands of Kramers and some of his pupils the method became a powerful tool."

The method is based upon Weyl's work on group theoretical methods in quantum mechanics (Weyl 1931) and uses the transformation properties of invariant expressions involving the components of a spinor, that is, a two-component function ξ, η.[15] It was first used by Kramers (1930c, 1931a) to derive formulæ for the intensities of atomic multiplets, an application extended by Brinkman in his doctoral thesis (Brinkman 1932; see also Brinkman 1956). In his textbook Kramers applies the method to the Dirac equation, and he also used it (Kramers 1943b) in his discussion of multipole radiation. We shall briefly sketch Kramers's treatment of the problem of relative intensities in the normal Zeeman effect in order to give some idea of the philosophy behind the method. Of necessity our discussion will be rather concise; we refer to Brinkman's thesis and monograph for more details.

Let ξ and η be the components of a spinor; they will be an irreducible representation of degree 2 of the rotation group under unitary transformations. Similarly, the quantities $\xi^{l+m}\eta^{l-m}$, $m = -l, -l+1, \cdots, +l$, will form an irreducible representation of degree $2l+1$, $l = \frac{1}{2}, 1, \frac{3}{2}, \cdots$, of the same group.

Consider now a free atom in a state with angular momentum l. Its state will be $2l+1$-fold degenerate and we can choose the corresponding $2l+1$ independent wavefunctions L_m in such a way that they transform as $\xi^{l+m}\eta^{l-m}$. Symbolically writing $L_m = \xi^{l+m}\eta^{l-m}$, we can form a generating function

$$Q^{2l} = (-b\xi + a\eta)^{2l}, \qquad (3.055)$$

where a, b is an arbitrary, constant spinor.

If $\int d\tau$ indicates integration over all spatial coordinates and summation over all spin coordinates, we find

---

[15] The name spinor is used because the vectors ξ, η used here transform under a rotation like the two independent components $S_+ \equiv S_x + iS_y$ and $S_- \equiv S_x - iS_y$ of the spin vector $\boldsymbol{\sigma}$.

$$\int Q^{*2l} Q^{2l} \, d\tau = \int (-b^*\xi^* + a^*\eta^*)^{2l}(-b\xi + a\eta)^{2l} \, d\tau$$
$$= C(a^*a + b^*b)^{2l}, \tag{3.056}$$

where we have used the invariance of the integral under rotation. By suitably choosing the normalisation of the L_m we can make $C = 1$ and it then follows from a comparison of the second and third members of Eq. (3.056) that

$$\varepsilon_{m'm} \equiv \int L_{m'}^* L_m \, d\tau = \binom{2l}{l+m}^{-1} \delta_{m'm}, \tag{3.057}$$

where $\binom{a}{b}$ is the binomial coefficient and δ_{kl} the Kronecker symbol.

In order to calculate the intensities of the lines of the normal Zeeman multiplet we need integrals of the kind $J_{km'm} = \int L_{m'}^* \Omega_k L_m \, d\tau$, where Ω_k ($k = 1, 2, 3$) is, say, a Cartesian coordinate, that is, a component of a vector Ω which transforms as the vector $\mathbf{r}$ with components $x + iy$, $-x + iy$, or z.[16] The important fact now is that these three components of $\mathbf{r}$ transform as X^2, Y^2, XY, where X, Y is a spinor. Symbolically we can combine the three components of Ω into the symbol $\Omega = (-BX + AY)^2$, where A, B is another constant spinor like a, b.

The matrix elements of the Ω_k follow from the following three integrals:

$$\Omega_{l,l} = \int Q^{*2l} \Omega Q^{2l} \, d\tau$$
$$= C_0 (a^*a + b^*b)^{2l-1}(a^*A + b^*B)(-bA + aB), \tag{3.058}$$

$$\Omega_{l+1,l} = \int Q^{*2(l+1)} \Omega Q^{2l} \, d\tau$$
$$= C_{+1}(a^*a + b^*b)^{2l}(a^*A + b^*B)^2, \tag{3.059}$$

$$\Omega_{l-1,l} = \int Q^{*2(l-1)} \Omega Q^{2l} \, d\tau$$
$$= C_{-1}(a^*a + b^*b)^{2l-2}(-bA + aB)^2. \tag{3.060}$$

The values of these three integrals follow from the fact that they are invariant under rotation and that the only invariants that can be constructed from the two constant spinors a, b and A, B are $a^*a + b^*b$, $a^*A + b^*B$, and $-bA + aB$. Moreover, it is clear that $\Omega_{l',l} = 0$, if $|l' - l| > 1$.

By comparing the coefficients of the various powers of a^*, a, b^*, b, A, and B in the different members of Eqs. (3.058) to (3.060) and using the

---

[16]It is important to choose here $x + iy$, $-x + iy$, z rather than x, y, z. This, of course, is well known from the classical theory of the normal Zeeman effect.

normalisation constants from Eq. (3.057). one obtains the intensities of the Zeeman components.

In the above discussion we mentioned that it is essential to use $x \pm iy$, z rather than x, y, z as the components of the spatial vector, in order to connect its transformation properties with those of a spinor. This had an important consequence for Kramers's discussion of multipole radiation (Kramers 1943b). Rather than working with electrical and magnetic multipoles, the symbolic method works with right- handed and left-handed circularly polarised multipoles, which thus turn out to be more suitable entities for describing multipole radiation.

3.4. Applications of Quantum Mechanics to Specific Problems

We now turn to applications of quantum mechanics to some specific problems. We first mention applications of the WKB method to a few particular problems. We refer to Zwaan's thesis (Zwaan 1929) where the theory is applied to the Ca arc spectrum and where one can also find a general discussion of the method, to a paper with van Engers (van Engers and Kramers 1933) where the method is used in a discussion of the hydrogen molecule ion H_2^+, and to a paper with Coenen (Coenen and Kramers 1936) discussing the potassium spectrum.

A second group of papers is concerned with the so-called sum rules. Thomas (1925) and Kuhn (1925) had proved that if $f_{i \to f}$ is the oscillator strength corresponding to a transition from an initial state i to a final state f,

$$f_{i \to f} = -\frac{4\pi m}{3\hbar e^2}\nu_{i \to f}\left[P_{x_{i \to f}}^2 + P_{y_{i \to f}}^2 + P_{z_{i \to f}}^2\right], \tag{3.061}$$

with $\nu_{i \to f}$ the frequency corresponding to the transition — negative for an absorption and positive for an emission process — and $P_{k_{i \to f}}$ the matrix elements of the components of the polarisation, which is related to the electric dipole moment, the sum of all oscillator strengths corresponding to a state i will satisfy the relation

$$\sum_f f_{i \to f} = Z, \tag{3.062}$$

where Z is the total number of electrons in the atom and where the sum is replaced by an integral for the states of the continuous spectrum.

Experimentally one might hope to test this relation by absorption measurements from which the oscilation strengths can be determined, but this is often very difficult because not all spectral regions are easily accessible. On the other hand, insofar as in most transitions only a single electron is involved, one might expect that a relation similar to Eq. (3.062) would be valid for separate groups in the atom. For instance, in the case of absorption processes in which only the K electrons change their state, that is, those corresponding to the K absorption bands, Eq. (3.062) would hold wih Z replaced by 2. However, it was found experimentally that the right-hand side has a value appreciably less than 2. This was explained by Kronig and Kramers (1928) by the fact that in deriving the value 2 for the right-hand side of Eq. (3.062) it is assumed that one may neglect the interactions between the electrons. Simple-mindedly one can take the interactions into account by assigning to the electrons in the various shells quantum numbers that differ from integers by appropriate quantum defects.[17] This change in the states of the electrons will mean that the matrix elements occurring in Eq. (3.062) will change and will in fact decrease, leading to a final result in good agreement with the experimental data.

Kronig and Kramers also point out that although the Pauli principle means that some of the oscillator strengths will not be present in the sum in Eq. (3.062), there will be as many positive as negative terms dropping out, and they will drop out in mutually cancelling pairs so that this will not affect the sum rule.

By analogy with Eq. (3.062) there are also sum rules relating not to all possible transitions but only to those corresponding to specific changes in quantum numbers. For instance, in the case of a single electron in a central field of force where the eigenfunctions are characterised by two quantum numbers, n and l, one has for electric dipole transitions the selection rule $\Delta l = \pm 1$ and the following sum rules:

$$\left.\begin{aligned} f_1 &\equiv \sum_{n'} f_{n,l \to n',l+1} = \frac{(l+1)(2l+3)}{3(2l+1)}, \\ f_2 &\equiv \sum_{n'} f_{n,l \to n',l-1} = -\frac{l(2l-1)}{3(2l+1)}, \end{aligned}\right\} \quad (3.063)$$

where, in accordance with Eq. (3.062), $f_1 + f_2 = 1$.

Similarly, in the case of a diatomic molecule where one is concerned with single-electron transitions and the case where the molecular orbital quantum number λ, which replaces l, is a good quantum number[18] the following sum rules hold:[19]

---

[17]For a discussion of how interactions lead to quantum defects, see, for instance, Kramers and Holst 1922, Ch. VII for a general discussion or ter Haar 1967, Ch. V for a more detailed discussion.

[18]The states corresponding to $\lambda = 0, 1, 2, \ldots$ correspond to the $\sigma, \pi, \delta, \ldots$ states, and the cases of single-electron transitions occur, for instance, in the case of the H_2 molecule (one of the $1s\sigma$ electrons), the He_2^+ molecule ion (the $2s\sigma$ electron), or the CH molecule, which is of interest to astrophysicists (the $2p\pi$ electron). We refer to a review article by Mulliken (1932) or to Herzberg's textbook (Herzberg 1950) for details about molecular spectra.

[19]As will be clear from the proof given below, this result holds for any axially symmetric single-electron system.

$$f_0 \equiv \sum_{n'} f_{n,\lambda \to n',\lambda} = \tfrac{1}{3}, \quad f_\pm \equiv \sum_{n'} f_{n,\lambda \to n',\lambda\pm 1} = \tfrac{1}{3}(1\pm\lambda), \quad (3.064)$$

and again we see that $f_0 + f_+ + f_- = 1$, as should be the case.

Kramers has given an elegant proof of these partial sum rules. The proof of Eqs. (3.060) is given in a paper with Jonker and Koopmans (Kramers, Jonker, and Koopmans 1933) while Eqs. (3.064) were used, without proof, in a paper on the formation of molecules in interstellar space (Kramers and ter Haar 1946). I shall give here the proof of Eqs. (3.064) that was sketched to me by Kramers when we were writing this paper.

Consider an axially symmetric system with Hamiltonian

$$\mathcal{H} = \Omega(\varrho, z) - \frac{C}{\varrho^2}\frac{\partial^2}{\partial\chi^2}, \qquad (3.065)$$

with

$$\Omega = -C\left[\frac{\partial^2}{\partial z^2} + \frac{\partial^2}{\partial\varrho^2} + \frac{1}{\varrho}\frac{\partial}{\partial\varrho}\right] + V(\varrho, z), \qquad C = \frac{\hbar^2}{2m}, \qquad (3.066)$$

where $V(\varrho, z)$ is the potential energy and ϱ, z, and χ are cylindrical coordinates.

The stationary states of the system are characterised by two quantum numbers, n and λ, and the corresponding normalised eigenfunctions $\phi_{n,\lambda}(\varrho, z, \chi)$ are of the form

$$\phi_{n,\lambda}(\varrho, z, \chi) = \frac{1}{\sqrt{2\pi}}\psi_{n,\lambda}(\varrho, z)e^{i\lambda\chi}, \qquad (3.067)$$

and satisfy the eigenvalue equation

$$\mathcal{H}\phi_{n,\lambda}(\varrho, z) = E_n\phi_{n,\lambda}(\varrho, z), \qquad (3.068)$$

and the completeness relation (see, for instance, QM §36)

$$\sum_n \psi^*_{n,\lambda}(\varrho, z)\psi_{n,\lambda}(\varrho', z') = \frac{1}{\sqrt{\varrho\varrho'}}\delta(\varrho - \varrho')\delta(z - z'). \qquad (3.069)$$

To get the first of the sum rules (3.064) we need the following expression:

$$f_0 = -\sum_{n'} \frac{E_{n'} - E_n}{3C}\left|\int \phi^*_{n,\lambda} z \phi_{n',\lambda}\, \varrho\, d\varrho\, dz\, d\chi\right|^2, \qquad (3.070)$$

where we have used the Bohr relation to replace the frequency by a difference in energies and where we have used the fact that in this case the polarisation operator is just equal to ez.

Using Eq. (3.068) to replace one of the energies by the Hamiltonian, using Eq. (3.069), integrating by parts, and combining Eqs. (3.067) and (3.068), we find ($d^2\omega \equiv \varrho \, d\varrho \, dz$)[20]

$$f_0 = -\frac{1}{3C} \sum_{n'} \int d^2\omega \, d\chi \, d^2\omega' \, d\chi' \, \phi^*_{n,\lambda} \phi'_{n,\lambda} zz' \phi'^*_{n',\lambda} (\mathcal{H} - E_n) \, \phi_{n',\lambda}$$

$$= -\frac{1}{3C} \int d^2\omega \, d^2\omega' \psi^*_{n,\lambda} \psi'_{n,\lambda} zz' \left[\sum_{n'} \psi'^*_{n',\lambda} \left(\Omega + \frac{C\lambda^2}{\varrho^2} - E_n \right) \psi_{n',\lambda} \right]$$

$$= -\frac{1}{3C} \int d^2\omega \, d^2\omega' \psi^*_{n,\lambda} \psi'_{n,\lambda} zz' \left[\Omega + \frac{C\lambda^2}{\varrho^2} - E_n \right] \frac{\delta(\varrho - \varrho')\delta(z - z')}{\sqrt{\varrho\varrho'}}$$

$$= -\frac{1}{3C} \int d^2\omega z \psi_{n,\lambda} \left[\Omega + \frac{C\lambda^2}{\varrho^2} - E_n \right] z \psi^*_{n,\lambda}$$

$$= -\frac{1}{3} \int d^2\omega \frac{\partial \psi^*_{n,\lambda}}{\partial z} z \psi_{n,\lambda}. \qquad (3.071)$$

Similarly one finds for $f_\pm$ the equations

$$f_\pm = -\frac{1}{6} \int d^2\omega \left[\mp 2\lambda \psi^*_{n,\lambda} \psi_{n,\lambda} + \frac{\partial \psi^*_{n,\lambda}}{\partial \varrho} \varrho \psi_{n,\lambda} \right]. \qquad (3.072)$$

Integrating by parts and using the fact that the $\psi_{n,\lambda}$ are normalised we obtain the sum rules (3.064).

In the paper with Jonker and Koopmans a proof was also given of the fact that for a system governed by the Dirac equations the sum rule (3.062) must be replaced by[21]

$$\sum_f f_{i \to f} \, (\equiv S) = 0. \qquad (3.073)$$

The proof is simple and goes as follows. The time-dependent Dirac equation can be written in the form

$$i\hbar \frac{\partial \psi}{\partial t} = \mathcal{H}_D \psi, \qquad (3.074)$$

with

$$\mathcal{H}_D = -eV - c\varrho_3 (\mathbf{p} \cdot \boldsymbol{\sigma}) - mc^2 \varrho_1. \qquad (3.075)$$

The left-hand side of Eq. (3.073) can — apart from a numerical constant — be written in the form

---

[20] The primes on a function indicate that its arguments are primed.
[21] This result only holds as long as the negative-energy states are not assumed to be filled.

$$S = \sum_n \sum_{k=1}^{3} \omega_{mn} |q_{mn}^k|^2, \qquad (3.076)$$

where q^k ($k = 1, 2, 3$) are the components of the coordinate **q**.
Using the commutation relation[22]

$$p^k q^l - q^l p^k = -i\hbar \delta_{kl}, \qquad (3.077)$$

one finds for the matrix elements q_{mn}^k the equation[23]

$$q_{mn}^k = \frac{\dot{q}_{mn}^k}{i\omega_{mn}} = \frac{\mathcal{H} q^k - q^k \mathcal{H}}{\hbar \omega_{mn}} = \frac{ic}{\omega_{mn}} (\varrho_3 \sigma^k)_{mn}. \qquad (3.078)$$

As q_{mn}^k is the complex conjugate of q_{nm}^k and $\omega_{mn} = -\omega_{nm}$ we have, on the one hand,

$$S = \sum_n \omega_{mn} q_{mn}^k q_{nm}^k = ic \sum_n (\varrho_3 \sigma^k)_{mn} q_{nm}^k$$
$$= ic \left(\varrho_3 \sigma^k q^k \right)_{mm}, \qquad (3.079)$$

and, on the other hand,

$$S = -\sum_n \omega_{nm} q_{mn}^k q_{nm}^k = -ic \sum_n q_{mn}^k (\varrho_3 \sigma^k)_{nm}$$
$$= -ic \left(q^k \varrho_3 \sigma^k \right)_{mm}. \qquad (3.080)$$

Since q^k and $\varrho_3 \sigma$ commute, Eq. (3.073) follows.

The reason for the vanishing of S is the existence of negative terms in the sum corresponding to transitions to negative-energy states.

The third group of papers is concerned with molecular spectra. Two of those (Kramers 1929a, b) are dealing with the spectrum of O_2.[24] The oxygen molecule is a special case of a diatomic molecule in that its ground state is a triplet state,[25] $^3\Sigma$, instead of the usual singlet state. The atmospheric rotational spectrum of oxygen was explained by Mulliken as being due to transitions to an excited $^1\Sigma$ state and in order to explain the experimental data it was necessary that the molecular states be split into very narrow doublets with $J = \Lambda \pm 1$ with their centre being at a distance from the

---

[22] Immediately after the advent of quantum mechanics Kramers (1925d) had pointed out that the Thomas-Kuhn sum rules were connected with these commutation rules.
[23] One assumes here that there exists a weak field which has lifted all degeneracies.
[24] The theory of molecular spectra in the form in which we know it nowadays was to a very large extent developed in the late twenties. For details of the experimental data and of the theory of molecular spectra we refer to Herzberg's monumental textbook (Herzberg 1950).
[25] In fact, it is a $^3\Sigma_g^-$ state; in 1928 the notation had not yet been agreed on and Kramers refers to it as a ^3S state and to the excited state as a ^1S state.

$J = \Lambda$ level, which was practically constant, that is, independent of Λ. This splitting was unexpected at the time and in the above-mentioned papers Kramers explained it by taking into account not only the coupling of the spin of an electron to both its own orbit and the orbits of the other electrons but also the coupling to the other spins — a coupling neglected until then. If this is done, excellent agreement with the experimental data is found.

The other papers in this group (Kramers and Ittmann 1929a, b, 1930) deal with the quantum theory of an asymmetric top. Casimir (1952) writes about these papers that "although he is dealing here with a concrete problem which was not without importance for the interpretation of molecular spectra, yet I think that, as the work proceeded, it was the general mathematical contexture rather than a preoccupation with the properties of molecules that became his main concern. As a matter of fact he had on that occasion to elaborate considerably the existing theory of Lamé functions." In fact, there is nowhere in these papers a comparison with experimental data.

In their first paper Kramers and Ittmann show that the problem of the asymmetric top can be solved by a separation of variables. Classically, the kinetic energy $\mathcal{T}_{\mathrm{cl}}$ of an asymmetric top is given by the equation

$$\mathcal{T}_{\mathrm{cl}} = \frac{(\dot{\psi}\sin\vartheta\sin\varphi + \dot{\vartheta}\cos\varphi)^2}{2A} + \frac{(\dot{\psi}\sin\vartheta\cos\varphi - \dot{\vartheta}\sin\varphi)^2}{2B} + \frac{(\dot{\psi}\cos\vartheta + \dot{\varphi})^2}{2C}, \quad (3.081)$$

where A, B, and C are the principal moments of inertia of the top, which we have ordered in such a way that $A > B > C$, and ϑ, ψ, and φ are the Euler angles.

One now introduces instead of ϑ and φ new coordinates λ and μ, which satisfy the condition

$$C \leqslant \lambda \leqslant B \leqslant \mu \leqslant A, \quad (3.082)$$

through the equations[26]

$$\left. \begin{aligned} \cos^2\vartheta &= \frac{(\lambda-C)(\mu-C)}{(A-C)(B-C)}, \\ \sin^2\vartheta\sin^2\varphi &= \frac{(A-\lambda)(A-\mu)}{(A-C)(A-B)}, \\ \sin^2\vartheta\cos^2\varphi &= \frac{(B-\lambda)(\mu-B)}{(A-B)(B-C)}. \end{aligned} \right\} \quad (3.083)$$

---

[26] Of course, these three equations are not independent, since the left-hand (or the right-hand) sides add up to 1.

If we write the T_{cl} in the form

$$T_{cl} = \tfrac{1}{2} \sum_{ij=1}^{3} g_{ij} \dot{q}_i \dot{q}_j, \tag{3.084}$$

where we have put $\lambda = q_1$, $\mu = q_2$, $\psi = q_3$, the Schrödinger equation will be of the form[27]

$$\tfrac{1}{2}\hbar^2 \sum_{ij} \frac{1}{\sqrt{g}} \frac{\partial}{\partial q_i} \left(\sqrt{g}\, g_{ij} \frac{\partial \Phi}{\partial q_j} \right) + E\Phi = 0; \tag{3.085}$$

here g is the determinant of the g_{ij} which is given by the equation

$$\sqrt{g} = \frac{\mu - \lambda}{4\sqrt{-f(\lambda)f(\mu)}\sqrt{ABC}}, \tag{3.086}$$

where $f(x)$ is the expression

$$f(x) = (A-x)(B-x)(C-x). \tag{3.087}$$

We shall look for the solution of Eq. (3.085) in the form

$$\Phi = V(\lambda)W(\mu)e^{im\psi}. \tag{3.088}$$

The integer m is the quantum number corresponding to the angular momentum around the z-axis. In their third paper Kramers and Ittmann show how the solutions with non-vanishing m can be derived from those with $m = 0$.[28] Because of the rotational invariance of the problem one can, in fact, choose the coordinate axes such that the angular momentum around the z-axis vanishes, which means that Eq. (3.086) simplifies to

$$\Phi = V(\lambda)W(\mu). \tag{3.089}$$

On substituting the expressions for the g_{ij} and g, the Schrödinger equation becomes

$$\frac{\mu}{\mu-\lambda}\left(\sqrt{-4f(\lambda)}\frac{\partial}{\partial\lambda}\right)^2 \Phi + \frac{\lambda}{\mu-\lambda}\left(\sqrt{4f(\mu)}\frac{\partial}{\partial\mu}\right)^2 \Phi + \frac{2E}{\hbar^2}\Phi = 0. \tag{3.090}$$

Multiplying this equation by $(\mu - \lambda)/\mu\lambda$ and substituting Eq. (3.089) for Φ we see that it can be separated, which leads to the two equations

---

[27] Compare the expression for the Laplacian in the case of non-orthogonal curvilinear coordinates given by Margenau and Murphy (1956).
[28] Instead of the Lamé equation which occurs in the case of $m = 0$, there is now a transformed Lamé equation.

$$\left.\begin{array}{rcl}\left(\sqrt{-4f(\lambda)}\dfrac{\partial}{\partial\lambda}\right)^2 V &=& \left(-\dfrac{2E}{\hbar^2}+A\lambda\right)V, \\[2mm] \left(\sqrt{4f(\mu)}\dfrac{\partial}{\partial\mu}\right)^2 U &=& \left(\dfrac{2E}{\hbar^2}-A\mu\right)U,\end{array}\right\} \qquad (3.091)$$

where A is a constant. Eqs. (3.091) are Lamé-type equations.

Kramers and Ittmann prove that the constant A must be equal to $j(j+1)$ with j an integer, determining the total angular momentum of the top. They discuss in detail the energy spectrum and especially the asymptotic behaviour of the eigenspectrum and the eigenfunctions for large quantum numbers, using the WKB approximation.

The last paper we want to mention in this subsection is a paper with Brinkman (Kramers and Brinkman 1930) in which they use the first Born approximation to calculate the cross-section for the capture of an electron from an atom by an α-particle passing by.

3.5. Kramers Degeneracy and Charge Conjugation

To conclude this chapter we discuss two papers, both of which are reprinted at the end of the book — as papers E and H. The first one proves that the energy states of systems with an odd number of electrons are at least twofold degenerate in the presence of purely electric fields, and the second one introduces charge-conjugation. Although they are only mentioned in passing by Kramers, both papers are concerned with important symmetry properties of quantum systems, namely the so-called T- and C-invariances, that is, invariance under time reversal and under charge conjugation.

The paper in which Kramers proves what is now called the Kramers degeneracy theorem is entitled "Théorie générale de la rotation paramagnétique dans les cristaux" (General theory of paramagnetic rotation in crystals). From the title it is clear that Kramers was not primarily interested in the theorem but in its consequences for the behaviour of paramagnetic systems. The paper is one of a series of papers on the rotation of the plane of polarisation in paramagnetic crystals, an effect studied experimentally by Becquerel.[29] We shall discuss these papers in some detail in §5.3; here we shall confine ourselves to giving a proof of the theorem.

Consider a system of N electrons with spatial coordinates $\mathbf{r}_i$, momenta $\mathbf{p}_i$, and spin coordinates $\mathbf{S}_i$ ($i = 1, \ldots, N$) governed by a Hamiltonian $\mathcal{H}(\mathbf{r}_1, \ldots, \mathbf{r}_N; \mathbf{p}_1, \ldots, \mathbf{p}_N; \mathbf{S}_1, \ldots, \mathbf{S}_N)$, where the operators corresponding to the components of the spin of the ith electron are, apart from a factor $\frac{1}{2}\hbar$, the Pauli matrices (3.036):

---

[29] We should mention that the degeneracy theorem was mentioned by Kramers in his first paper of the series (Kramers 1929f), where he states that the proof of the theorem would take up too much space.

$$S_{ix} = \tfrac{1}{2}\hbar\sigma_{1i}, \qquad S_{iy} = \tfrac{1}{2}\hbar\sigma_{2i}, \qquad S_{iz} = \tfrac{1}{2}\hbar\sigma_{3i}. \qquad (3.092)$$

Let $\varphi_{s_1,..,s_N}(\mathbf{r}_1, \ldots, \mathbf{r}_N)$ describe a state of the system in which the spin of the ith electron has a value $s_i \hbar$ ($s_i = \pm\tfrac{1}{2}$) in a chosen direction which we shall assume to be the z-direction.

We define the spin-conjugate wavefunction by the equation

$$\varphi^\dagger_{s_1,..,s_N}(\mathbf{r}_1, \ldots, \mathbf{r}_N) = \Sigma_y \varphi^*_{s_1,..,s_N}(\mathbf{r}_1, \ldots, \mathbf{r}_N), \qquad (3.093)$$

where

$$\Sigma_y = \sigma_{21}\sigma_{22} \cdots \sigma_{2N}. \qquad (3.094)$$

We next define the spin-conjugate Hamiltonian $\mathcal{H}^\dagger$ by the equation

$$\mathcal{H}^\dagger = \Sigma_y \mathcal{H}^* \Sigma_y. \qquad (3.095)$$

Taking the complex conjugate of $\mathcal{H}$ means that all $\mathbf{p}_i$ are replaced by $-\mathbf{p}_i$ and that all σ_{2i} are replaced by $-\sigma_{2i}$, while the action of the operators Σ_y from behind and from in front means that the σ_{1i} and the σ_{3i} change their signs; the final result therefore is

$$\mathcal{H}^\dagger = \mathcal{H}(\mathbf{r}_1, \ldots, \mathbf{r}_N; -\mathbf{p}_1, \ldots, -\mathbf{p}_N; -\mathbf{S}_1, \ldots, -\mathbf{S}_N). \qquad (3.096)$$

From Eqs. (3.093) and (3.094) it follows that

$$\mathcal{H}^\dagger \varphi^\dagger = \Sigma_y \mathcal{H}^* \Sigma_y \Sigma_y \varphi^* = \Sigma_y \mathcal{H}^* \varphi^* = E \Sigma_y \varphi^* = E \varphi^\dagger. \qquad (3.097)$$

Hence, if φ is an eigenfunction of $\mathcal{H}$ corresponding to an eigenvalue E, $\varphi^\dagger$ will be an eigenfunction of $\mathcal{H}^\dagger$ corresponding to the same eigenvalue. In order that the eigenvalue E will be non-degenerate in the case where $\mathcal{H}$ and $\mathcal{H}'$ are identical, $\varphi^\dagger$ and φ must be the same function, $\varphi^\dagger = a\varphi$. Since we have

$$(\varphi^\dagger)^\dagger = a^* \varphi^\dagger = a^* a \varphi, \qquad (3.098)$$

a necessary condition for non-degeneracy is $|a|^2 = 1$. However, it follows from Eq. (3.093) that $(\varphi^\dagger)^\dagger = (-1)^N \varphi$. Hence follows the theorem: Every eigenvalue of a Hamiltonian which is invariant under a reversal of the sign of all momenta and spins, that is, under time reversal, is for a system with an odd number of electrons degenerate with an even degree of degeneracy.

In the same paper Kramers also proves that the magnetic moment of a system with N electrons vanishes for all nondegenerate states. We refer to either paper E or to a paper by Klein (1952) for the proof of this theorem.

To conclude this chapter we discuss the paper in which Kramers (1937a) introduces charge conjugation. He shows that if Ψ satisfies the time-dependent Dirac equation (compare Eq. (3.037))

$$((\boldsymbol{\alpha} \cdot \boldsymbol{\pi}) + e\Phi + \beta mc)\Psi = i\hbar \frac{\partial \Psi}{\partial t}, \qquad (3.099)$$

where the 4×4 matrices α and β are defined by the equations[30]

$$\alpha = \varrho_3 \sigma, \qquad \beta = \varrho_1, \tag{3.100}$$

there is a *charge-conjugated* wavefunction Ψ^L which satisfies the same equation, but with the sign of the charge reversed:

$$((\alpha \cdot \pi^L) - e\Phi + \beta mc)\Psi^L = i\hbar \frac{\partial \Psi^L}{\partial t}, \tag{3.101}$$

with

$$\pi^L = \mathbf{p} + \frac{e\mathbf{A}}{c}. \tag{3.102}$$

It can be verified through direct calculation that if Ψ is given by Eq. (3.039), Ψ^L satisfies the equation

$$\Psi^L = \begin{pmatrix} \phi_{22}^* \\ -\phi_{21}^* \\ -\phi_{12}^* \\ \phi_{11}^* \end{pmatrix}. \tag{3.103}$$

The representation (3.100) of α and β is the one used preferentially by Kramers because of its properties under Lorentz transformations. An interesting different choice of representation is the one where

$$\alpha_1 = \varrho_3 \sigma_1, \qquad \alpha_2 = \varrho_3 \sigma_2, \qquad \alpha_3 = -\varrho_1, \qquad \beta = \varrho_3 \sigma_2, \tag{3.104}$$

when Ψ^L and Ψ turn out to be each other's complex conjugates.

It is clear from Kramers's paper, although not stated explicitly, that charge conjugation and especially invariance under charge conjugation — the so-called C-invariance — can only be consistently formulated when one promotes the wavefunction to be an operator.[31]

Kramers discusses in some detail what happens in hole theory. If Ψ is the "promoted" wavefunction operator it will satisfy the anticommutation relation

$$\Psi^\dagger(q)\Psi(q') + \Psi(q')\Psi^\dagger(q) = \delta(q, q'), \tag{3.105}$$

where the † indicates the Hermitean conjugate function, the q stand for the complete set of space and spin coordinates, and $\delta(q, q')$ is a

---

[30] Of course, β has a factor which is the unit matrix as far as the second index of the components ϕ_{rs} of Ψ is concerned.

[31] This is often called "second quantisation" which is a very bad misnomer, since the quantum of action enters nowhere. A better name is "occupation number representation" since usually the wavefunctions are expanded in terms of the Jordan-Klein and Jordan-Wigner matrices (see, for instance, QM § 72).

product of Dirac delta functions for all of the space coordinates and Kronecker delta functions for all the spin coordinates.

Any operator Ω acting upon the space and spin coordinates in its "promoted" form will become

$$\Omega = \int \boldsymbol{\Psi}^\dagger \Omega \boldsymbol{\Psi} \, dq, \tag{3.106}$$

where the integration is over all space and spin coordinates.

In hole theory one wants to have expressions that are symmetric in the electrons and the positrons (holes).[32] Kramers therefore introduces the charge-conjugated wavefunction operator — which also satisfies the anticommutation relation (3.105) — and instead of the "promoted" operator (3.106) the expression

$$\Omega = \tfrac{1}{2} \int \left(\boldsymbol{\Psi}^\dagger \Omega \boldsymbol{\Psi} + \boldsymbol{\Psi}^{L\dagger} \Omega^L \boldsymbol{\Psi}^L \right) dq, \tag{3.107}$$

where Ω^L is obtained from Ω by changing the sign of the charge.

Kramers uses the above formulæ to express various quantities, such as the energy, the number density, and the charge density, in terms of the Jordan-Wigner matrices, that is, the creation and annihilation operators of free electrons and positrons which are introduced by expanding $\boldsymbol{\Psi}$ as follows:[33]

$$\boldsymbol{\Psi} = \sum \mathbf{a}_k \varphi_k + \sum \mathbf{b}_k^\dagger \varphi_k^L, \tag{3.108}$$

where the φ_k are four-component free electron wavefunctions.

The charge-conjugate wavefunction $\boldsymbol{\Psi}^L$ corresponding to $\boldsymbol{\Psi}$ of Eq. (3.108) is of the form

$$\boldsymbol{\Psi}^L = \sum \mathbf{b}_k \varphi_k + \sum \mathbf{a}_k^\dagger \varphi_k^L, \tag{3.109}$$

and from the formulæ given by Kramers it follows that, indeed, the hole theory is completely symmetric under charge conjugation, that is, C-invariant.

In this paper Kramers considers both the case where there are no external fields and also the case where Φ and $\mathbf{A}$ are non-zero, including the case of hydrogen-like atoms. It is interesting to note that when considering the expression for the Coulomb interaction energy in hole theory he concludes with the statement "we expect that a *correction must be applied to the energy values of the stationary states of the hydrogen atom, as given by the Dirac theory of 1928.* (Kramers's italics)

---

[32] It is interesting to note that in this paper in 1937 Kramers uses for electrons and positrons the names negatons and positons.
[33] For details we refer to QM § 72.

4 Quantum Electrodynamics

The main content of the present chapter will be a discussion of Kramers's research on the fundamental problems raised by the interaction between electrons and electromagnetic radiation, although not such a detailed one as given by Dresden (1987, Ch. 16). By far the most extensive discussion of his work can be found in Chapter VIII of the second part of his own textbook (Kramers 1938e); other accounts are found in his contributions to the 1937 Galvani Congress (Kramers 1938c) and the 1948 Solvay Congress (Kramers 1950) as well as in a lecture given to the Dutch Physical Society in April 1944 (Kramers 1944d); an English translation of this talk was published in the *Collected Scientific Papers* and is reprinted as paper K at the end of this book.

Kramers was the first to introduce mass renormalisation in quantum mechanics[1] by constructing a theory in which the experimental rather than the inertial mass appears. That it was, in fact, Kramers who introduced the renormalisation concept was well known to people like Bethe and Schwinger, but their acknowledgement of this fact did not appear in their early papers and as a result later generations have as often as not been oblivious of the true facts, and attribute the renormalisation idea to theorists such as Schwinger or Tomonaga.[2]

We must also briefly mention Kramers's unpublished contribution to S-matrix theory, which is discussed in some detail by Dresden (1987, § 17IIC). In his talk to the Dutch Physical Society in April 1944 (see paper K at the end of this book) Kramers lists three attempts to get rid of the divergences of quantum electrodynamics as it was at that time. One is that by Heisenberg. When he introduced quantum mechanics, he had followed Kramers's idea that one should concentrate on quantities which allow of a direct physical interpretation. In the early forties he tried the same approach in developing a theory in which the Hamiltonian is replaced by the so-called S-matrix or scattering matrix, which contains the scattering data. When Kramers in his talk points out that the S-matrix is in principle able to answer the question

---

[1] Lorentz had introduced a similar idea in his theory of the electron (see, e.g., Kramers 1944d).

[2] Dresden (1987, § 16IVB) has described in detail the belated recognition of Kramers's seminal ideas, which unfortunately occurred after his death.

about bound states of particles, since those are related to the zeroes and poles of the scattering matrix, considered as an analytic function of its arguments, he fails to mention that it was he himself who pointed out this fundamental property of the scattering matrix to Heisenberg.[3] This is one more instance of Kramers's unpublished contributions to some of the basic principles of modern theoretical physics.

Another approach mentioned in Kramers's talk is the one where one considers classical particle theory and uses the correspondence principle to develop an improved quantum theory. This was the way Kramers himself approached the problem. Pais (1986, p. 449; 1995) has related that Kramers felt that since there is no correct relativistic theory, one should start by trying to construct a correct non-relativistic classical theory, quantise it, and afterwards go over to the relativistic theory. This is in contrast to most other theorists, who approach the problem by starting from existing relativistic quantum theories, attempting to improve those. As he states in his contribution to the 1948 Solvay Congress, one should construct a classical electron theory in canonical form, since one will then have obtained "a trustworthy basis for a quantum theory of interaction between particles and field, that is, for a quantum theory which shows the necessary correspondence with that part of the classical electron theory which is of physical importance."

Apart from Kramers's own book and his three papers, there are only four other publications using his approach: two papers by Serpe (1940, 1941), who applied Kramers's theory as presented in his book to a harmonic oscillator and showed that in Kramers's theory some of the divergences of Dirac's theory had disappeared; a paper by Opechowski (1941), who obtains — in the dipole approximation — a Hamiltonian which is correct to first order in the electron charge; and the 1952 Leiden thesis of van Kampen (1951), who used the Hamiltonian obtained by Kramers, which is correct in the dipole approximation to all orders in the electronic charge, to discuss the scattering of light. To some extent van Kampen's account is more systematic than Kramers's own in his talk to the Solvay Congress, and we have therefore largely followed van Kampen.[4]

Kramers proceeds essentially as follows. He starts with the classical Hamiltonian of the system consisting of an electron and the electromagnetic field. In this Hamiltonian one identifies that part which depends on the structure of the electron, and through a canonical transformation eliminates it. One is then left with a structure-independent Hamiltonian which can be quantised. It is not surprising that this ambitious programme cannot be carried out in

---

[3] Kramers's son has pointed out to me that it would have been undiplomatic to mention this fact, indicating contact with a German physicist, during the occupation of the Netherlands. Hilgevoord (1996) states that it was, in fact, Wouthuysen, a research student of Kramers, who found this property; as Wouthuysen was a Jew and at that time in hiding his name could also not be mentioned.

[4] Dresden presents both these accounts in detail and we refer to him for some of those details.

all details and it is necessary to make various approximations, but Kramers's hope was that even such an approximate theory would contain all physically relevant features. The advantage of this approach, as he saw it, is that at all times one knows what one is doing, whereas one has no control of the errors introduced in the approach starting from the Dirac theory as it sets out from flawed foundations.

In order to obtain the structure-independent Hamiltonian of the system consisting of a single electron together with the radiation field, Kramers (1950) starts from the (non-relativistic, classical) equations of motion and transforms those into a structure-independent form. He then constructs his Hamiltonian. Van Kampen (1951) starts directly from the classical non-relativistic Hamiltonian of the system and through a number of canonical transformations produces essentially the same Hamiltonian.

We shall now follow van Kampen. Since the radiation field is a transverse field, we start from the Hamiltonian of an electron in a transverse field. If one considers the Maxwell equations as the canonical equations of a continuous field, the vector potential $\mathbf{A}$ will be the canonical coordinate and $-c\mathbf{E}/4\pi$ its canonically conjugate momentum; this means that in the Hamiltonian of our system we must express the magnetic field in terms of the vector potential. It is convenient to introduce, instead of the continuous fields $\mathbf{A}$ and $\mathbf{E}$, an infinite set of discrete coordinates and momenta, $\mathbf{q}_n$ and $\mathbf{p}_n$. This can be done by enclosing the system in a finite volume, which we shall choose to be in the shape of a sphere with the electron at its centre. The Hamiltonian that one obtains in this way is given by Eq. (4.12), where the canonical coordinates are the $\mathbf{q}_n$ and the position $\mathbf{R}$ of the electron with, as the canonically conjugate momenta, the $\mathbf{p}_n$ and the electron momentum $\mathbf{P}$.

At this point we are still working with an electron that has an unspecified, extended structure characterised by a charge density ϱ, which we shall assume to be spherically symmetric. Our Hamiltonian therefore still depends on the structure of the electron. The next step consists of two consecutive canonical transformations, the first one to remove the term which is linear in the electron momentum and the second to reduce the dependence on the transformed field coordinates to a sum of squares. The final result is the Hamiltonian (4.16).

We start with an electron, at a position $\mathbf{R}$ and with a momentum $\mathbf{P}$. If it is interacting with a radiation field we start from the following Hamiltonian:

$$\mathcal{H} = \frac{\boldsymbol{\Pi}^2}{2m_0} + U(\mathbf{R}) + \int \frac{\mathbf{E}^2 + [\nabla \wedge \mathbf{A}]^2}{8\pi} \, d^3\mathbf{r}, \qquad (4.01)$$

where (compare Eq. (3.038))

$$\boldsymbol{\Pi} = \mathbf{P} - \frac{e\mathbf{A}_{\mathrm{av}}}{c}, \qquad (4.02)$$

and where $U(\mathbf{R})$ is the static potential after the longitudinal field has been eliminated. In Eqs. (4.01) and (4.02) $\mathbf{A}$ is the vector potential which is related to the electric field $\mathbf{E}$ and the magnetic field $\mathbf{H}$ through the relations[5]

$$\mathbf{E} = -\frac{1}{c}\frac{\partial \mathbf{A}}{\partial t}, \qquad \mathbf{H} = \text{curl } \mathbf{A}, \tag{4.03}$$

and we have chosen a gauge in which $\text{div} \mathbf{A} = 0$. The quantity $\mathbf{A}_{\text{av}}$ is an average over the extended charge:

$$e\mathbf{A}_{\text{av}} = \int \mathbf{A}(\mathbf{r})\varrho(|\mathbf{R}+\mathbf{r}|)\,d^3\mathbf{r} = \int \mathbf{A}(\mathbf{r}-\mathbf{R})\varrho(r)\,d^3\mathbf{r}. \tag{4.04}$$

If we now use the dipole approximation, that is, assume that all relevant wavelengths are long as compared to the size of the electron, Eq. (4.04) reduces to

$$e\mathbf{A}_{\text{av}} = \int \mathbf{A}(\mathbf{r})\varrho(r)\,d^3\mathbf{r}. \tag{4.05}$$

The canonical coordinates and momenta corresponding to the Hamiltonian (4.01), in which $\mathbf{H}$ is replaced by curl $\mathbf{A}$, are $\mathbf{R}$ and $\mathbf{P}$ and $\mathbf{A}$ and $-c\mathbf{E}/4\pi$, respectively.

We next assume that the system is contained in a sphere of radius L. In that case we have in the dipole approximation the following expansion for $\mathbf{A}$:

$$\mathbf{A}(\mathbf{r}) = \text{Trv}\left\{\sum_{n=1}^{\infty}\sqrt{\frac{3c^2}{L}}\,\mathbf{q}_n\,\frac{\sin \nu_n r}{r}\right\}, \tag{4.06}$$

where

$$\nu_n = \frac{n\pi}{L}, \tag{4.07}$$

and Trv $\{\cdot\cdot\}$ indicates that one must take the transverse part of $\{\cdot\cdot\}$; the right-hand side of Eq. (4.06) is thus defined as follows:

$$\text{Trv}\left\{\sum_{n=1}^{\infty}\mathbf{q}_n\,\frac{\sin \nu_n r}{r}\right\} \equiv \sum_{n=1}^{\infty}\left\{\mathbf{q}_n + \frac{1}{\nu_n^2}(\mathbf{q}_n\cdot\nabla)\nabla\right\}\frac{\sin \nu_n r}{r}. \tag{4.08}$$

A similar expansion for the electric field can be written in the form

$$\mathbf{E} = -\text{Trv}\left\{\sum_{n=1}^{\infty}\sqrt{\frac{3}{L}}\,\mathbf{P}_n\,\frac{\sin \nu_n r}{r}\right\}, \tag{4.09}$$

---

[5] We have assumed here that the longitudinal part of the electromagnetic field has already been eliminated so that there is no scalar potential. The radiation field itself is, of course, a transverse field.

where the expansion coefficient is chosen in such a way that now $\mathbf{p}_n$ is the canonically conjugate of $\mathbf{q}_n$.

Substituting the expansion (4.06) into Eq. (4.05) we find

$$e\mathbf{A}_{\text{av}} = \sqrt{\frac{4c^2}{3L}} \sum_n \mathbf{q}_n \int \sin \nu_n r\, \varrho(r)\, 4\pi\, dr = \sum_n \varepsilon_n c \mathbf{q}_n, \qquad (4.10)$$

where the ε_n, which are defined through Eq. (4.10), depend on the structure of the electron. If we write

$$\varepsilon_n = \sqrt{\frac{4e^2}{3L}}\, \delta_n \nu_n, \qquad (4.11)$$

the structure-dependent quantities δ_n become equal to unity in the limit of a point electron.

Substituting Eqs. (4.06), (4.08), and (4.11) into Eq. (4.01), we find for the Hamiltonian the equation

$$\mathcal{H} = \frac{\mathbf{P}^2}{2m_0} + U(\mathbf{R}) - \frac{(\mathbf{P} \cdot \sum \varepsilon_n \mathbf{q}_n)}{m_0} + \frac{(\sum \varepsilon_n \mathbf{q}_n)^2}{2m_0}$$

$$+ \sum_n \tfrac{1}{2}\left(\mathbf{p}_n^2 + c^2 \nu_n^2 \mathbf{q}_n^2\right). \qquad (4.12)$$

We now apply the canonical transformation

$$\mathbf{P} = \mathbf{P}', \qquad \mathbf{p}_n = \mathbf{p}'_n,$$

$$\mathbf{R} = \mathbf{R}' + \sum \frac{\varepsilon_n}{m\nu_n^2} \mathbf{p}'_n, \quad \mathbf{q}_n = \mathbf{q}'_n + \frac{\varepsilon_n}{m\nu_n^2} \mathbf{P}', \qquad (4.13)$$

where

$$m = m_0 + \sum \frac{\varepsilon_n^2}{\nu_n^2}. \qquad (4.14)$$

This leads to the transformed Hamiltonian

$$\mathcal{H} = \frac{\mathbf{P}'^2}{2m} + U\!\left(\mathbf{R}' + \sum \frac{\varepsilon_n}{m\nu_n^2} \mathbf{p}'_n\right)$$

$$+ \frac{(\sum \varepsilon_n \mathbf{q}'_n)^2}{2m_0} + \sum_n \tfrac{1}{2}\left(\mathbf{p}'^{\,2}_n + c^2 \nu_n^2 \mathbf{q}'^{\,2}_n\right). \qquad (4.15)$$

We next apply a second canonical transformation to bring the quadratic expression in the $\mathbf{q}'_n$ to principal axes. This finally leads to the Hamiltonian

$$\mathcal{H} = \frac{\mathbf{P}'^2}{2m} + U\!\left(\mathbf{R}' + \frac{e}{m}\sum \sqrt{\frac{4}{3L_n}}\, \frac{\cos \eta_n}{k_n} \mathbf{p}''_n\right)$$

$$+ \sum_n \tfrac{1}{2}\left(\mathbf{p}''^{\,2}_n + c^2 k_n^2 \mathbf{q}''^{\,2}_n\right), \qquad (4.16)$$

where the k_n are the roots of a certain characteristic equation[6] while the η_n and L_n are defined by the equations

$$\left. \begin{array}{ll} Lk_n = \eta_n + n\pi, & 0 < \eta_n < \tfrac{1}{2}\pi; \\ L_n = L - r_0 \cos^2 \eta_n, & r_0 = \dfrac{2e^2}{3mc^2}. \end{array} \right\} \quad (4.17)$$

Note that the experimental mass m rather than the inertial mass m_0 enters into the definition of r_0. We also note that r_0 is essentially the classical electron radius, defined using the experimental electron mass.

The structure of the electron enters into Eq. (4.16) through the k_n and the hope is that its effect on the physics of the system is small. In fact, as long as we are dealing with wavelengths that are long as compared to r_0 so that as first approximation we have $kr_0 \approx 0$, the structure-dependent effects will be small. In that case we may assume that we are dealing with a point electron and the characteristic equation becomes

$$\tan Lk = kr_0, \qquad (4.18)$$

or, equivalently,

$$\tan \eta = kr_0. \qquad (4.19)$$

Insofar as the structure-dependent effects can, indeed, be neglected we have reached Kramers's goal and obtained a structure-independent Hamiltonian, albeit only in the non-relativistic and dipole approximations. This Hamiltonian is written in terms of canonically conjugate variables so that it can be quantised in a straightforward way. The second term on the right-hand side of Eq. (4.14) is the electromagnetic mass of the electron.[7] Although it

---

[6] For details see either van Kampen's thesis or Dresden's account of that thesis.

There is one amusing detail that has not attracted as much attention as it deserves. Provided the electron is sufficiently small, the characteristic equation has one imaginary solution, $k_* = i/r_0$, corresponding to a term $\tfrac{1}{2}(\mathbf{p}_*^2 - \mathbf{q}_*^2/r_0^2)$ in the Hamiltonian, as well as an extra term in the argument of U. The electromagnetic field corresponding to this solution has a factor $e^{-r/r_0}/r$, which means that the field vanishes outside a region of the size of the classical electron radius and hence does not contribute to the radiation field. The time-dependent solution will contain a factor e^{ct/r_0}; this means that we have here a "runaway" solution that is well known from Dirac's classical electron theory (Dirac 1938). The reason why this solution has not been found by other people is that they use an expansion in terms of e^2 while the charge here occurs in the denominator of the exponent. Van Kampen (private communication) has pointed out that the appearance of this runaway solution may be an important clue to the fundamental problems connected with the concept of point particles.

[7] Note the similarity between Eq. (4.14) and Lorentz's Eq. (3) in paper K.

diverges in the limit of a point electron this is immaterial, since we should only consider the experimental mass and forget about how it is made up.

From Eq. (4.16) we can see why the renormalised Hamiltonian will lead to changes in the energy levels and hence to effects like the Lamb shift. This arises through the change in the argument of the potential energy U. Classically one would expect that this change will affect different orbits differently, and since the Lamb shift is the difference in energy between the $2s_{1/2}$ and the $2p_{1/2}$ levels of the hydrogen atom, which according to the Dirac theory should be zero, its appearance is to some extent to be expected — and was, indeed, found by Kramers (1950) using his renormalised Hamiltonian.

5 Statistical Mechanics, Solid-State Physics, and Low-Temperature Physics

In the present chapter we shall first deal with Kramers's contributions to equilibrium statistical mechanics. He always had a special affection for statistical mechanics, possibly because he was a pupil of Ehrenfest, who himself was a pupil of Boltzmann, the founder of kinetic theory. However, whereas Ehrenfest preferred Boltzmann's kinetic theory approach to Gibbs's method, using ensembles, Kramers whenever possible used the Gibbs approach.

As in the fields we discussed in the preceding three chapters, Kramers made important contributions to statistical mechanics. Of some of those it is debatable whether they are pure statistical mechanics or fall into other categories. This is especially true of the papers dealing with non-equilibrium statistical mechanics. I have restricted myself in the present chapter to the papers dealing with systems in statistical equilibrium, whereas Dresden (1988) considers in his discussion of Kramers's contributions to statistical mechanics both equilibrium and non-equilibrium topics. As a result I am postponing to the next chapter the discussion of two of Kramers's most influential and seminal papers — the paper on the behaviour of polymer molecules in inhomogeneous flow (Kramers 1946b) and his paper on the theory of Brownian motion in a field of force (Kramers 1940a) — as well as his paper with Kistemaker (Kramers and Kistemaker 1943) and a related paper (Kramers 1949b) on the behaviour of a gas near a wall.

Second, we shall consider in the present chapter Kramers's papers dealing with solid-state and low-temperature physics. His two papers on ferromagnetism could be discussed under either the heading of statistical mechanics or the heading of solid-state physics, since the question of phase transitions plays an important role in both of them, while the second paper considers quite general methods for evaluating partition functions.

As far as Kramers's contributions to solid-state physics are concerned, apart from the above-mentioned papers on ferromagnetism and a short note on antiferromagnetism, the remaining papers in this category are all concerned with the theory of paramagnetic salts, which play such an important role in the attaining of low temperatures through adiabatic demagnetisation. The last paper to be discussed in the present chapter is one on the elementary excitations of liquid helium.[1]

---

[1] This paper was published posthumously.

68   Chapter 5

In the case of statistical mechanics, my restriction to equilibrium topics means that in that category there are only seven papers to be considered in the present chapter, whereas there are ten papers on paramagnetism. We remind ourselves that parts of one of these were discussed in Chapter 3, since the Kramers degeneracy was introduced in it. The papers to be discussed in this chapter will be split into three sections:

i. In §5.1 I shall discuss an early paper on the free energy of a mixture of ions (Kramers 1927a), a paper with Ornstein (Ornstein and Kramers 1927) on the Fermi distribution formula, and a paper (Kramers 1938a) discussing the theory of canonical grand ensembles. In that section I shall also discuss Kramers's proof of the Nernst Theorem — the Third Law of Thermodynamics.

ii. In §5.2 we consider the two papers on ferromagnetism (Heller and Kramers 1934, Kramers 1936a), and the two papers with Wannier (Kramers and Wannier 1941a, b) on the two-dimensional Ising model, the first of which is reprinted at the end of this book as paper J. It is interesting to mention that in the second paper on ferromagnetism Kramers was probably the first to use combinatorial methods to find series expansions.

iii. In §5.3 we shall discuss Kramers's contributions to the theory of paramagnetic rotation (Kramers 1929f, 1930d, 1932a, 1933b, h; Kramers and Becquerel 1929; Becquerel, de Haas, and Kramers 1929; Becquerel, van den Handel, and Kramers 1951), a paper on the interaction of magnetic atoms in paramagnetic crystals (Kramers 1934a), and the work with de Haas and Wiersma (de Haas, Wiersma, and Kramers 1934) on the adiabatic cooling method.

We conclude §5.3 with a brief discussion of a short note on antiferromagnetism (Kramers 1952a) and of a paper on phonons and rotons in liquid helium (Kramers 1952b).

5.1. Kramers's Contributions to the General Theory of Equilibrium Statistical Mechanics

One often finds the statement (see, for instance, Pathria 1972, p. 64, or Lifshitz and Pitaevskii 1980, p. 69) that Nernst's theorem, stating that the entropy of a system at equilibrium will vanish at absolute zero, is a result of the fact that at absolute zero a system will be in its ground state, which is non-degenerate, so that the entropy being proportional to the logarithm of the statistical weight of the state of the system will be equal to zero.[2]

---

[2] Even if the ground state were degenerate, its degree of degeneracy g would be independent of the number of constituents, N, so that the entropy, being equal to $k_B \ln g$ (k_B is Boltzmann's constant) would be negligibly small compared to the entropy at finite temperatures which is of the order of Nk_B.

It was pointed out by Kramers[3] that for macroscopic systems — for which Nernst's theorem is experimentally found to be satisfied — the spacing of the energy levels will be of the order of, say, 10^{-16} K, and there is therefore no question of the system being in its ground state at any experimentally accessible temperatures.[4] The reason that Nernst's theorem is satisfied must therefore be found elsewhere. Kramers pointed out that it is a consequence of the behaviour of the energy level density. The entropy S of a system can be expressed in terms of the energy level density as follows:[5]

$$S = k_B \ln \varrho(\varepsilon), \tag{5.001}$$

where $\varrho(\varepsilon)$ is the energy level density, that is, $\varrho(\varepsilon)\,d\varepsilon$ is the number of states for which the energy of the system ε lies between ε and $\varepsilon + d\varepsilon$; this energy is at the temperature considered a solution of the equation[6]

$$\frac{\partial}{\partial \varepsilon} \ln \varrho(\varepsilon) - \beta = 0, \tag{5.002}$$

where β is proportional to the inverse temperature:

$$\beta = \frac{1}{k_B T}. \tag{5.003}$$

Equation (5.001) can be derived as follows. The partition function Z of the system is given by the equation

$$Z = \sum_k e^{-\beta_k \varepsilon}, \tag{5.004}$$

where the sum is over all energy states of the system. If, as we are assuming, the distances between energy states are small as compared to $k_B T$, the sum in Eq. (5.004) can be replaced by an integral so that we have

$$Z = \int e^{-\beta \varepsilon} \varrho(\varepsilon)\, d\varepsilon, \tag{5.005}$$

or, taking into account that integrands of the kind we have here have a steep maximum,

$$Z = e^{-\beta \varepsilon} \varrho(\varepsilon) \Delta \varepsilon, \tag{5.006}$$

---

[3] This is mentioned by Casimir (1963) in a paper dedicated to Max Born on the occasion of his eightiethth birthday. I am indebted to Professor Casimir for first drawing my attention to Kramers's arguments, as long ago as in 1955.

[4] In fact, there are a very large number of states involved at these temperatures, as was pointed out by van Kampen (1954).

[5] We refer to Casimir's paper (Casimir 1963) and to a discussion I have given elsewhere (ter Haar 1966, Ch. 9) for more details of the arguments given in what follows.

[6] Equation (5.002) is essentially the thermodynamic equation $\beta = \partial[S/k_B]/\partial \varepsilon$.

where $\Delta\varepsilon$ is the width of the peak and where ε in Eq. (5.006) is the value of the energy at the maximum which is also the value of the energy at the temperature considered.[7]

At the same time we bear in mind that the entropy S is related to the partition function through the equation

$$S = k_B \ln Z + \frac{\varepsilon}{T} + C, \qquad (5.007)$$

where C is a constant. Neglecting C and $\ln \Delta\varepsilon$ we have Eq. (5.001). Eq. (5.002) is the condition for ε being the value of the energy at the maximum of the integrand.

The reason why Nernst's theorem can be proved to be satisfied for all models for which $\varrho(\varepsilon)$ can be calculated is that for all those models the energy level density satisfies either the relation

$$\ln \varrho(\varepsilon) = CN\varepsilon^p \qquad (5.008)$$

or the relation

$$\varrho(\varepsilon) = C\frac{\varepsilon}{\Delta}e^{\varepsilon/\Delta}, \qquad (5.009)$$

with C being a constant of order unity and the exponent p in Eq. (5.008) satisfying the inequality

$$\tfrac{1}{2} \leqslant p < 1. \qquad (5.010)$$

Using Eqs. (5.001) and (5.002) one can prove that, indeed, $S \to 0$ as $T \to 0$.

The proofs proceed as follows. From Eqs. (5.008), (5.002), and (5.001) we find

$$\varepsilon = C_1 \beta^{1/(p-1)}, \qquad S = C_2 \beta^{p/(p-1)}, \qquad (5.011)$$

where C_1 and C_2 are temperature-independent constants. Nernst's theorem follows from Eqs. (5.011) and (5.010).

Similarly, from Eqs. (5.009), (5.002), and (5.001) we have

$$\varepsilon = C_3 \Delta e^{-\beta\Delta}, \qquad S = C_4 \beta \Delta e^{-\beta\Delta}, \qquad (5.012)$$

where C_3 and C_4 are temperature-independent constants. Again, Nernst's theorem is satisfied.

It remains a curious — and so far unexplained — fact that all physical systems apparently have energy level densities that lead to Nernst's theorem being satisfied.[8]

---

[7] These relations all follow from the usual Gibbs relations; see, for instance, ter Haar 1995, Chs. 5 and 6.

[8] Casimir (1963) suggests that Kramers may, in fact, have intended to prove that this is the case.

Kramers's first paper (Kramers 1927a) in the field of statistical mechanics was concerned with the theory of strong electrolytes. Klein and he had studied these systems and essentially derived, but not published, the Debye-Hückel expression (1923) for the free energy of mixtures of ions. This expression is valid only in the limit of low concentrations, and was derived by Debye and Hückel by a clever trick which, however, prevented them from studying corrections to their result.

Kramers followed Gibbs and used ensemble theory to evaluate the free energy. Using a number of calculational tricks which nowadays are well known and widely used, but were relatively new in the 1920s, such as looking at the limit for large numbers of particles, he derived an expression for the free energy which in the limit of low concentrations reduced to the Debye-Hückel result but which showed for higher concentrations deviations from that result. He also showed that at those higher concentrations one must take into account the fact that the ions have a finite size.

In the same year Ornstein and Kramers (1927) used detailed balancing arguments to derive the expression for the Fermi-Dirac distribution function that had been found the previous year by Fermi and Dirac, who had used probability arguments to derive this formula.

Although Gibbs introduced grand ensembles in the last chapter of his famous monograph (Gibbs 1902), they have been widely applied only since the end of the war, when statistical mechanics became a more popular subject. We have mentioned that Kramers preferred the ensemble approach to the kinetic methods used by Boltzmann and preferred by Ehrenfest. In 1938 he published a paper (Kramers 1938a) essentially elaborating on Gibbs's treatment of grand ensembles and at the same time introducing some new concepts and showing how grand ensemble theory could be extended to cover quantum mechanical systems. The contents of this paper are now standard topics in statistical mechanics textbooks, so I shall not discuss them in any detail but only mention that it was Kramers who showed that the natural thermodynamic potential in the case of the grand ensembles is the so-called grand potential q which is the logarithm of the grand partition function,[9] $\mathcal{Z}$, and which in the case of a classical system is determined by the equation

$$e^q = \mathcal{Z} = \sum_{\nu_1} \cdots \sum_{\nu_r} \frac{e^{\alpha_1 \nu_1 + \cdots \alpha_r \nu_r}}{\nu_1! \cdots \nu_r!} \int_\Gamma e^{-\beta \varepsilon} \, d\Omega, \qquad (5.013)$$

where we are considering a system containing r kinds of particles, where ε is the energy of the system and $d\Omega$ an element of the phase space of the system containing ν_1 particles of kind 1, ..., ν_r particles of kind r, where the α_i are the logarithms of the absolute activities, where β is again given by Eq. (5.003), and where the integration is over the whole of Γ-space, that is,

---
[9] For details of grand ensemble theory see, for instance, ter Haar 1995, §§ 5.09 - 5.12 and 6.4.

phase space. In his paper Kramers proved that the grand potential satisfies the equation[10]

$$q = \frac{pV}{k_B T},\qquad(5.014)$$

where p is the pressure of the system represented by the grand ensemble and V and T are its volume and temperature.

5.2. Phase Transitions

Kramers was the first to realise the importance of the so-called thermodynamic limit in the theory of phase transitions. If one accepts that the partition function (5.013) will describe the equilibrium behaviour of a system, the question arises how this function, which looks like a continuous function of the parameters such as the volume V or the temperature T, can give rise to discontinuities in the thermodynamic functions such as the free energy. The answer to this question is hidden in Kramers's 1936 paper on the theory of ferromagnetism:[11] The discontinuities arise only in the thermodynamic limit, that is, when one takes the limit in which the number of particles in the system as well as its volume go to infinity while at the same time their ratio is kept constant. A year later, at the van der Waals Congress in Amsterdam, Kramers suggested this solution, and as there was no immediate acceptance (or rejection) of his statement he put it to a vote. Unfortunately there does not seem to be a record of the outcome of the vote—but nowadays, of course, Kramers's solution is universally accepted.

In the first of his two papers on ferromagnetism (Heller and Kramers 1934) Kramers is interested in constructing a classical model of a ferromagnet and in determining whether such a model will have a non-vanishing magnetic moment at low temperatures. He starts from the Heisenberg Hamiltonian of a ferromagnet,

$$\mathcal{H} = -\mu_B B \sum_i \sigma_{iZ} - \tfrac{1}{2} I \sum_{i<j} (\boldsymbol{\sigma}_i \cdot \boldsymbol{\sigma}_j),\qquad(5.015)$$

where B is the external magnetic field which we have assumed to be along the Z-axis, where the first sum is over all atoms in the system, which are assumed to have an angular momentum $\tfrac{1}{2}\hbar\boldsymbol{\sigma}_i$ and a magnetic moment $\mu_B \boldsymbol{\sigma}_i$

---

[10] In the paper discussed here, Kramers only considered the quantity $\Omega = k_B T q$, which is the negative of the quantity Ω introduced by Gibbs. The grand potential q itself was introduced by Kramers in lectures in the early 1940s.

[11] The crucial remarks can be found on pp. 18 and 20 of this paper, where Kramers states: "Our calculation requires only that we know the limit of $\ln V_m(\tau)/N$ as $N \to \infty$" ($V_m(\tau)$ is a function which directly determines the free energy of the system) and "The function $\ln V_m(\tau)/N$ thus consists of two analytically different parts."

with μ_B the Bohr magneton, and the second sum over all nearest neighbour pairs. We have assumed that only nearest neighbours interact and that the exchange integrals I which occur in their interaction energy are all the same.

The magnetic moment $\mathcal{M}$ of the system is given by the equation

$$\mathcal{M} = \frac{1}{\beta} \frac{\partial \ln Z}{\partial B}, \tag{5.016}$$

where Z is the partition function.

The three systems considered by Heller and Kramers are the linear chain, the square lattice, and the simple cubic lattice. They introduce their classical model by replacing the spin operators σ_i by c-number vectors $\mathbf{s}_i$ with length n and direction cosines X_i, Y_i, and Z_i. The Hamiltonian then becomes

$$\mathcal{H} = -\mu_B n B \sum_i Z_i - \tfrac{1}{2} I n^2 \mathcal{Q}, \tag{5.017}$$

where $\mathcal{Q}$ is a quadratic expression in the X_i, Y_i, and Z_i. The partition function Z is obtained by integrating $e^{-\beta \mathcal{H}}$ over the canonical variables of the system.

Even for the linear chain, the evaluation of the partition function is very complicated, and Heller and Kramers do not tackle the general problem but restrict themselves to the case of low temperatures, when we have $\beta I \gg 1$. In that case we may assume that almost all spins are practically along the Z-axis so that we can expand $\mathcal{H}$ in powers of the X_i and the Y_i and retain only the quadratic terms in those quantities. Moreover, the partition function can now in the $\beta I \to \infty$ limit be written as an integral over all the X_i and Y_i from $-\infty$ to $+\infty$ and be evaluated.

The important result obtained by Heller and Kramers is that in the case of a simple cubic lattice the magnetisation tends to a finite limit as $B \to 0$, which means that even a classical model can show ferromagnetism. On the other hand, they find that there is no spontaneous magnetisation in the cases of the linear chain or the square lattice.

The evaluation of the partition function proceeds in very much the same way for all three lattices considered by Heller and Kramers. We shall give some details for the case of the square lattice. If we assume it to lie in the x, y-plane,[12] we have for the quadratic function $\mathcal{Q}$

$$\begin{aligned}\mathcal{Q} = {} & X_{k,l} X_{k+1,l} + X_{k,l} X_{k,l+1} + Y_{k,l} Y_{k+1,l} \\ & + Y_{k,l} Y_{k,l+1} + Z_{k,l} Z_{k+1,l} + Z_{k,l} Z_{k,l+1},\end{aligned} \tag{5.018}$$

where the atoms are characterised by the pair of indices k, l which number their place in the x- and y-direction, respectively.

Using the fact that $X_i^2 + Y_i^2 + Z_i^2 = 1$ we find for $\mathcal{H}$

---

[12] The x-, y-, and z-axes may or may not coincide with the X-, Y-, and Z-axes.

74   Chapter 5

$$\mathcal{H} = -N^2 n(\tfrac{1}{2}nI + \mu_B B) + \tfrac{1}{2}n^2 I \sum_{k,l=1}^{N} [(2+\alpha)(X_{k,l}^2 + Y_{k,l}^2)$$
$$- X_{k,l}X_{k+1,l} - X_{k,l}X_{k,l+1} - Y_{k,l}Y_{k+1,l} - Y_{k,l}Y_{k,l+1}], \quad (5.019)$$

where

$$\alpha = \frac{\mu_B B}{nI}, \quad (5.020)$$

where we have assumed that there are N^2 atoms with N in both the x- and the y-direction, and where we have introduced periodic boundary conditions, that is, "closed" the lattice by assuming that

$$\left.\begin{array}{ll} X_{N+1,l} = X_{1,l}, & X_{k,N+1} = X_{k,1}, \\ Y_{N+1,l} = Y_{1,l}, & Y_{k,N+1} = Y_{k,1}. \end{array}\right\} \quad (5.021)$$

The partition function Z is given by the expression

$$Z = (\tfrac{1}{2}n\hbar)^{N^2} \int e^{-\beta\mathcal{H}} dX_{1,1} \cdots dX_{N,N} \, dY_{1,1} \cdots dY_{N,N}. \quad (5.022)$$

As for large values of β only small values of the X_i and Y_i contribute significantly to the integral, we may assume the integral to be from $-\infty$ to $+\infty$ for all X_i and Y_i.

Since $\mathcal{H}$ is, apart from a constant term, homogeneous quadratic in the X_i and Y_i, we can use the formula (Heller and Kramers 1934, Eq. (7))

$$\int e^{-\sum \alpha_{ij} x_i x_j} dx_1 \cdots dx_M = \prod_{\lambda=1}^{M} \sqrt{\frac{\pi}{\Delta_\lambda}}, \quad (5.023)$$

where the α_{ij} form a symmetric matrix with eigenvalues Δ_λ.

In the case of the square lattice the Δ_λ can, like the X_i and the Y_i, be characterised by a pair of indices, $\Delta_{\mu,\nu}$, and the eigenvalue problem in this case is of the form

$$\left(u_{k-1,l}^{\mu,\nu} + u_{k,l-1}^{\mu,\nu}\right) - 2(2+\alpha-\Delta_{\mu,\nu}) + \left(u_{k+1,l}^{\mu,\nu} + u_{k,l+1}^{\mu,\nu}\right) = 0, \quad (5.024)$$

where the $u_{i,j}^{\mu,\nu}$ are the components of the eigenvectors. The solution of the eigenvalue problem (5.024) is

$$\Delta_{\mu,\nu} = 2 + \alpha - \cos\frac{2\pi\mu}{N} - \cos\frac{2\pi\mu}{N}. \quad (5.025)$$

Using Eq. (5.023) we get for the logarithm of the partition function the expression

$$\ln Z = \beta n N^2 (nI + \mu_B B) + N^2 \ln \frac{\pi \beta \hbar}{nI}$$

$$- \left(\frac{N}{2\pi}\right)^2 \int\int \ln(2 + \alpha - \cos\xi - \cos\eta)\, d\xi\, d\eta, \qquad (5.026)$$

where we have replaced the sums over the μ and the ν by integrals over $\xi = 2\pi\mu/N$ and $\eta = 2\pi\nu/N$, which is justified in the case of large N.[13]

For the magnetisation we get from Eqs. (5.016), (5.020), and (5.026) for the square lattice

$$\mathcal{M} = N^2 n\mu_B - \left(\frac{N}{2\pi}\right)^2 \frac{\beta\mu_B}{nI} \int_0^{2\pi}\int_0^{2\pi} \frac{d\xi\, d\eta}{2 + \alpha - \cos\xi - \cos\eta}, \qquad (5.027)$$

while for the simple cubic lattice the result, obtained by similar calculations, is

$$\mathcal{M} = N^3 n\mu_B - \left(\frac{N}{2\pi}\right)^3 \frac{\beta\mu_B}{nI} \int \frac{d\xi\, d\eta\, d\zeta}{3 + \alpha - \cos\xi - \cos\eta - \cos\zeta}. \qquad (5.028)$$

To see whether there is ferromagnetism we must take the limit as $\alpha \to 0$ and see whether $\mathcal{M}$ tends to a finite positive value in that limit. In the case of the square lattice—and of the linear chain—$\mathcal{M}$ tends to $-\infty$, indicating the absence of ferromagnetism,[14] while in the case of the simple cubic lattice we find for $\mathcal{M}$ up to the first non-vanishing correction in the temperature T (a is a constant)

$$\mathcal{M} \cong N^3 n\mu_B (1 - aT), \qquad (5.029)$$

which must be compared with the expression

$$\mathcal{M} \cong N^3 n\mu_B (1 - a'T^{1/2}) \qquad (5.030)$$

for the Heisenberg ferromagnet.

In an interesting last section Heller and Kramers introduce quantisation for their model at low temperatures where the Hamiltonian (5.019) holds. In that case $\sqrt{\frac{1}{2}n\hbar}X_i$ and $\sqrt{\frac{1}{2}n\hbar}Y_i$ are canonically conjugate and the Hamiltonian is that of a set of harmonic oscillators which can easily be quantised. The eigenvalues correspond to the eigenfrequencies of the oscillators and the oscillators themselves are the so-called spin waves.

Kramers's second paper on ferromagnetism (Kramers 1936a) is well hidden in the *Communications of the Kamerlingh Onnes Laboratorium*. In it

---

[13] There are small misprints in the formulae for $\Delta_{\mu,\nu}$ and ξ and η in the original paper.

[14] In his second paper on ferromagnetism Kramers uses a similar argument to reach the conclusion that the one- or two-dimensional Heisenberg ferromagnets show no magnetisation in the absence of a magnetic field.

he considers the partition function of the Heisenberg ferromagnet for the case of simple, body-centered, and face-centered cubic lattices. In this paper Kramers proposes a method to evaluate the partition function of a system, the energy of which is bounded both from below and from above. He then applies this method to the Heisenberg model of ferromagnetism. In fact, his main interest seems to be to give series expansions of the free energy of the Heisenberg ferromagnet at low and at high temperatures, both when there is no magnetic field present and in the presence of a magnetic field.

Kramers starts this paper with a general discussion of the partition function of a homogeneous system of a very large number, N, of particles:

$$Z = e^{-\beta NF} = \int e^{-\beta \varepsilon}, \tag{5.031}$$

where F and ε are, respectively, the free energy per particle and the energy (Hamiltonian) of the system, and where the integration in the case of a quantum mechanical system must be replaced by a summation over all energy eigenvalues.

If we expand the exponential we have

$$e^{-\beta NF} = \sum_n \frac{\beta^n}{n!} A_N(n), \tag{5.032}$$

with the positive quantities $A_N(n)$ given by the equation

$$A_N(n) = \int (-\varepsilon)^n. \tag{5.033}$$

We know from the general theory of statistical mechanics that the free energy and the other properties of the system at equilibrium are determined by the largest term in the series (5.032).

If we write the largest term in the series (5.032) in the form

$$\ln A_N(n) = N(K(\nu) + \nu \ln N), \tag{5.034}$$

with

$$n = \nu N, \tag{5.035}$$

and from now on assume all quantities to refer to the largest term in (5.032), we find for the free energy the expression

$$-\beta F = K(\nu) - \nu(\ln \nu - 1) + \nu \ln \beta, \tag{5.036}$$

while the condition that we are dealing with the largest term gives us for K the relation

$$\frac{\partial K}{\partial \nu} - \ln \nu = -\ln \beta. \tag{5.037}$$

From Eqs. (5.036) and (5.037) and the fact that F is independent of N it follows that K is independent of N so that the value of ν corresponding to the largest term is also independent of N.

If E is the energy of the system per particle, it follows from the Gibbs-Helmholtz equation

$$E = \frac{\partial \beta F}{\partial \beta} \tag{5.038}$$

that ν is satisfies the relation

$$\nu = -\beta E. \tag{5.039}$$

The free energy per particle, F, and entropy per particle, S, can be expressed in terms of K and its derivative K' with respect to ν:

$$-\beta F = K - \nu K' + \nu, \tag{5.040}$$

and

$$S = \beta(E - F) = K - \nu K'. \tag{5.041}$$

Reversely, K and $A_N(n)$ can be expressed in terms of E and S:

$$K = S - \beta E \ln(-E), \tag{5.042}$$

and

$$\ln A_N(n) = NS - \beta NE \ln(-NE). \tag{5.043}$$

In the remainder of the paper Kramers considers the following cases of a Heisenberg ferromagnet with spin-$\frac{1}{2}$ particles on a cubic lattice: high and low temperatures both without and with a magnetic field being present. In each case the relevant $A_N(n)$ is evaluated using combinatorial methods, thus applying this technique before it became such a common practice in the hands of Mayer, Kirkwood, and countless other researchers, especially once Feynman diagrams became the vogue after the war. The Hamiltonian is again given by Eq. (5.015), which through a particular choice of units and of the origin of the energy can be written in the form

$$\varepsilon = -a \sum_k \sigma_{kz} - \sum_{kl} (kl), \tag{5.044}$$

with

$$a = \frac{\mu_B B}{I} \tag{5.045}$$

and

$$(kl) = \tfrac{1}{2}(1 + (\sigma_k \cdot \sigma_l)). \tag{5.046}$$

78   Chapter 5

The second sum in Eq. (5.044) is again over nearest neighbour pairs and the components of the σ_k are now the Pauli matrices (3.036).

The task now is to find a general expression for the $A_N(n)$, and then to find its value corresponding to the largest term. The $A_N(n)$ are given by the expression

$$A_N(n) = \sum_\varphi \int \varphi^*(-\varepsilon)^n \varphi, \qquad (5.047)$$

where ε is here the operator given by Eq. (5.044), where φ is a wavefunction describing a particular spin distribution, and where the sum is over the 2^N possible spin distributions.[15] One can prove that Eq. (5.047) leads to the following expression for $A_N(n)$:

$$A_N(n) = \sum_\varphi \sum_l \{N'_\varphi + (N - 2m)a\}^{n-l} \binom{n}{l} s_\varphi^{(l)}, \qquad (5.048)$$

where m is the number of down-spins in φ, N'_φ is the number of parallel nearest neighbours in φ, and $s_i^{(l)}$ is the number of states, obtained from φ by replacing one down-spin l times, which are the same as φ.

For the case where there is no magnetic field ($a = 0$) we can prove Eq. (5.048) as follows. If in the φ configuration the kth and the lth spins are parallel ($\uparrow\uparrow$ or $\downarrow\downarrow$) we have

$$(kl)\varphi = \varphi, \qquad (5.049)$$

whereas if they are antiparallel ($\uparrow\downarrow$ or $\downarrow\uparrow$) we have

$$(kl)\varphi = \varphi^{(1)}, \qquad (5.050)$$

where $\varphi^{(1)}$ is obtained from φ by the kth and lth spins having changed their directions.

Hence it follows that

$$-\varepsilon\varphi = N'_\varphi \varphi + \sum \varphi^{(1)}, \qquad (5.051)$$

and

$$(-\varepsilon)^2 \varphi = {N'_\varphi}^2 \varphi + (N'_\varphi + N''_\varphi) \sum \varphi^{(1)} + \sum \varphi^{(2)}, \qquad (5.052)$$

where N''_φ is the average number of parallel nearest neighbours in $\varphi^{(1)}$ which we may put equal to N'_φ, and $\varphi^{(2)}$ is obtained from φ by a down-spin twice having exchanged directions with a nearest neighbour. Finally we get

---

[15] The integral is over the arguments of φ, that is, it is a symbolic indication that $\int \varphi^* \varphi'$ vanishes unless the spin distributions corresponding to φ and φ' are the same.

$$(-\varepsilon)^n \varphi = \sum_l N_\varphi'^{\,n-l} \binom{n}{l} \sum \varphi^{(l)}; \tag{5.053}$$

the last sum contains $\overline{N_\varphi}^{\,l}$ terms where $\overline{N_\varphi}$ is the number of antiparallel pairs in the state φ, which in the case when there is no magnetic field is equal to $\frac{1}{2}zN - N_\varphi'$ with z the coordination number, that is, the number of nearest neighbours per spin. Equation (5.048) follows directly from Eq. (5.053), if we further bear in mind that in the case where there is a magnetic field present we must add a term $(N-2m)a\varphi$ to the right-hand side of Eq. (5.051).

Kramers uses Eq. (5.048) to find approximations for the thermodynamic functions at high and at low temperatures. At high temperatures when there is no magnetic field we find for the free energy F and the magnetisation σ $(=1-2m/N)$

$$F = -\tfrac{1}{4}z - \frac{\ln 2}{\beta} - \frac{3z}{16}\beta + \cdots, \qquad \sigma = 0, \tag{5.054}$$

and we find in the case when there is a magnetic field at high temperatures for the magnetisation the implicit equation

$$\sigma = \tanh\left[\beta(a + \tfrac{1}{2}z\sigma) - \tfrac{4}{z}\beta^2\sigma(2-\sigma^2) + \cdots\right]. \tag{5.055}$$

At low temperatures we find for the magnetisation when there is no magnetic field

$$\sigma = 1 - CT^{3/2} + \cdots, \tag{5.056}$$

where C is a constant, and when there is a magnetic field

$$\sigma = 1 - CT^{3/2}e^{-2\beta a} + \cdots. \tag{5.057}$$

We shall sketch some of the details of the calculations for the case when there is no magnetic field. The case when there is a magnetic field present is similar, albeit more complicated. In the case of high temperatures we are dealing with states φ in which the number of up- and down-spins is practically the same and the value of ν corresponding to the largest term is small. The main contribution to $s_\varphi^{(l)}$ comes from those cases where a down-spin moves its site $\tfrac{1}{2}l$ times and then retraces its path. This leads to the following expression for $s_\varphi^{(l)}$:

$$s_\varphi^{(l)} = \binom{\tfrac{1}{2}zN - N_\varphi'}{\tfrac{1}{2}l} \frac{l!}{2^{l/2}}. \tag{5.058}$$

Moreover, we may replace the sum over φ by 2^N. To find the largest term in the series for the free energy we still need to find the mean value of $N_\varphi'^{\,l}$. The probability to find N_φ' pairs of parallel spins is

$2^{-zN/2}\binom{\frac{1}{2}zN}{N'_\varphi}$ so that the average value of N'^{p}_{φ} will be given by the equation

$$\langle N'^{p}_{\varphi}\rangle = 2^{-zN/2}\sum_{N'_\varphi}\binom{\frac{1}{2}zN}{N'_\varphi}N'^{p}_{\varphi}. \qquad (5.059)$$

The sum in this equation can be replaced by its largest term. Moreover, in the sum over l in Eq. (5.048) the largest term will occur for relatively small values of l, which makes the calculation of $N'^{\,l}_{\varphi}$ easier. Taking the largest term in the sum in Eq. (5.059) and the largest term in the sum in Eq. (5.048), one finally finds

$$A_N(n) = 2^N \left(\tfrac{1}{4}zN\right)^n e^{3n^2/zN}, \qquad (5.060)$$

or

$$\ln A_N(n) = N\left(\ln 2 + \nu \ln(\tfrac{1}{4}zN) + \frac{3\nu^2}{z}\right). \qquad (5.061)$$

Using Eqs. (5.061), (5.034), and (5.040) we then find Eq. (5.054).

At low temperatures we must consider states for which the number m of down-spins is very small compared to N, which means that in general the distance between down-spins will be large and hence that N'_φ can be put equal to $\tfrac{1}{2}zN - zm$ in the expression for $A_N(n)$. Since there are $\binom{N}{m}$ ways of putting the m down-spins in the lattice we must look for

$$A_N(n) = \text{Max}\left\{\binom{N}{m}\left(\tfrac{1}{2}zN - zm\right)^{n-l}\binom{n}{l}s^{(l)}_\varphi\right\}, \qquad (5.062)$$

where the maximum is with respect to m and l for fixed n.

To find $s^{(l)}_\varphi$ we note that of l shifts on average l/m shifts refer to down-spins and one must therefore find the probability that after l/m shifts a down-spin has returned to its original position. This probability can be found from the theory of the random walk problem and is, for a simple cubic lattice, equal to $(2\pi l/3m)^{-3/2}$.[16] We thus find for $s^{(l)}_\varphi$ the expression[17]

$$s^{(l)}_\varphi = (zm)^l \left(\frac{2\pi l}{3m}\right)^{-3m/2} \qquad (5.063)$$

---

[16] For the other cubic lattices the probability is the same, except for a numerical factor of order unity.

[17] We have neglected here those $\varphi^{(l)}$, which are identical with the original φ because some down-spins have swapped places. These terms are important and Kramers considers them in detail. In fact, if one neglects them one would come to the erroneous conclusion that the one- and two-dimensional Heisenberg ferromagnets would show spontaneous magnetisation at low temperatures.

Substituting Eq. (5.063) into expression (5.062), minimising with respect to l and m, and taking into account that $l/n \ll 1$, we find for the maximum of $A_N(n)$, in the case of the simple cubic lattice,

$$A_N(n) = (\tfrac{1}{2}zN)^{N\nu} e^{N(4\pi\nu/3)^{-3/2}}, \qquad (5.064)$$

whence follows Eq. (5.056).

Kramers's final contribution to equilibrium statistical mechanics, and arguably his most important one, are the two papers with Wannier on the two-dimensional Ising problem (Kramers and Wannier 1941a, b), the first of which is reprinted at the end of this book as paper J. In this case instead of expression (5.015) we have the Hamiltonian

$$\mathcal{H} = -\mu_B B \sum_i \mu_i - \tfrac{1}{2} I \sum_{i,j} \mu_i \mu_j, \qquad (5.065)$$

where the spins are either up, that is, parallel to the magnetic field, corresponding to $\mu_i = 1$, or down, corresponding to $\mu_i = -1$, and where the second sum is over all nearest-neighbour pairs.

The partition function can be written in the form

$$Z = \sum_{\mu_i = \pm 1} e^{K \sum_{i,j} \mu_i \mu_j + C \sum_i \mu_i}, \qquad (5.066)$$

where the first sum is over all 2^N possible spin configurations (N is the total number of spins in the lattice), and where K and C are given by the equations

$$K = \tfrac{1}{2}\beta I, \qquad C = \beta \mu_B B. \qquad (5.067)$$

The magnetisation $\mathcal{M}$ and the energy E are given by the equations

$$\mathcal{M} = m \frac{\partial \ln Z}{\partial C}, \qquad (5.068)$$

and

$$E = -\mathcal{M}B - \tfrac{1}{2} I \frac{\partial \ln Z}{\partial K}. \qquad (5.069)$$

The method used by Kramers and Wannier to find the partition function consists in building up the lattice step by step. In their first paper they show how the so-called transfer matrix method works in the cases of the linear chain and of the square lattice. It is this method which is the basis of practically all later research in the field of cooperative phenomena, including Onsager's exact solution (Onsager 1944) of the two-dimensional Ising model.[18]

---

[18]It seems (see, for instance, Newell and Montroll 1953, §2.1) that the transfer matrix method was discovered independently by Kramers and Wannier, Montroll (1941), Lassettre and Howe (1941), and Kubo (1943), but it was Kramers and Wannier who set up a form of operator algebra that was used by Onsager (1944) to find his exact solution of the two-dimensional Ising problem.

In the case of the square lattice, the method starts by successively adding one more column (Fig. 5.1): consider a two-dimensional lattice where each column contains a spins and where columns are added until the lattice contains $a \times b \, (= N)$ spins.

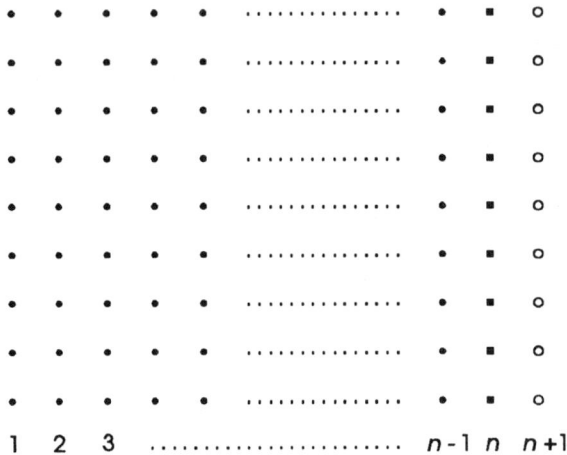

Fig. 5.1. Adding one more column to a square lattice.

We denote the spins of the nth column by μ_i $(i = 1, 2, \ldots, a)$ and those of the $n-1$st column by μ'_i $(i = 1, 2, \ldots, a)$. Let $P(\mu'_i)$ be the probability that the a spins in the $n-1$st column and $P(\mu_i)$ that the spins in the nth column have given values. These probabilities will be proportional to the Boltzmann factor $e^{-\beta \mathcal{H}}$. Using Eq. (5.065) for $\mathcal{H}$ we then find that

$$P(\mu'_i) = p' \exp\left(K \sum_{i,j} \mu_i \mu_j\right), \tag{5.070}$$

where p' is a normalising constant and where the summation is over all nearest-neighbour pairs in the lattice with $n-1$ columns. Including one more column (see Fig. 5.1) we have

$$P(\mu'_i, \mu_i) = \frac{p}{p'} P(\mu'_i) \exp\left[K \sum_{i=1}^{a-1} \mu'_i \mu'_{i+1} + K \sum_{i=1}^{a} \mu_i \mu'_i\right], \tag{5.071}$$

where $P(\mu'_i, \mu_i)$ is the probability that the spins in the $n-1$st and the nth column all have given values.

From Eq. (5.071) we get, by summing over all possible configurations of the $n-1$st column, the following equation for $P(\mu_i)$

$$P(\mu_i) = \sum_{\mu'_i = \pm 1} \frac{p}{p'} P(\mu'_i) \exp\left[K \sum_{i=1}^{a-1} \mu'_i \mu'_{i+1} + K \sum_{i=1}^{a} \mu_i \mu'_i\right]; \tag{5.072}$$

if we take into account that adding an extra column does not alter the physical situation, so that $P(\mu_i)$ and $P(\mu'_i)$ are the same functions of their arguments, we get an equation relating $P(\mu_n)$ to $P(\mu_{n-1})$, or, since these two probabilities should for a long chain be the same functions of their arguments, a matrix equation for $P(\mu)$:

$$\rho P(\mu_i) = \sum_{\mu'_i = \pm 1} \mathbf{K}(\mu_i, \mu'_i) P(\mu'_i), \tag{5.073}$$

where ρ appears due to normalisation and where $\mathbf{K}(\mu_i, \mu'_i)$ is given by the equation

$$\mathbf{K}(\mu_i, \mu'_i) = \exp\left[K \sum_{i=1}^{a-1} \mu'_i \mu'_{i+1} + K \sum_{i=1}^{a} \mu_i \mu'_i\right]. \tag{5.074}$$

Introducing $A(\mu_i)$ by the equation

$$A(\mu_i) = P(\mu_i) \exp\left(\frac{1}{2} K \sum \mu_i \mu_{i+1}\right), \tag{5.075}$$

we have can bring Eq. (5.073) into symmetric form:

$$\rho A(\mu_i) = \sum_{\mu'_i = \pm 1} \mathbf{T}(\mu_i, \mu'_i) A(\mu'_i), \tag{5.076}$$

where we have now introduced the so-called *transfer matrix*

$$\mathbf{T}(\mu_i, \mu'_i) = \exp\left\{K\left[\sum \mu_i \mu'_i + \frac{1}{2}\sum \mu_i \mu_{i+1} + \sum \mu'_i \mu'_{i+1}\right]\right\}. \tag{5.077}$$

The transfer matrix $\mathbf{T}$ is of order 2^a and if we expand it in terms of its eigenvectors one can show that the partition function Z is determined by the eigenvalues ρ_j of $\mathbf{T}$ in the following way:

$$Z (= \mathrm{Tr}\, \mathbf{T}^b) = \sum_{j=1}^{2^a} \rho_j^b, \tag{5.078}$$

where we have imposed a periodicity condition $(\mu_i^{(1)} = \mu_i^{(b+1)})$ on the lattice.

If b is sufficiently large, only the largest eigenvalue will contribute and we have

$$Z = \rho_{\max}^b. \tag{5.079}$$

To see in detail how the transfer matrix method works we shall consider the case of a linear chain in some detail although, of course, the partition function in that case can be evaluated by much simpler methods. In this case we shall assume that there is a magnetic field present.

The probability $P(\mu_1, \ldots, \mu_{n-1})$ that the spins $\mu_1, \mu_2, \ldots, \mu_{n-1}$ of a chain of $n-1$ members will have given values will be proportional to the Boltzmann factor $e^{-\beta\mathcal{H}}$. Using Eq. (5.065) for $\mathcal{H}$ we then find that

$$P(\mu_1, \ldots, \mu_{n-1}) = p_{n-1}\, e^{K(\mu_1\mu_2 + \cdots + \mu_{n-2}\mu_{n-1}) + C(\mu_1 + \cdots + \mu_{n-1})}, \qquad (5.080)$$

where p_{n-1} is a normalising constant. Including one more spin in the chain (see Fig. 5.2) we have

$$P(\mu_1, \mu_2, \ldots, \mu_n) = \frac{p_n}{p_{n-1}}\, P(\mu_1, \mu_2, \ldots, \mu_{n-1})\, e^{K\mu_{n-1}\mu_n + C\mu_n}. \qquad (5.081)$$

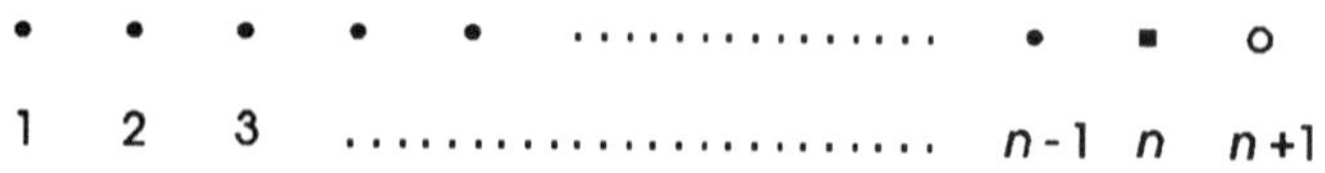

Fig. 5.2. Adding one more spin to a linear chain.

From Eqs. (5.080) and (5.081) we can obtain the probabilities $P(\mu_{n-1})$ and $P(\mu_{n-1}, \mu_n)$ that, respectively, μ_{n-1} and both μ_{n-1} and μ_n have given values, independent of the values of the $n-2$ other spins. For those probabilities we have

$$P(\mu_{n-1}) = \sum_{\mu_1=\pm 1}\cdots\sum_{\mu_{n-2}=\pm 1} P(\mu_1, \mu_2, \ldots, \mu_{n-1}), \qquad (5.082)$$

and

$$P(\mu_{n-1}, \mu_n) = \sum_{\mu_1=\pm 1}\cdots\sum_{\mu_{n-2}=\pm 1} \frac{p_n}{p_{n-1}}$$
$$\times P(\mu_1, \ldots, \mu_{n-1})\, e^{K\mu_{n-1}\mu_n + C\mu_n}, \qquad (5.083)$$

or

$$\lambda P(\mu_{n-1}, \mu_n) = P(\mu_{n-1})\, e^{K\mu_{n-1}\mu_n + C\mu_n}, \qquad (5.084)$$

with $\lambda = p_{n-1}/p_n$.

Summing both sides of Eq. (5.084) over the two possible values of μ_{n-1}, we get the probability $P(\mu_n)$ that μ_n has a given value irrespective of the values of the $n-1$ preceding spins, that is,

$$\lambda P(\mu_n) = \sum_{\mu_{n-1}=\pm 1} P(\mu_{n-1})\, e^{K\mu_{n-1}\mu_n + C\mu_n}. \qquad (5.085)$$

If our chain is sufficiently long $P(\mu_n)$ and $P(\mu_{n-1})$ should be the same function of their argument. We can consider the $P(\mu)$ to be spinors, that is, one by two matrices. Equation (5.085) is then a matrix equation and, introducing the new spinor $A(\mu)$ through the relation

$$A(\mu) = P(\mu)\,e^{-\frac{1}{2}C\mu}, \tag{5.086}$$

we can write it in the form

$$\lambda A(\mu) = \sum_{\mu'=\pm 1} \mathbf{T}(\mu,\mu')\,A(\mu'), \tag{5.087}$$

where the transfer matrix $\mathbf{T}(\mu,\mu')$ is a two by two matrix whose elements are given by the equation

$$\mathbf{T}(\mu,\mu') = e^{K\mu\mu' + \frac{1}{2}C(\mu+\mu')}. \tag{5.088}$$

Written out in all detail, Eq. (5.087) looks as follows:

$$\lambda \begin{pmatrix} A(+1) \\ A(-1) \end{pmatrix} = \begin{pmatrix} e^{K+C} & e^{-K} \\ e^{-K} & e^{K-C} \end{pmatrix} \begin{pmatrix} A(+1) \\ A(-1) \end{pmatrix}. \tag{5.089}$$

Equation (5.087) has the form of a matrix eigenvalue equation. Let us examine the importance of the eigenvalues of $\mathbf{T}$. We denote the orthonormalised eigenvectors of $\mathbf{T}$ by $A_1(\mu)$ and $A_2(\mu)$ and the corresponding eigenvalues by λ_1 and λ_2. Since $\mathbf{T}(\mu,\mu')$ is a function of both μ and μ', we can expand it in the form

$$\mathbf{T}(\mu,\mu') = \sum_{i,j=1}^{2} c_{ij} A_i(\mu) A_j(\mu'), \tag{5.090}$$

where the c_{ij} are given by the equations

$$c_{ij} = \sum_{\mu=\pm 1} \sum_{\mu'=\pm 1} \mathbf{T}(\mu,\mu') A_i(\mu) A_j(\mu'). \tag{5.091}$$

Since the $A_i(\mu)$ are the orthonormalised eigenvectors of $\mathbf{T}$ we have

$$\sum_{\mu'=\pm 1} \mathbf{T}(\mu,\mu')\,A_j(\mu') = \lambda_j A_j(\mu), \tag{5.092}$$

and

$$\sum_{\mu=\pm 1} A_i(\mu)\,A_j(\mu) = \delta_{ij}, \tag{5.093}$$

where δ_{ij} is the Kronecker symbol.

Using these relations we find from Eq. (5.091) that $c_{ij} = \lambda_i\,\delta_{ij}$ and hence that

$$\mathbf{T}(\mu,\mu') = \lambda_1 A_1(\mu)A_1(\mu') + \lambda_2 A_2(\mu)A_2(\mu'). \tag{5.094}$$

It now easily follows from the orthonormality relation (5.093) and Eq. (5.094) that we have

$$\sum_{\mu_2=\pm 1} \mathbf{T}(\mu_1,\mu_2)\,\mathbf{T}(\mu_2,\mu_3)$$
$$= \lambda_1^2 A_1(\mu_1)A_1(\mu_3) + \lambda_2^2 A_2(\mu_1)A_2(\mu_3), \tag{5.095}$$

$$\sum_{\mu_2=\pm 1}\sum_{\mu_3=\pm 1} \mathbf{T}(\mu_1,\mu_2)\,\mathbf{T}(\mu_2,\mu_3)\,\mathbf{T}(\mu_3,\mu_4)$$
$$= \lambda_1^3 A_1(\mu_1)A_1(\mu_4) + \lambda_2^3 A_2(\mu_1)A_2(\mu_4), \tag{5.096}$$

and so on, until

$$\sum_{\mu_2,\ldots,\mu_N=\pm 1} \mathbf{T}(\mu_1,\mu_2)\,\mathbf{T}(\mu_2,\mu_3)\cdots\mathbf{T}(\mu_N,\mu_{N+1})$$
$$= \lambda_1^N A_1(\mu_1)A_1(\mu_{N+1}) + \lambda_2^N A_2(\mu_1)A_2(\mu_{N+1}). \tag{5.097}$$

Assuming that $\mu_1 = \mu_{N+1}$, which corresponds either to closing the chain to form a ring or to imposing a periodicity condition on the chain, we can sum Eq. (5.097) over μ_1 and obtain

$$\sum_{\mu_1,\ldots,\mu_N=\pm 1} \mathbf{T}(\mu_1,\mu_2)\,\mathbf{T}(\mu_2,\mu_3)\cdots\mathbf{T}(\mu_N,\mu_1)$$
$$\left(= \operatorname{Tr}\mathbf{T}^N\right) = \lambda_1^N + \lambda_2^N. \tag{5.098}$$

From Eqs. (5.089) and (5.066) it follows that the left-hand side of Eq. (5.098) is just the partition function Z of a chain of N spins.

We have thus proven that the partition function of the one-dimensional Ising ferromagnet of N spins in a magnetic field is given by the equation

$$Z = \lambda_1^N + \lambda_2^N, \tag{5.099}$$

where the λ_i are the eigenvalues of the matrix (5.088). If N is sufficiently large and if $|\lambda_2| < |\lambda_1|$ we can write Eq. (5.099) in the form[19]

$$Z = \lambda_1^N. \tag{5.100}$$

The eigenvalues of $\mathbf{T}$ are found from the equation

---

[19] The various thermodynamic quantities that one calculates using the partition function involve $\ln Z$, and in going over from Eq. (5.099) to (5.100) one neglects terms of order $(\lambda_2/\lambda_1)^N$ as compared to the main term $N\ln\lambda_1$.

$$\begin{vmatrix} e^{K+C} - \lambda & e^{-K} \\ e^{-K} & e^{K-C} - \lambda \end{vmatrix} = 0, \tag{5.101}$$

or

$$\lambda_{1,2} = e^K \cosh C \pm \sqrt{e^{2K} \sinh^2 C + e^{-2K}}. \tag{5.102}$$

In the case when there is no magnetic field present, so that $C = 0$, we find for Z

$$Z = (2 \cosh K)^N, \tag{5.103}$$

a result which could also have been obtained from directly summing the partition function of the one-dimensional Ising ferromagnet.

Using Eqs. (5.100), (5.102), and (5.068) we see that if there is no magnetic field the magnetisation is always equal to zero. This corresponds to the well-known fact that a one-dimensional lattice will not show ferromagnetism. The reason is that one wrong spin will completely upset any tendency for long-range order, since there is no way for the later spins to find out whether or not they are in step with the earlier parts of the chain. The situation is different in the two-dimensional case, since then each spin is connected with all the other spins in the lattice in a multitude of ways and not only through a single nearest neighbour.

In their first paper Kramers and Wannier also determined the exact position of the transition temperature using the symmetry properties of the transfer matrix. If we introduce a variable K^* through the equation

$$\sinh 2K \sinh 2K^* = 1, \tag{5.104}$$

and put

$$K^* = \tfrac{1}{2}\beta^* I, \tag{5.105}$$

we have associated with each temperature T (or β) another temperature T^* (or β^*). We see that as T increases from 0 to ∞, T^* decreases from ∞ to 0.

For the case of the two-dimensional Ising ferromagnet on a square lattice without a magnetic field, Kramers and Wannier show that

$$\frac{Z(T)}{(\cosh 2K)^N} = \frac{Z(T^*)}{(\cosh 2K^*)^N}, \tag{5.106}$$

which relates the partition function at temperature T to its value at temperature T^*. This means that critical temperatures must occur in pairs, or, if there is just one critical temperature, T_c, it must occur for a value of K_c determined by the equation

$$\sinh 2K_c = 1, \tag{5.107}$$

which fixes the position of T_c. Of course, this does not prove that there exists, in fact, a transition temperature.

Using the symmetry properties of the transfer matrix Kramers and Wannier also find that there can be no jump in the specific heat at the transition temperature. In their second paper they develop a powerful, variational approximation method to find power series solutions at high and at low temperatures. They also conclude from numerical solutions that the specific heat is, in fact, infinite at the transition temperature.

5.3. Solid-State and Low-Temperature Physics

Becquerel showed that uniaxial crystals with rare earth ions, when placed in a magnetic field along the crystal axis, showed at low temperatures a strong rotation of the plane of polarisation of light propagating along the magnetic field due to the paramagnetic ions. The effect increases strongly with decreasing temperature and shows saturation. For tysonite (a lanthanum and cerium fluoride) the angle of rotation ϱ can be described by the formula

$$\varrho = \varrho_\infty(T) \tanh(n\beta\mu_B B), \tag{5.108}$$

where n is a constant which for tysonite is equal to 1 within the experimental accuracy, and where the saturation value $\varrho_\infty(T)$ is temperature-dependent.

In the case of xenotime (an yttrium, erbium, and gadolinium phosphate) the experimental data obtained by Becquerel and de Haas show deviations from Eq. (5.108) and it was originally suggested that they might be represented by a formula of the kind

$$\varrho = \varrho_\infty(T) \frac{n\mu_B B}{\sqrt{(n\mu_B B)^2 + K^2}} \tanh\left(\beta\sqrt{(n\mu_B B)^2 + K^2}\right), \tag{5.109}$$

with $n = 7$.

From the behaviour of absorption lines produced by rare earth crystals, Becquerel concluded that the metallic ions in those crystals were subject to a strong internal axially symmetric electric field with its axis along the crystal axis. Becquerel asked Kramers to explain the experimental data, and this resulted in a series of papers in which the theory was gradually developed, and in which Kramers considered the paramagnetic susceptibility of rare-earth crystals as well as their paramagnetic rotation. The first three papers in this series (Kramers 1929f, Kramers and Becquerel 1929, Becquerel, de Haas, and Kramers 1929) belong together. In the first of these Kramers announced the Kramers degeneracy theorem without proving it, and considered the case of an axially symmetric electric field resulting in a quadratic potential energy for the $4d$ electrons of the trivalent rare-earth ions. This potential will split the degenerate energy level of the ground state of the free ion, with the lowest level of the resulting multiplet being split further by the applied magnetic field. One of the two components of the Zeeman doublet will produce left-handed and the other right-handed rotation of the polarisation of light propagating along the magnetic field, and since their relative populations will be governed by the Boltzmann factors $e^{\pm\alpha}$ ($\alpha = \beta g|m|\mu_B B$, where g is the Landé factor and m the magnetic quantum number), we get for the resulting rotation of the polarisation plane

$$\varrho = \varrho_\infty \frac{e^\alpha - e^{-\alpha}}{e^\alpha + e^{-\alpha}} = \varrho_\infty \tanh(\beta g|m|\mu_B B), \tag{5.110}$$

in agreement with Eq. (5.108). In the second paper of the series Kramers and Becquerel compare this result with the experimental data on tysonite and, assuming that the paramagnetic rotation is due to the Ce^{+++} ions, find good agreement between theory and experiment.

As to the experimental xenotime data, reported in the third paper by Becquerel, de Haas, Kramers, in his first paper Kramers suggests that the difference between Eqs. (5.108) and (5.109) is due to the fact that the potential is more complicated than a simple quadratic one. In the second paper Kramers and Becquerel find good agreement with the theory if they assume that the rotation is due to the Gd^{+++} ions and Eq. (5.109) is used.

A year later Kramers returns to this class of problems. As the earlier work still left a number of features unexplained, Kramers develops in his fourth paper on the subject (Kramers 1930d, reprinted as paper E at the end of the book) a general phenomenological theory of paramagnetic rotation in crystals. He gives the proof of the degeneracy theorem and of the fact that non-degenerate stationary states are always non-magnetic. Concentrating on atoms with an odd number of electrons placed in an internal electric field the ground state of which shows the smallest possible degeneracy, that is, a two-fold degeneracy, he calculates the magnetisation produced by an external magnetic field and derives again Eq. (5.108) for the case of a weak magnetic field. He now rejects the earlier explanation of Eq. (5.109) for xenotime and suggests that the deviation from Eq. (5.108) may be due to magnetic interactions between the atoms, leading to an extra magnetic field, proportional to the magnetisation, acting on the ions. This will lead to the following implicit formula for ϱ/ϱ_∞:

$$\varrho = \varrho_\infty(T) \tanh\left(n\beta\mu_B \left[B - a\frac{\varrho}{\varrho_\infty}\right]\right), \tag{5.111}$$

which Kramers suggests may equally well represent the experimental data for xenotime.

Further experiments on rare-earth crystals showed that the theory in the form proposed by Kramers in his first paper was deficient, especially in its assumption that the potential could be represented by a quadratic term. This assumption had not been made in the third paper, but in that paper there was no comparison between theory and experimental data. Moreover, in deriving a formula for the paramagnetic rotation, Kramers had, in first instance, restricted himself to terms linear in B, and Van Vleck (1932) had pointed out that in certain cases the B^2 terms may be predominant. In a pair of papers (Kramers 1932a, 1933b)[20] further generalises the theory to take into account the possible magnetic-field dependence of ϱ_∞ and at the same time he studies in detail the problem of the paramagnetic susceptibility of rare-earth crystals. It is interesting to note that in these papers he uses the symbolic method (see § 3.3).

In the *Leipziger Vorträge* on magnetism Kramers (1933h) presents a slightly less technical account of his theory, once again changing his interpretation of the data on the paramagnetc rotation in xenotime, which he now ascribes to the Er^{+++} ions. He also devotes a section to the theory of the Debye-Giauque demagnetisation effect. This theory was tested in the adiabatic cooling experiments using ceriumethylsulfate with which temperatures of 0.08 K were reached (de Haas, Wiersma, and Kramers 1934).

In 1951 Kramers (Becquerel, van den Handel, and Kramers 1951) returned to this subject and applied — though without great success — his theory to measurements of the magnetisation and paramagnetic rotation of nickelsulphate.

Kramers's last paper (Kramers 1934a) on paramagnetism is a discussion of the interaction between magnetic atoms in a paramagnetic crystal. Kramers considers a crystal containing magnetic atoms in doubly degenerate states and non-magnetic atoms in non-degenerate states. He shows that in such a crystal the interaction between two magnetic atoms will be an indirect exchange interaction, involving a third, non-magnetic, atom.

In the spirit of the Hartree approach to atoms, one can describe the wavefunction Φ of the system as a sum of products of single-atom wavefunctions:

$$\Phi = \Phi_{t_1, t_2, \ldots} = C \sum_P \pm \prod_k \varphi_k^{t_k} \prod_n \chi_n, \qquad (5.112)$$

where the first product is over all magnetic atoms with the superscript t_k ($= \pm\frac{1}{2}$ indicating the two substates, for which one can choose two spin-conjugated states, of the doubly degenerate state). The second product is over all non-magnetic atoms and the sum is over all possible permutations of all the electrons in the system with $\pm$ included to take account of the Pauli principle. To zeroth approximation the wavefunction of the system will be a sum of wavefunctions of the form (5.112) with the φ's and χ's being the ground-state wavefunctions of

---

[20]These papers appeared in successive issues of the *Proceedings of the Dutch Academy of Sciences* due to this academy's policy against papers with more than a certain length. However, in reprints of these papers — and as a result in the *Collected Scientific Papers* volume — the papers are reunited; this also explains the footnote on p. 545 of the *Collected Scientific Papers*.

the atoms and the different Φ differing only in the set of t_k-values—which means that this state will be 2^K-fold degenerate, if K is the total number of magnetic atoms.

In the next approximation we can put

$$\Psi^r = \sum_{t_1,t_2,\cdots} a^r_{t_1,t_2,\ldots} \Phi_{t_1,t_2,\ldots} + \psi^r, \quad (r = 1, 2, \cdots, 2^K) \quad (5.113)$$

where we assume that ψ^r is a small correction, that is, we assume that the coupling between the atoms is weak.

In the secular problem we have to solve to find the splitting of the degenerate zero-approximation ground state we have a Hermitian perturbing Hamiltonian that can be expanded in terms of Pauli matrices (3.033), referring to each of the magnetic atoms as follows:

$$\mathcal{H} = b^{(0)} + \sum_{k,p} b^{(1)}_{k,p} \sigma^k_p + \sum_{kl,pq} b^{(2)} \sigma^k_p \sigma^l_q + \cdots, \quad (5.114)$$

where $k, l = 1, 2, \cdots, K$ and $p, q = 1, 2, 3$. Kramers proves that, in fact, all terms with an odd number of Pauli matrices vanish so that in first instance we are dealing with a Hamiltonian of the form

$$\mathcal{H} = b^{(0)} + \sum_{kl} B_{kl}, \quad B_{kl} = b^{kl}_{pq} \sigma^k_p \sigma^l_q. \quad (5.115)$$

The B_{kl} indicate a coupling between the atoms k and l, which is partly a direct magnetic interaction and partly an exchange coupling. However, in paramagnetic salts the magnetic atoms will not be nearest neighbours and the direct magnetic and the exchange forces will be negligibly small.

In fact, Kramers reports that calculations carried out with Bloch showed that there will still be an interaction between magnetic atoms, but the energy will be of third order in the perturbing Hamiltonian. The interaction is mediated by a non-magnetic atom and we have the so-called indirect exchange interaction.

In a short note on the quantum theory of antiferromagnetism, Kramers (1952a) points out that the simple method of dividing the lattice into two, or more, sublattices, each of which produces its own magnetisation which then acts on the whole lattice and which reaches saturation at absolute zero, needs to be treated with care in the case of one- and two-dimensional antiferromagnets.

During Kramers's stay at the Institute for Advanced Studies in Princeton in the autumn before his death he wrote a paper on "Some Reflections of Phonons and Rotons," which was published posthumously (Kramers 1952b). There are clear indications that this is a first, rather than a final draft of a paper on the subject.

Kramers studies in this paper liquid helium at low temperatures and introduces quantised excitations, obeying Bose-Einstein statistics, which he calls excitons. The phonons and Landau's rotons are special cases of these excitons. In common with other authors he introduces the mass density ϱ_n of the exciton excitations — in the two-fluid theory called the normal density — through the momentum density $\mathfrak{P}(=\varrho_n \mathbf{u}$, where $\mathbf{u}$ is the velocity of the exciton gas with respect to the underlying liquid). He derives both the London formula for the fountain effect and Landau's formula for the second-sound velocity.

It is interesting to quote Kramers when he comments on the fact that he does not have a theory about the interactions between excitons — which means that his considerations only hold for a dilute exciton gas, that is, small values of ϱ_n. He states: "we are as yet utterly unable to approach the problem of the existence of the λ-point, and its possible connection with the Bose-Einstein statistics of the ^{4}He atoms" — a statement that can still be made more than forty years later.

Kramers also speculates about the nature of the excitons and tentatively suggests that they may be linked to the possible states of individual helium atoms inside the box formed by the surrounding atoms; excitons might be the propagation of such states through the liquid.

6 The Kramers Problem and Polymer Physics

In common with the other giants of twentieth-century theoretical physics Kramers's interests ranged over the whole of physics and not just over a specialised small domain of it, as is the modern trend. So far we have discussed his work in quantum mechanics, quantum electrodynamics, solid state physics, and equilibrium statistical mechanics. In the present chapter we shall consider the so-called Kramers problem as well as his paper on the flow of polymers, leaving for the next chapter the remainder of his work, ranging from topics in non-equilibrium statistical mechanics to astrophysics and relativity.

It is a matter of taste what one considers to be Kramers's most important contribution to physics. Is it his derivation of the dispersion formulæ which led to quantum mechanics; is it his application of the transfer matrix method to the two-dimensional Ising model, which led to Onsager's exact solution; is it his derivation of a structure-independent renormalised Hamiltonian; or is it his introduction of the semi-classical WKB approximation, which is still daily applied? High on the list, surely, must come the so-called Kramers equation or what Mel'nikov (1991) calls the Kramers problem. This is the following. Consider a particle trapped in a potential well and subject to the irregular forces of a surrounding medium in temperature equilibrium (Brownian motion). What is the probability that the particle may escape by passing over a potential barrier? Kramers studied this problem in 1940 and solved it for various limiting cases. Since then it has been studied by many people and applied to a great variety of physical situations. Finally, as far as polymer physics is concerned, its practitioners might well consider Kramers's paper on the behaviour of macromolecules in a streaming liquid as his most important one (see, for instance, Bird 1989 or Isihara 1989).

We shall discuss these two papers in the present chapter, the plan of which is the following:
i. In §6.1 we shall discuss the paper on the Brownian motion in a field of force (Kramers 1940a; reprinted as paper I) as well as the related paper with Christiansen (Christiansen and Kramers 1923) on chemical reaction rates.
ii. In §6.2 we consider the paper on the behaviour of macromolecules in inhomogeneous flow (Kramers 1946b; reprinted as paper L).

6.1. The So-Called Kramers Problem

In 1923 Kramers studied with Christiansen (Christiansen and Kramers 1923) the rate of chemical reactions. He returned to the problem of chemical reactions in 1940 (Kramers 1940a; reprinted as paper I at the end of this book) in a paper that deals besides with a large number of other possible applications. We shall discuss these two papers in the present section.

Christiansen and Kramers discuss the theory of exothermal unimolecular reactions of the form

$$g \to g' + g'', \tag{6.01}$$

where g, g', and g'' indicate molecules in their ground state. The idea is that these reactions take place after the molecule g is excited into an "activated" state a through either collisions or radiation. They show, first, that radiation can be neglected in the activation process and, second, that this simple picture will not work since the experimental data require that the activation is sufficiently rapid to lead to a temperature equilibrium between the g and a states, whereas their calculations show that the decay rate leading to the reaction (6.01) is much faster than the activation rate.

However, they show that the following chain of reactions will lead to the same expression for the reaction constant:

$$g \leftrightarrows a \to a' + a'' \to g' + g'', \tag{6.02}$$

where a' and a'' are activated states of the reaction products g' and g''.

In the 1940 paper Kramers studies the probability density in phase space of a Brownian particle in a field of force for the case where this particle is trapped in a potential well. Van Kampen (1991) suggests that although the apparent motivation was a discussion of chemical dissociation and, possibly, nuclear fission,[1] the real reason for this research was that Kramers was fascinated by the intricacies of the mathematical problem. The problem is the following. Consider a particle moving in an external field of force, but in addition subject to the irregular forces of a surrounding medium in temperature equilibrium; the particle is originally caught in a potential well but may escape in the course of time by passing over a potential barrier. By constructing a diffusion equation obeyed by a density distribution of particles in phase space, one can calculate the probability of escape as a function of the temperature and viscosity of the medium.

Van Kampen in his after-dinner speech (van Kampen 1991) to the "50 years after Kramers" Discussion Meeting has given a masterly description of the main features of the paper, and we cannot do better in what follows than quote large sections of this description, interspersed with a more detailed

---
[1] Not fusion, as a misprint in van Kampen's paper suggests.

description of some of the steps.² As there is no model known which can be treated exactly, approximations must be made and "it seems as if almost everything that can be done in that direction has been done by Kramers." As van Kampen says, "this is a typical Kramers paper, containing many gems, but as a whole somewhat confusing. It needs a careful perusal and that may well be the reason that it was not well known for many years."

Consider the one-dimensional motion of a particle of unit mass subject to an external field of force $K(q) = -\partial U/\partial q$ and the random force $X(t)$ due to the medium. Its equations of motion are

$$\dot{p} = K(q) + X(t), \qquad \dot{q} = p, \tag{6.03}$$

where q and p are the coordinate and momentum of the particle and the dot indicates the time derivative.

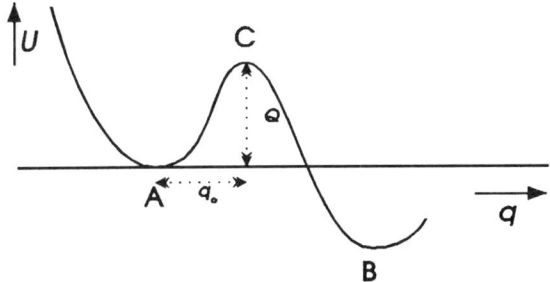

Fig. 6.1. Potential field with smooth barrier.

Kramers first derives the following Fokker-Planck type of diffusion equation for the particles in phase space:

$$\frac{\partial \varrho}{\partial t} = -K(q)\frac{\partial \varrho}{\partial p} - p\frac{\partial \varrho}{\partial q} + \eta\frac{\partial}{\partial p}\left(p\varrho + T\frac{\partial \varrho}{\partial p}\right), \tag{6.04}$$

where $\varrho(p,q)$ is the probability distribution in phase space — the ensemble density — and η and T are, respectively, the viscosity and the temperature, which is here expressed in energy units, that is, Boltzmann's constant is put equal to unity. Equation (6.04) is generally known as the Kramers equation although it had been derived originally by Klein (1922).

Kramers used Eq. (6.04) to study the escape of a particle over a potential barrier, either a smooth one, as shown in Fig. 6.1, or a sharp one, as shown in Fig. 6.2, with barrier heights which we assume to be large compared to T. "The *first*³ gem is the discovery that it is sufficient to investigate the stationary case, even though one is interested in a decay rate. Note that this is not

---
[2] I am indebted to Professor van Kampen for his permission to quote him extensively; quotation marks indicate where I have done so.
[3] The italics in the quotations are van Kampen's.

96   Chapter 6

an approximation, but is precisely correct within the margin of uncertainty inherent in the very concept of escape time." If we look at Figs. 6.1 and 6.2 we see that we are interested in the passage of particles from the bound state at A to another bound state at B. If the system were in thermodynamic equilibrium, the ensemble density would be proportional to $e^{-E/T}$ and the net number of particles passing from A to B would vanish. If, however, initially the number at A would be larger than that corresponding to thermodynamic equilibrium, diffusion will occur, tending to establish equilibrium. However, since the barrier height is large compared to T this diffusion process will be slow, and we may assume that the process corresponds to a stationary diffusion process and that the particle distribution near A (and also near B) will correspond to thermodynamic equilibrium.

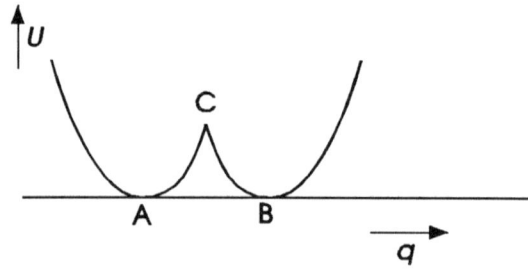

Fig. 6.2. Potential field with edge-shaped barrier.

As Kramers did not manage to solve the general escape problem for arbitrary values of the viscosity, he considered the limiting cases of large and small values of η. He first considers the case of large values of the viscosity. "The *second* gem is that in *the limit of large friction* the equation reduces to a diffusion equation in coordinate space alone. This requires the elimination of the momentum of the particle, which Kramers achieved by means of an ingenious step. This has now become an industry under the title 'adiabatic elimination of fast variables.' " The resulting equation has the form

$$\frac{\partial \sigma}{\partial t} = -\frac{\partial}{\partial q}\left(\frac{K}{\eta}\sigma - \frac{T}{\eta}\frac{\partial \sigma}{\partial q}\right), \tag{6.05}$$

where we have written ϱ in the form

$$\varrho(q,p,t) = \sigma(q,t)e^{-p^2/2T}. \tag{6.06}$$

In the stationary case the solution of Eq. (6.05) is

$$w \equiv \frac{K}{\eta}\sigma - \frac{T}{\eta}\frac{\partial \sigma}{\partial q} = -\frac{T}{\eta}e^{-U/T}\frac{\partial}{\partial q}\left(\sigma e^{U/T}\right) = \text{constant}, \tag{6.07}$$

where U is the potential of the external force K.

Integrating this solution between two values, P and Q, of q we find

$$w = \left[T|\sigma\, e^{U/T}\Big|_Q^P\right] \Big/ \left[\int_P^Q \eta\, e^{U/T}\, dq\right]. \tag{6.08}$$

If the viscosity is large, the effect of the random forces on the motion of the particle will be much larger than that of the external force $K(q)$. Hence, we may, provided K does not change much over distances of the order $\sqrt{T}/\eta$, expect that after a time of the order $1/\eta$ a Maxwell velocity distribution will be established for every value of q, that is, that Eq. (6.06) will hold.[4]

Equation (6.05) follows from Eq. (6.04) by integrating both sides of that equation along a straight line $q + p/\eta = \text{const.} = q_0$ from $p = -\infty$ to $+\infty$.[5]

We now come to the calculation of the mean first-passage time, that is, the calculation of the probability that a particle originally trapped at A escapes to B. To quote van Kampen again: "*Thirdly*, having obtained the one-dimensional diffusion equation Kramers found the mean first-passage time by deriving a formula for it, which is now common knowledge, although it is occasionally rediscovered."

Let us continue considering the large viscosity case, assuming a quasi-stationary state in which practically no particles have arrived at B, while at A thermal equilibrium has practically been reached. In that case we find the escape probability r from the equation

$$r = \frac{w}{n_A}, \tag{6.09}$$

where w is given by Eq. (6.08) and n_A is the number of particles near A.

Kramers now "used a *fourth* ingredient: he decomposed the range of the coordinate into one region around the top of the potential region, and another region covering the potential well. This has now become a standard trick of singular perturbation theory. It enables one to apply different expansions in both regions, provided one can fit them smoothly together so as to get an approximation that covers the whole range." Assuming that the potential near A can be represented by the harmonic oscillator potential $\frac{1}{2}(2\pi\omega)^2 q^2$, Kramers finds in the case of the potential of Fig. 6.1

$$r \approx \frac{2\pi\omega\omega'}{\eta} e^{-Q/T}, \tag{6.10}$$

where ω' is found by approximating the potential near the top of the barrier by

---

[4] Kramers points out the interesting fact that whereas the velocities that a given particle will have during its motion follow a Maxwell distribution, the velocity distribution of the particles at a given point q will not be Maxwellian — otherwise the diffusion current $\int \varrho p\, dp$ would vanish.

[5] A detailed justification of the derivation of Eq. (6.05) can be found in a paper by Wycoff and Balasz (1987).

98   Chapter 6

$$Q_{\text{near C}} = Q - \tfrac{1}{2}(2\pi\omega')^2(q-q_C)^2, \qquad (6.11)$$

while for the potential of Fig. 6.2 we have

$$r \approx \frac{2\pi\omega^2}{\eta}\sqrt{\frac{\pi Q}{T}}\,e^{-Q/T}, \qquad (6.12)$$

where Q is again the barrier height.

For the case we are considering we find from Eq.(6.08)

$$w = \frac{T}{\eta}\sigma_A \left\{\int_A^B e^{U/T}\,dq\right\}^{-1}, \qquad (6.13)$$

where $\sigma_A = \left(\sigma\,e^{U/T}\right)_{\text{near A}}$. For n_A we have

$$n_A = \int_{-\infty}^{+\infty} \sigma_A e^{-(2\pi\omega)^2 q^2/2T}\,dq = \frac{\sigma_A}{\omega}\sqrt{\frac{T}{2\pi}}. \qquad (6.14)$$

In the case of the potential of Fig. 6.1 the integral in Eq. (6.13) can be evaluated by taking into account that the main contribution comes from near the top of the barrier so that we can change the limits from A to B to from $-\infty$ to $+\infty$ and Eq. (6.10) follows.

If we assume that the potential of Fig. 6.2 can to the left of C be represented exactly by $\tfrac{1}{2}(2\pi\omega)^2 q^2$ and to the right of C by the mirror image of the potential to the left of C, we can again, extending the limits as above, evaluate the integral in Eq. (6.20), and Eq. (6.19) follows.

"The *fifth* ingredient is a real gem:[6] the very ingenious construction of a solution in the barrier region. It is true that this is only one special solution, but it is precisely the one he needs: no incident particles from infinity, and thermal equilibrium on the side of the well. Hence it can be attached smoothly to the equilibrium distribution inside the well." Kramers had noted that, if one appoximates the potential of Fig. 6.1 near the barrier by Eq. (6.11), one can in that region solve Eq. (6.04) exactly in the stationary case for all values of the viscosity. The special solution mentioned by van Kampen leads for large values of the viscosity ($\eta \gg \omega'$) to expression (6.10) for the escape probability, while in the case of small viscosities ($\eta \ll \omega'$) one finds the transition state method value

$$r \approx \omega e^{-Q/T} \equiv r_{\text{tr}}. \qquad (6.15)$$

Using Eq. (6.11) for the potential, writing $q' = q - q_C$ and

$$\varrho = \zeta\,e^{-[p^2-(2\pi\omega')^2(q')^2]/2T}, \qquad (6.16)$$

---

[6] In fact, reading that part of the paper one is reminded of a Mozart string quartet (footnote by D.t.H.).

we find from Eq. (6.04) the following equation for ζ:

$$(2\pi\omega')^2 q' \frac{\partial \zeta}{\partial p} - p \frac{\partial \zeta}{\partial q'} - \eta p \frac{\partial \zeta}{\partial p} + \eta T \frac{\partial^2 \zeta}{\partial p^2} = 0. \tag{6.17}$$

This equation allows a solution of the form

$$\zeta = \zeta(u), \quad \text{where} \quad u = p - aq', \tag{6.18}$$

provided a satisfies the equation

$$a(a - \eta) = (2\pi\omega')^2, \tag{6.19}$$

and ζ the equation

$$(a - \eta)u \frac{d\zeta}{du} + \eta T \frac{d^2\zeta}{du^2}. \tag{6.20}$$

The solution $\zeta = $ const. corresponds to thermal equilibrium on both sides of the barrier and is therefore not of interest for our purpose. However, there is also the solution

$$\zeta = C \int^u e^{-(a-\eta)u^2/2\eta T} du, \tag{6.21}$$

where C is a constant.

Taking $-\infty$ as the lower limit of the integral, and as the solution of Eq. (6.19)

$$a = \tfrac{1}{2}\eta + \sqrt{\tfrac{1}{4}\eta^2 + (2\pi\omega')^2}, \tag{6.22}$$

so that $a - \eta$ is positive, this solution represents a situation where there are practically no particles to the right of the barrier while to the left of the barrier we have

$$\zeta = C \sqrt{\frac{2\pi\eta T}{a - \eta}}, \tag{6.23}$$

so that we get for the density in phase space near A the equilibrium expression

$$\varrho = C \sqrt{\frac{2\pi\eta T}{a - \eta}} e^{Q/T} e^{-[p^2 + (2\pi\omega')^2 q^2]/2T}. \tag{6.24}$$

The number of particles n_A near A is obtained by integrating expression (6.24) over the whole of phase space, and the number of particles w passing per unit time over the barrier is obtained by integrating $p\varrho$ over p from ∞ to $+\infty$ for $q' = 0$. This then leads to the following expression for r:

$$r = \frac{w}{n_A} = w\sqrt{\frac{a-\eta}{a}}\,e^{-Q/T}, \tag{6.25}$$

which leads to the results mentioned earlier for large and small values of η.

We now must consider the case of small viscosities. In the transition state method one assumes that the particles at A are in perfect thermodynamic equilibrium with the particles at B so that the number of particles passing the barrier per unit time is obtained by integrating $p\varrho_0$ over p, where $\varrho_0 = e^{-E/T}$ is the equilibrium Gibbs distribution, while the number of particles at A is given by integrating the Boltzmann distribution over the whole of phase space. This leads to expression (6.15). However, "Kramers also realised that *for very small friction* the fluctuations might not be able to maintain the equilibrium in the well; rather the leakage across the barrier would deplete the high energy tail of the distribution. For this case he introduced his *sixth device*. In the limit of low friction the particle in the well oscillates roughly with a constant energy. It is therefore possible to average out the phase. This leads to a one-dimensional diffusion equation in the energy scale, from which it is easy to compute the average time needed to reach the energy of the top of the barrier." Instead of the energy E of the particle Kramers introduced the action $I(E)$, that is, the area in phase space within a curve of constant energy:

$$I(E) = \oint p\,dq. \tag{6.26}$$

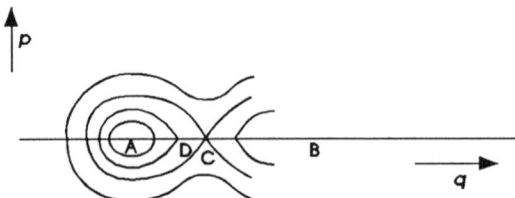

Fig. 6.3. Phase space for smooth potential barrier.

In the two cases of Figs. 6.1 and 6.2 the phase space of the particle has the form shown in Figs. 6.3 and 6.4. Instead of the diffusion equation (6.05) we now find a diffusion equation for the ensemble density itself:

$$\frac{\partial\varrho}{\partial t} = \eta\,\frac{\partial}{\partial I}\left(I\varrho + TI\,\frac{\partial\varrho}{\partial E}\right), \tag{6.27}$$

and solving this for the stationary case we get instead of Eq. (6.08) the equation

The Kramers Problem and Polymer Physics

$$w = \left[\eta T \left| \varrho e^{E/T} \right|_Q^P \right] \Big/ \left[\int_P^Q I^{-1} e^{E/T} dE \right]. \tag{6.28}$$

Equations (6.27) and (6.28) can be derived as follows. We first note that for the barriers of Figs. 6.1 and 6.2 the motion of the particle in the potential well will be oscillatory. Moreover, if the viscosity is small, the Brownian motion will only lead to small changes in the particle energy during a single oscillation. Let $\bar{\varrho}\, dI$ denote the fraction of the ensemble corresponding to values of I lying between I and $I + dI$. We then get an equation for $\bar{\varrho}$ by averaging Eq. (6.04) over the ring in phase space corresponding to those I-values.

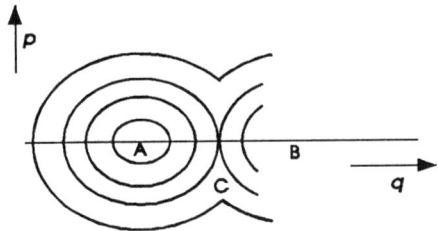

Fig. 6.4. Phase space for edge-shaped potential barrier.

Since the ensemble density is unchanged when there are no Brownian forces, the averaging of the first two terms on the right-hand side of Eq. (6.04) will give zero and as $\partial E/\partial p = p$ and $\overline{p^2} = I/\omega$, where ω is the period of the oscillation which satisfies the equation

$$\omega = \frac{\partial E}{\partial I}, \tag{6.29}$$

we find Eq. (6.27) when we take into account that in the small viscosity case the energy during an oscillation is hardly changed so that $\bar{\varrho} \approx \varrho$.

In the stationary state the solution of Eq. (6.27) is

$$w \equiv -\eta \left(I\varrho + TI \frac{\partial \varrho}{\partial E} \right)$$
$$= -\eta T I\, e^{-E/T} \frac{\partial}{\partial E} \left(\varrho\, e^{E/T} \right) = \text{constant}, \tag{6.30}$$

whence follows Eq. (6.28).

Applying Eq. (6.28), we finally get for the escape probability the following expression in the small viscosity case:

$$r \approx \eta \frac{Q}{T} e^{-Q/T}. \tag{6.31}$$

We can derive Eq. (6.31) as follows. We first of all note that the integral in Eq. (6.28) diverges at $E = I = 0$ and we therefore take as the lower limit of the integral $E = T$. Moreover, denoting the value of $\varrho e^{E/T}$ at A by ϱ_A and taking into account that there will practically no particles at C and to the right of it, we find

$$w \approx \eta T \varrho_A \left\{ \int_T^Q I^{-1} e^{E/T}\, dE \right\}^{-1} \tag{6.32}$$

As the main contribution to the integral will come from values of E that differ from Q by amounts of the order of T, we can put I equal to the value I_C corresponding to the energy curve through C. Since this quantity is of the order of Q/ω, we find

$$w \approx \frac{\eta \varrho_A Q}{\omega} e^{-Q/T}, \tag{6.33}$$

and since we have again $n_A = \varrho_A T/\omega$, Eq. (6.31) follows.

There are many applications of the theory developed by Kramers. He himself discusses several of them and for a more up-to-date discussion we refer to the review papers by Hänggi and collaborators (Hänggi, Talkner, and Borkovec 1990) and Mel'nikov (1991).

6.2. Flow of Polymers

Another of Kramers's seminal papers is the one on the behaviour of macromolecules in inhomogeneous flow (Kramers 1944a, 1946b; reprinted as paper L at the end of this book). Because of wartime restrictions, this paper was originally published in Dutch in 1944, but after the war it was translated and published in English.

The paper is a more sophisticated and generalised version of the theory by Hermans (1943) who characterised the behavior of a macromolecule in a flowing liquid by the diffusion of its endpoints, while being bound together by a fictitious elastic attraction. The distribution of the vector from the one end to the other is changed by the flow, and Hermans showed that as a result the changes in the viscosity and in the flow birefringence of the liquid due to the macromolecules are both proportional to the square of the number of the elementary links in the molecule, in accordance with experiment (Staudinger's rule)

Kramers's generalisation of Hermans's method consists in considering the statistical behaviour of the individual links of the macromolecules. We write the force exerted by the flow on the ith particle as $-\zeta_i \mathbf{v}_i$ where $\mathbf{v}_i$ is the velocity of the particle relative to the fluid,

$$\mathbf{v}_i = -\mathbf{v}'_i + \mathbf{v}''_i, \tag{6.34}$$

where $\mathbf{v}'_i$ is the velocity of the fluid and $\mathbf{v}''_i$ that of the particle. If we assume that the flow is irrotational and stationary so that it can be derived from a velocity potential Ψ,

$$\mathbf{v}'_i = -\nabla_i \Psi, \tag{6.35}$$

where ∇_i indicates that we take the value of the gradient at the position of the ith particle, we can write down the following equation for the total force $\mathbf{F}_i$ acting on the ith particle:

$$\mathbf{F}_i = -\nabla_i U - \zeta_i \dot{\mathbf{r}}_i + \mathbf{F}'_i, \qquad \dot{\mathbf{r}}_i = \mathbf{i}, \tag{6.36}$$

where $\mathbf{F}'_i$ is a random (Brownian) force that would lead to thermal equilibrium, if no other forces were acting, and where U is a potential given by the equation

$$U = \sum_i \zeta_i \Psi(\mathbf{r}_i). \tag{6.37}$$

We note that the equations of motion are the same as for a particle moving in a stationary fluid and in a potential field U. This means that although we are considering a non-equilibrium situation of a particle in a moving fluid, the situation is analogous to the equilibrium situation of a particle moving in a potential field. Hence, for the calculation of averages we can use a Boltzmann distribution corresponding to the temperature T.

Kramers now introduces the pearl-necklace model of a polymer, which is a freely jointed bead-rod chain. The molecule consists of N particles of mass m at positions $\mathbf{r}_i$, connected by a weightless rod of fixed length L between each two consecutive ones. We assume that the quantity ζ_i that measures the friction between the particles and the fluid (see Eq. (6.37)) is the same ($=\zeta$) for all the particles. Let $\boldsymbol{\omega}_i$ be the unit vector along the direction of the rod connecting the ith and the $i+1$st particle, and let $\mathbf{r}$ be the coordinate of the centre of mass of the molecule. We then have

$$\mathbf{r}_i = \mathbf{r} + \frac{L}{N}(\boldsymbol{\omega}_1 + 2\boldsymbol{\omega}_2 + \cdots + i\boldsymbol{\omega}_i$$
$$-[N-i-1]\boldsymbol{\omega}_{i+1} - [N-i-2]\boldsymbol{\omega}_{i+2} - \boldsymbol{\omega}_{N-1}), \tag{6.38}$$

and the kinetic energy $\mathcal{T}$ of a molecule will therefore be

$$\mathcal{T} = \frac{mL^2}{2N} \sum_{\mu\nu} g_{\mu\nu} (\dot{\boldsymbol{\omega}}_\mu \cdot \dot{\boldsymbol{\omega}}_\nu) + \tfrac{1}{2} Nm (\dot{\mathbf{r}} \cdot \dot{\mathbf{r}}), \tag{6.39}$$

where

$$g_{\mu\nu} = \begin{cases} \mu(N-\nu) & \text{when } \nu \geq \mu, \\ \nu(N-\mu) & \text{when } \nu \leq \mu. \end{cases} \tag{6.40}$$

Because we have assumed that the distance between two consecutive particles in the molecule is fixed, there are constraints and the number of degrees of freedom is reduced; we must therefore introduce generalised coordinates. The simplest model is the one where the rods meeting at a particle can rotate freely.[7] In that case we can introduce as the $2N+1$ generalised coordinates the three centre-of-mass coordinates and for each rod two polar angles ϑ_μ and φ_μ such that the three components of $\boldsymbol{\omega}_\mu$ are given by the relations

$$\left.\begin{aligned} \omega_{\mu x} &= \cos\vartheta_\mu, \\ \omega_{\mu y} &= \sin\vartheta_\mu \cos\varphi_\mu, \\ \omega_{\mu z} &= \sin\vartheta_\mu \sin\varphi_\mu. \end{aligned}\right\} \tag{6.41}$$

Let us now calculate the change in the viscosity coefficient η due to the presence of the macromolecules. One usually considers a flow such that the x-, y-, and z-components of the fluid velocity are given by the equations

$$u' = \varkappa y, \qquad v' = 0, \qquad w' = 0; \tag{6.42}$$

however, this flow is not irrotational and we shall instead consider a velocity potential given by the equation

$$\Psi(\mathbf{r}) = -\tfrac{1}{2}\varkappa xy \tag{6.43}$$

so that the fluid velocity has the components

$$u' = \tfrac{1}{2}\varkappa y, \qquad v' = -\tfrac{1}{2}\varkappa x, \qquad w' = 0. \tag{6.44}$$

This flow differs from the flow given by Eqs. (6.42) in that there is an additional uniform rotation superimposed on it. Kramers, in fact, showed that one may neglect the effect of this rotation, provided $\varkappa$ is sufficiently small.

The viscosity η is related to the xy-component P_{xy} of the stress tensor in the usual way:

$$P_{xy} = \eta\left(\frac{\partial u'}{\partial y} + \frac{\partial v'}{\partial x}\right) = \eta\varkappa. \tag{6.45}$$

This means that we can find the change in η by calculating the contribution the macromolecules make to P_{xy}. There are two contributions to the momentum in the x-direction transported per unit time through a unit area with a normal parallel to the y-direction: the first one comes from the motion of the particles and the second from the tension in the links.

For the first contribution $\delta_1\eta (= \delta_1 P_{xy}/\varkappa)$ we have

$$\delta_1\eta = -\frac{m}{\varkappa}\left\langle \sum_i \dot{x}_i \dot{y}_i \right\rangle_{\mathrm{Av}}, \tag{6.46}$$

---

[7] Such a link was called a statistical chain element by W. Kuhn.

where the average must be taken over the Boltzmann distribution with the potential energy U corresponding to expression (6.43), that is,

$$U = -\tfrac{1}{2}\zeta\varkappa \sum_{i=1}^{N} x_i y_i, \tag{6.47}$$

which, in terms of the components of the ω_μ's, becomes

$$U = -\frac{\varkappa\zeta L^2}{2N} \sum_{\mu\nu} g_{\mu\nu}\omega_{\mu x}\omega_{\mu y}. \tag{6.48}$$

The second contribution can be written in the form

$$\delta_{\mathrm{II}}\eta = \frac{L}{\varkappa}\left\langle \sum_\mu S_\mu \omega_{\mu x}\omega_{\mu y} \right\rangle_{\mathrm{Av}}, \tag{6.49}$$

where the summation is over all links intersecting the area concerned and S_μ is the tension in the μth link.

Equation (6.46) follows by bearing in mind that the probability that the ith particle crosses per unit time a unit area at right angles to the y-axis in the direction of the negative y-axis is equal to $-\dot{y}_i$ while the momentum in the x-direction it carries along is $m\dot{x}_i$

Equation (6.49) follows when we bear in mind that the tension in a link intersecting the area concerned contributes $S_\mu \omega_{\mu x}$ to P_{xy} while $L\omega_{m\mu y}$ is the probability that the intersection takes place.

After one has evaluated the averages in Eqs. (6.46) and (6.49), the final result for the change in the viscosity becomes in the case of small $\varkappa$:

$$\delta\eta = C\zeta L^2 N^2, \tag{6.50}$$

with C a constant which depends on whether or not there are any restrictions on the rotation of the links with respect to one another. We see that we are, indeed, led to Staudinger's rule, that is, a change in the viscosity proportional to the square of the number of links.

Equation (6.50) can be derived as follows. We can rewrite expression (6.49) as follows:

$$\delta_{\mathrm{II}}\eta = -\frac{1}{\varkappa}\left\langle \sum_i F''_{ix} y_i \right\rangle_{\mathrm{Av}}, \tag{6.51}$$

where $\mathbf{F}''$ is the total force on the ith particle due to the tensions in the links,

$$\mathbf{F}''_i = \sum_j S_i^{(j)} \omega_{ix}^{(j)}; \tag{6.52}$$

the summation is over all links j meeting in the ith particle.[8]

The equality of the right-hand sides of Eqs. (6.49) and (6.51) follows from the fact that for each link there will be two terms corresponding to the same $S_i^{(j)}$ for the two particles at the end of the link with opposite signs of the $\omega_{ix}^{(j)}$, while we can write $L\omega_{\mu y} = y_k - y_l$, where k and l are the two particles at the end of the link.

For $\mathbf{F}_i''$ we have the equation

$$\mathbf{F}_i'' = m\ddot{\mathbf{r}}_i - \mathbf{F}_i = m\ddot{\mathbf{r}}_i + \nabla_i U + \zeta_i \dot{\mathbf{r}}_i - \mathbf{F}_i', \tag{6.53}$$

where we have used Eq. (6.36) for $\mathbf{F}_i$. Substituting expression (6.53) into Eq. (6.51) we see that the last two terms on the right-hand side of Eq. (6.53) do not give a contribution to the average so that combining Eqs. (6.46), (6.51), and (6.53) we finally find

$$\delta\eta = \delta_I\eta + \delta_{II}\eta$$

$$= -\frac{m}{\varkappa}\left\langle \sum_i (\dot{x}_i \dot{y}_i + \ddot{x}_i y_i) \right\rangle_{\text{Av}} - \frac{1}{\varkappa}\left\langle \sum_i \frac{\partial U}{\partial x_i} y_i \right\rangle_{\text{Av}} \tag{6.54}$$

As the first sum on the right-hand side of Eq. (6.54) is a time derivative, its average will be zero. Using Eq. (6.47) for the potential, we finally find for the change in the viscosity

$$\delta\eta = \tfrac{1}{2}\zeta \sum_i \langle y_i^2 \rangle_{\text{Av}} = \tfrac{1}{6}\zeta \sum_i \langle (\mathbf{r}_i \cdot \mathbf{r}_i) \rangle_{\text{Av}}$$

$$= \frac{\zeta L^2}{6N} \sum_{\mu\nu} g_{\mu\nu} (\boldsymbol{\omega}_\mu \cdot \boldsymbol{\omega}_\nu). \tag{6.55}$$

To evaluate the right-hand side of Eq. (6.55), Kramers points out that the correlation of the directions of two links, μ and ν, decreases steeply with increasing distance between the links and in the cases where it has been calculated the average of $(\boldsymbol{\omega}_\mu \cdot \boldsymbol{\omega}_\nu)$ vanishes. Therefore we can limit ourselves in the sum over μ and ν to the terms with $\mu = \nu$ and in that case we have $(\boldsymbol{\omega}_\mu \cdot \boldsymbol{\omega}_\mu) = 1$ so that Eq. (6.55) leads to the result

$$\delta\eta = \frac{\zeta L^2}{6N} \sum_{\mu=1}^{N-1} g_{\mu\mu} = \frac{\zeta L^2}{6N} \sum_\mu \mu(N-\mu)$$

$$= \frac{\zeta L^2}{6N} \frac{(N-1)N(N+1)}{6} \approx \frac{\zeta L^2 N^2}{36} \tag{6.56}$$

in agreement with Eq. (6.50).

---

[8] In the pearl-necklace model there are one or two of such links, but there may be more in the case of branched polymers.

The calculation of the change in the flow birefringence is analogous to the one we have just sketched for the change in the viscosity.

Kramers discusses the changes which have to be made if one is dealing with branched or ringlike polymers. He also discusses how to extend the theory to the case of finite $\varkappa$. For a discussion of the influence of Kramers's paper on the statistical mechanics of macromolecules we refer to a book by Bird, Curtiss, Armstrong, and Hassager (1987).

7 Miscellaneous Topics

In this last chapter we shall discuss those papers from the Collected Scientific Papers which do not fall into any of the categories covered so far. The plan of the present chapter is the following:

i. In §7.1 we shall discuss Kramers's papers on problems in kinetic theory and gas dynamics. There are three of those: two papers deal with the slipping of a gas along a wall — the first with Jaap Kistemaker[1] (Kramers and Kistemaker 1943) deals with the so-called diffusion-slip and the second with the viscosity-slip (Kramers 1949b) — and the third one deals with vibrations of a gas column.

ii. In the final section, §7.2, we consider the six papers that could not be put in any of the categories of the previous chapters or sections. These is an early paper on general relativity (Kramers 1921), a paper on the vibrations of the acetylene molecule (Olsen and Kramers 1932), a paper on the tension in the cornea (Kramers and ter Haar 1942), a paper which van Dishoeck (1996) mentions as containing the first of interstellar chemistry models (Kramers and ter Haar 1946), a paper on the stopping power of α-particles in metals (Kramers 1947a), and a paper on column ionisation (Kramers 1952c).

7.1. Kinetic Theory of Gases and Gas Dynamics

Kramers was very much interested in transport phenomena, especially in gases, and he regularly lectured on the theory of non-uniform gases following the approach by Chapman and Cowling (1939). Two of his papers deal with this topic, and especially the problem of the behaviour of a gas near a wall. The first paper on this subject is the one with Kistemaker (Kramers and Kistemaker 1943), and the second one (Kramers 1949b) is his contribution to the Florence Statistical Mechanics Conference.

In hydrodynamics and aerodynamics one usually assumes that at a wall the velocity of the fluid is zero. However, experiments with gases show that this condition is only approximately satisfied and one must account for a certain amount of slipping along the wall, which we shall assume to be stationary. There are, in general, three sources of this slip: a gradient, $\partial u/\partial z$,

---
[1] Not Jan Kistemaker, as wrongly stated by Dresden (1988).

of the fluid velocity u along the normal to the wall, assumed to be parallel to the positive z-axis, which leads to the so-called viscosity slip; a gradient along the wall, $\partial T/\partial x$, of the temperature T, which gives rise to the so-called thermal slip; and in the case of a gas consisting of two diffusing components, a concentration gradient $\partial n_1/\partial x$ along the wall, the source of the so-called diffusion slip. This means that we can write for the slip velocity u_0

$$u_0 = \zeta \frac{\partial u}{\partial z} + \zeta' \frac{\partial T}{\partial x} + \zeta'' \frac{\partial n_1}{\partial x}. \tag{7.01}$$

The above-mentioned papers deal with the first and the third of these slips: the paper with Kistemaker with the diffusion slip, an effect predicted and experimentally observed by Kistemaker,[2] and the contribution to the Florence Conference with the viscosity slip.

In both those papers Kramers discusses in some detail why one would expect that u_0 should be equal to zero. The reason is that one expects that the molecules on hitting the wall will be diffusely reflected so that they will leave the wall with an average zero momentum along the wall. On colliding with other molecules of the gas this will lead to a distribution function with an average zero parallel momentum.[3] However, the collisions between the gas molecules and the molecules which are reflected from the wall will take place, on average, at a distance of the order of the mean free path, l, from the wall and one would therefore expect a complicated type of distribution in a transition layer near the wall with a thickness of the order of l. This means that one would expect corrections of the relative order of l/L where L is a characteristic length of the system considered. Since in ordinary hydrodynamics one assumes this parameter to be vanishingly small, the usual boundary condition ensues. However, in gases where the parameter may not be negligibly small one must expect that, as Kramers puts it, there will be a compromise between the deviations from the stationary Maxwell distribution in the interior of the gas — which will lead to a non-zero slip velocity u_0 when extrapolated through the transition layer to the wall — and the purely stationary (or near-stationary) Maxwell distribution of the reflected molecules.

---

[2] I am indebted to Professor Kistemaker for telling me about the circumstances that led to the Kramers-Kistemaker paper. Kistemaker had been asked in 1943 by Keesom to measure the deviations from the perfect gas law in gaseous helium below the λ-point. This meant that pressures had to be measured with an accuracy of better than 0.001 mm Hg, so that many small corrections became important. In looking at this problem Kistemaker considered the effect of the temperature variation along the connecting tube between the helium at low temperatures and the mercury meniscus at room temperature. He then saw that the interdiffusion of the heavy mercury and the light helium atoms would lead to momentum transfer to the wall of the connecting tube. When Kramers was shown Kistemaker's rough estimates of the effect, he produced the detailed theory that is given in their joint paper.

[3] Although quantum theory predicts a certain amount of specular (de Broglie) reflection, this will be significant only for specially prepared walls.

Maxwell (1879) has suggested the following (approximate) calculation of the slip velocity. Consider a gas in a non-uniform state and a mathematical surface S. Compare, on the one hand, the case where this surface is part of the wall so that no molecules can pass through it and, on the other hand, the case where this surface lies in the gas so that molecules can freely pass through it, although there is no net transport of molecules through it. In the first case the molecules will be reflected from S and there will be a net transfer of momentum parallel to S corresponding to the wall velocity, u_W. In the second case there will be a transfer of momentum corresponding to the mass velocity parallel to S, u_M, of the gas at S. Since, in general, for a non-uniform gas the velocity distribution of the molecules which in the second case passed through S will be different from the distribution of the reflected molecules in the first case, we expect that the slip velocity $u_M - u_W \equiv u_0$ will be non-vanishing.

Let us now consider the diffusion slip. We consider the case of a light gas with number density n_1, mass m_1, and average velocity u_1 parallel to the x-axis, which itself is parallel to the wall and to the concentration gradient, and a heavy gas with number density n_2, mass m_2, and average velocity in the x-direction u_2. One can express the slip velocity u_0 in terms of the diffusion coefficient D, the number densities n_1 and n_2, and the concentration gradient dn_1/dx as follows:

$$u_0 = \left[\frac{m_2 - m_1}{n_1 m_1 + n_1 m_2} - \frac{\sqrt{m_2} - \sqrt{m_1}}{n_1 \sqrt{m_1} + n_1 \sqrt{m_2}} \right] D \frac{dn_1}{dx}. \tag{7.02}$$

Equation (7.02) can be proved as follows for the case of a Maxwell gas, that is, a gas where the forces between the molecules are inversely proportional to the fifth power of their distance apart.[4] Maxwell has shown that in that case the distribution functions of the two gases are simply Maxwell distributions moving with constant velocities u_1 and u_2, that is, the interdiffusion can be described as the motion of two gases in thermal equilibrium moving through one another with a relative velocity $u_{\rm rel} = u_1 - u_2$. In the second of the cases considered above, where the molecules pass through the surface S the momentum transferred per unit time per unit area to S by the first gas will be given by the well known expression from kinetic theory, $\frac{1}{4} n_1 c_1 \cdot m_1 u_1$, where c_1 is the mean velocity in the light gas with a similar expression for the heavy gas. If now the surface S is replaced by a wall, the momentum carried by the molecules reflected from the wall will be

$$\left(\tfrac{1}{4} n_1 c_1 m_1 + \tfrac{1}{4} n_2 c_2 m_2 \right) u_W. \tag{7.03}$$

We find thus the following relation for u_W:

---

[4] For the case where we are dealing not with a Maxwell gas but with gases that obey different laws for their interaction potentials, we must correct Eq. (7.02). However, one hopes that the corrections are small.

$$\tfrac{1}{4}\left(n_1 m_1 c_1 u_1 + n_2 m_2 c_2 u_2\right) = \tfrac{1}{4}\left(n_1 m_1 c_1 + n_2 m_2 c_2\right) u_W. \tag{7.04}$$

On the other hand, from the definition of the diffusion coefficient it follows that

$$n_1 u_1 = -D\frac{dn_1}{dx} = D\frac{dn_2}{dx} = -n_2 u_2, \tag{7.05}$$

and hence we have for the mass velocity

$$u_M = \frac{n_1 m_1 u_1 + n_2 m_2 u_2}{n_1 m_1 + n_2 m_2} = \frac{m_2 - m_1}{n_1 m_1 + n_2 m_2} D\frac{dn_1}{dx}. \tag{7.06}$$

If we now use the fact that in thermal equilibrium we have $m_1 c_1^2 = m_2 c_2^2$, we obtain Eq. (7.03) by combining Eqs. (7.04) to (7.06).

This effect was measured by studying a mixture of air and hydrogen in a circular tube along which the relative concentrations changed. Due to diffusion there would be a mass flow with velocity u_M unless a pressure gradient is maintained, setting up a Poiseuille flow sufficient to produce a steady state. The necessary pressure gradient will be given by the usual Poiseuille formula, corrected for the fact that there is a slip velocity u_0 so that we now have the equation

$$\frac{dp}{dx} = -\frac{8\eta}{R^2}\left(u_M - u_0\right), \tag{7.07}$$

where η is the viscosity and R the tube radius. We see that, as one would expect, the effect of the diffusion slip is to decrease the pressure gradient.

The pressure difference Δp between the ends of a tube of length L will be given by the formula

$$\Delta p = \frac{8}{R^2} \int_0^L \eta\left(u_M - u_0\right) dx. \tag{7.08}$$

We can substitute for u_M and u_0 from Eqs. (7.06) and (7.02) and if we assume that we have only the light component at $x = 0$ ($n_2 = 0$) and only the heavy one at $x = L$ ($n_1 = 0$) we find

$$\Delta p = \frac{4\overline{\eta D}}{R^2} \ln \frac{m_2}{m_1}, \tag{7.09}$$

where $\overline{\eta D}$ is an appropriate average of ηD. If we had neglected the slip, that is, put $u_0 = 0$, Eq. (7.08) would have led to

$$\Delta p = \frac{8\overline{\eta D}}{R^2} \ln \frac{m_2}{m_1}, \tag{7.10}$$

and we see that the slip leads to an extra factor $\tfrac{1}{2}$. We notice that as both η and D are proportional to the mean free path l, the effect is of order $(l/R)^2$. The experiments showed, indeed, a decrease in the pressure head — which was even slightly larger than predicted by the simple theory. It is interesting to note that in most of their experiments with a capillary of 0.46 mm radius they were working with l/R ratios of between 0.01 and 0.04.

In his contribution to the Florence Statistical Mechanics Conference, Kramers considered the viscosity slip, that is, the slip occurring when there is a velocity gradient in the direction of the normal to the wall — along which we take the z-axis of our coordinate system. The slip coefficient ζ in Eq. (7.01) has the dimensions of a length and was shown by Maxwell (1879) to be of the order of the mean free path:

$$\zeta = kl, \tag{7.11}$$

where k is a dimensionless constant, called the slip number by Kramers, of the order of unity for which experiments by Lignac (1949) gave values of between 1.14 and 1.20 for hydrogen, helium, nitrogen, oxygen, and methane. The mean free path l in Eq. (7.11) is defined from the relation

$$\eta = \tfrac{1}{2}nmlc \tag{7.12}$$

for the viscosity coefficient η; in Eq.(7.12) n is the number density of the gas. We emphasise that once again the slip disappears in the limit as $l \to 0$.

Maxwell found by his approximate treatment $k = 1$ for a plane wall with cosine-law reflection, that is, for the case where a molecule, independent of its speed and direction of incidence, will on leaving the wall have a probability $(\cos \alpha/\pi) \, d^2\omega$ that its direction will make an angle α with the normal to the wall and lie within an element of solid angle $d^2\omega$. Kramers's aim was to obtain the exact theoretical value of k for this case. In his paper, which presents a programme for a method to find a solution rather than the solution itself, he showed, by slightly refining Maxwell's method and combining this with known results for the viscosity coefficient, that indeed k would be greater than unity and that under not too unreasonable assumptions a value of the order of 1.2 would be found.

Kramers approached the problem by considering the Boltzmann transport equation for the case of a stationary laminar flow of a gas parallel to a plane wall and looking for a solution of the form

$$f = f^{(0)} + f^{(1)}, \tag{7.13}$$

for the distribution function $f(\mathbf{c},\mathbf{r})$ which is a function of both the velocity $\mathbf{c}$ (components u, v, and w) and the position $\mathbf{r}$ of the molecules; here $f^{(0)}$ is the Maxwell distribution corresponding to a stationary laminar flow with a velocity gradient $\partial \bar{u}/\partial z = \varkappa$ (the bar indicates an average value),

$$f^{(0)} = n\left(\frac{2\pi}{m\beta}\right)^{3/2} e^{-\tfrac{1}{2}\beta m[(u-z\varkappa)^2 + v^2 + w^2]}, \tag{7.14}$$

where $\beta = 1/k_B T$ with k_B the Boltzmann constant and T the absolute temperature.

Using Enskog's solution (Chapman and Cowling 1939) for the case where there is a velocity gradient, and taking into account that in the case of slip the mean velocity of the gas at a position z will be $\varkappa(z+\zeta)$ Kramers finds the following expression for $f^{(1)}$ at distances from the wall which are large as compared to l:

$$f^{(1)} = \frac{f^{(0)}}{n} \left[-\varkappa(u - \varkappa z)wB + \beta m(u - \varkappa z)\zeta n \right], \qquad (7.15)$$

where B is a function of the speed, independent of n.

In a layer near the wall with a thickness of the order of l the distribution function will change significantly, especially for those molecules which are moving away from the wall, that is, with $w > 0$. We can therefore, in general, write for the distribution function

$$f = f^{(0)} \left[1 - \frac{\varkappa w B}{n}(u - \varkappa z) + \beta \varkappa (u - \varkappa z) m\zeta + X \right], \qquad (7.16)$$

where the function X will tend to zero at large distances from the wall.

At the wall we must have for $w > 0$:

$$X = \frac{\varkappa w B u}{n} - \beta m u \varkappa \zeta, \qquad (7.17)$$

whereas to a first approximation we may assume that Eq. (7.15) will continue to hold right down to the wall for $w < 0$ so that we have $X = 0$ for $w < 0$.

Kramers then rephrased the problem as follows: Consider a gas in thermal equilibrium and at rest with respect to the wall. Let the law of reflection at the wall be changed in such a way that the reflected molecules, that is, the molecules for which $w > 0$, have at the wall a distribution which rather than the Maxwell distribution $f^{(0)}$ is given by the equation

$$f = f^{(0)}(1 + X) = f^{(0)} \left[1 + \frac{\varkappa w B u}{n} - \beta m u \varkappa \zeta \right], \quad w > 0. \quad (7.18)$$

How can we now choose ζ in such a way that the gas at large distances from the wall stays at rest?

If we accept that X stays zero for $w < 0$ right down to the wall, we must determine ζ in such a way that the molecules leaving the wall with the distribution (7.18) do not give any momentum parallel to the wall to the gas. This means that we have

$$\int_{w>0} \frac{\varkappa m}{n} u^2 w^2 B f^{(0)} \, d^3\mathbf{c} = \beta \varkappa m^2 \zeta \int_{w>0} u^2 w f^{(0)} \, d^3\mathbf{c}. \qquad (7.19)$$

The integral on the left-hand side, if extended over the whole of velocity space, would represent the transport of momentum parallel to the direction of flow in laminar flow and thus, by definition, be equal to $\varkappa\eta$ where η is the viscocity. The integral on the right-hand side can be evaluated exactly and the final result is

$$\tfrac{1}{2}\varkappa\eta = \tfrac{1}{4}\varkappa nm\zeta c, \qquad (7.20)$$

or, if we use Eq. (7.12) for η:

$$\zeta = l, \quad \text{or} \quad k = 1. \qquad (7.21)$$

The reason why k is, in fact, larger than unity is, as Kramers points out, the circumstance that the distribution of the molecules with $w < 0$ which arrive at the wall is altered by collisions with reflected molecules, and one should therefore require that the tangential momentum given off by the wall be equal to that received by it. Considering all possible collisions this leads, indeed, to an increase in the value of k.

The last paper to be discussed in this section is the one in which Kramers considers vibrations in a gas column along which the temperature varies. It had been known (Keesom 1942) in Leiden for a long time that under certain circumstances spontaneous vibrations will occur in a helium cryostat and that these vibrations may cause such a large heat transport that it is impossible to fill the cryostat. To treat this problem Kramers started from the exact solution Kirchhoff (1868) gave of gas vibrations in a tube along which the temperature is constant. This solution describes three possible waves, called by Kramers the main wave, which is essentially a free wave, practically independent of viscosity or heat conduction; the friction wave, which depends strongly on the value of the viscosity; and the heat conduction wave, which depends on the magnitude of the thermal conductivity.

If the temperature is no longer constant everywhere the solutions become more complicated, and Kramers studied the case of wide tubes, which in practice correspond to tubes with a diameter of about 0.6 cm. He found that in the case where the vibrations of a gas column enclosed between a closed end and a liquid surface are forced by the oscillations of the liquid, these vibrations may become unstable. However, he did not find instability in the case of free gas vibrations. He felt that the fact that he was unable to account for the observed spontaneous vibrations was probably due to the linearisation of the problem.[5]

---

[5] It is also possible that the restriction to the wide tube case may be important, as there are indications (ter Haar 1955) that vibrations in narrow tubes are less stable than those in wide tubes. Unfortunately the case of most practical interest is the one where the tube is neither wide nor narrow.

7.2. Other Problems

It used to be a general rule that in order to obtain the doctor's degree in the Netherlands one should submit — and defend — not only a thesis but also a set of "stellingen" (propositions),[6] many of which were usually directly related to the subject matter of the thesis but some of which dealt with other subjects — mainly in the discipline of the candidate, but sometimes ranging over the whole gamut of human life. The last of Kramers's propositions reads:

In Einstein's theory consider a gravitational field with line element $ds^2 = \sum g_{\mu\nu} dx_\mu dx_\nu$ where x_1, x_2, x_3 are space coordinates and x_4 is the time coordinate and where the gravitational potentials $g_{\mu\nu}$ are independent of x_4. The most general coordinate transformation such that the transformed gravitational potentials are again independent of the time and that a point at rest again is at rest is given by the relations

$$x'_k = \phi_k(x_1, x_2, x_3), \quad k = 1, 2, 3, \quad x'_4 = ax_4 + \psi(x_1, x_2, x_3), \quad (7.22)$$

where ϕ_k and ψ are arbitrary functions and a is a constant. With regards to this group of transformations the quantities

$$R_{kl} = \tfrac{1}{2}\sqrt{g_{44}}\left[\frac{\partial}{\partial x_k}\left(\frac{g_{l4}}{g_{44}}\right) - \frac{\partial}{\partial x_l}\left(\frac{g_{k4}}{g_{44}}\right)\right], \quad k, l = 1, 2, 3, \quad (7.23)$$

and

$$G_{kl} = g_{kl} - \frac{g_{k4}g_{l4}}{g_{44}}, \quad k, l = 1, 2, 3, \quad (7.24)$$

have the properties of a tensor; the R_{kl} define an antisymmetric and the G_{kl} a symmetric tensor. One can consider these tensors to express the "rotatory" properties of the stationary gravity field considered, which are connected with the Coriolis force acting on every point of a moving mass. The absolute magnitude of the angular velocity corresponding at every point to the rotation is given by a scalar quantity Ω defined by

$$2\Omega^2 = \sum_{k,l,a,b=1,2,3} G^{ka} G^{lb} R_{kl} R_{ab}, \quad (7.25)$$

where, as usual, the G^{kl} are defined as the algebraic complements of the G_{kl}.

---

[6] At one time it was even possible to obtain the doctor's degree in certain subjects, such as law, solely by submitting a set of such propositions.

Kramers (1921)[7] returned to the axial vector defined by the antisymmetric tensor R_{kl}, which he calls the rotation vector, in 1920, when he discussed the effect of the sun's gravitational field on the precession of the earth's axis. He found that, in confirmation of a suggestion by Schouten (1918), Einstein's theory of general relativity leads to the result that, if one neglects the effect of the moon and of the mass of the earth, there will be a non-Newtonian contribution to the earth's precession, independent of the constitution of the body of the earth, amounting to a progressive precession of 0.019 arc seconds a year.

In 1932 Olson and Kramers (1932) published a short note in which they used a linear model of the acetylene molecule (C_2H_2) and applied the theory of small vibrations of classical mechanics to derive the frequencies of its five normal modes of an acetylene molecule.

In the early 1940s an eye-specialist friend of Kramers had asked him whether it would be possible to find out anything about the anisotropy of the tangential tensions in the cornea, if one knows the shape of its surface. This is a straightforward classical problem of equating the resultant force deriving from the tensions to the pressure inside the eye. From the experimental data about the shape of the eye—which for normal eyes is not spherical—one can, indeed, calculate (Kramers and ter Haar 1942) the anisotropy which may, at the edge of the cornea, be as high as 18%.[8]

In 1941 the University of Leiden offered a prize for a solution of the problem whether, in the time of about 10^9 years that our stellar system has probably existed, an appreciable number of solid particles could have been formed in the interstellar clouds. In one of the answers submitted (ter Haar 1943) the formation of solid grains was likened to the formation of liquid drops in a supersaturated vapour. In that case the rate at which such grains can be formed depends crucially on the rate at which di- and triatomic molecules are formed. Under laboratory conditions diatomic molecules are formed in three-body collisions, the third atom taking away the excess energy. In interstellar space, however, the density is so low—typically one atom per cm^3—that

---

[7] This paper was presented by Lorentz and Ehrenfest to the Royal Dutch Academy of Sciences on 25 September 1920 and printed in Dutch in the "Verslagen." A correction to the paper was presented on 18 December 1920; Fokker had drawn Kramers's attention to a mistake which this addition corrected. The paper, in English, published in the *Proceedings of the Dutch Academy* and reprinted in the *Collected Scientific Papers*, is the corrected version of the original Dutch paper in the Verslagen.

[8] After the paper was written and submitted for publication, we decided to see whether the thickness of the cornea would affect the final result. This thickness increases toward the edge and its effect is such that the anisotropy is greatly diminished—so much so that it is not excluded that the whole effect is completely negligible. I am convinced that, if the actual numerical calculations had not been made by a young research student, whose first publication this paper was to be, Kramers would never have published this purely academic exercise.

such three-body collisions are extremely rare and other mechanisms must be considered.

It is interesting to note that at that time spectral lines of only three interstellar molecules, CH, CH^+, and CN, had been observed in the ultraviolet, whereas at the present time the number of interstellar molecules of which lines have been observed is legion, thanks to the power of modern radio and infrared telescopes. Although Shklovskii (1946, 1952, 1953) and Townes (1957) suggested that radio telescopes might observe several molecular lines at radiofrequencies, it was not until 1963 that the 18-cm OH line was observed — the beginning of an ever more flourishing branch of interstellar chemistry. The explanation of the presence of CH, CH^+, and CN — and especially that of CH and CH^+ — in the interstellar space was the subject of a paper by Kramers and ter Haar (1946) in which all possible processes involving carbon and hydrogen atoms and ions were considered. They found that under the conditions of interstellar space an equilibrium between the following four processes leads to reasonable agreement between theoretical and observed densities of the CH molecule and the CH^+ molecular ion: radiation capture of a hydrogen atom by a carbon ion, electron capture by the CH^+ molecular ion, photoionisation of the CH molecule, and photodissociation of that molecule. The presence of the CN molecule was explained through a chain starting by the radiation capture of a nitrogen atom by a carbon ion.[9]

In contrast to predictions by von Weizsäcker (1933) that the stopping power of a metal should increase when the temperature is lowered, due to the increasing electrical conductivity leading to an increasing influence of the conduction electrons, experiments by Gerritsen (1946) found no such increase. Kramers (1947a) explained this by using Bohr's classical theory, taking the polarisation due to the conduction electrons into account.

In the last of his published papers[10] Kramers (1952c) explained the results of some other experiments by Gerritsen (1948, 1949; Gerritsen and Koolhaas 1943) on the passage of ions through liquid nitrogen, hydrogen, and helium, which showed a behaviour different from the one predicted by Jaffé's theory (1913) of column ionisation. He showed that a better agreement with the experimental data could be obtained if in the equation of motion one omits the diffusion rather than the recombination term — the correct procedure at low temperatures.

---

[9] It was assumed that under interstellar conditions most of the carbon is ionised and most of the nitrogen neutral.

[10] According to a note by Gerritsen at the end of the paper this work was done four years earlier but through accidental circumstances (Gerritsen's words) had not been published.

Reprinted Papers

A. On the Scattering of Radiation by Atoms

Abstract. If an atom is subjected to external radiation of frequency ν it not only emits secondary monochromatic spherical waves of frequency ν, which are coherent with the incident radiation, but the correspondence principle requires that, in general, also spherical waves of different frequencies are emitted. These frequencies are all of the form $|\nu \pm \nu^*|$ where $h\nu^*$ is the difference between the energies of the state in which the atom finds itself and another state. This incoherent scattered radiation corresponds in part to certain processes recently envisaged by Smekal in his considerations connected with the concept of light quanta. We use the correspondence principle in this paper to show how one can give a natural and apparently unique wave analysis of the scattering action of the atom. The discussion throughout extends the idea of a connection between the wave emission by an atom and stationary states which is given in a recent paper by Bohr, Kramers, and Slater; if the conclusions of the present paper are confirmed they would be an interesting support for that idea.

1. Introduction

It is well known that one can use an atomic picture to interpret the optical phenomena of dispersion and absorption occurring when monochromatic light passes through a gas as the emission by the irradiated atom of secondary spherical waves with the same frequency as the incident light and coherent with it. The conclusion that according to this picture weak scattered radiation is emitted in all directions has been beautifully confirmed by Rayleigh's theory of the blue sky and also directly in laboratory experiments. An extremely strong support for the theory is the fact that the appropriate observations allow us to determine Avagadro's number.

It has been possible to use electron theory based upon classical electrodynamics to give a closer theoretical description of the scattering action of atoms. For instance, the idea that quasi-elastically bound electrons in the atom can perform harmonic oscillations about an equilibrium position, and that these electrons are made to oscillate through the electric forces in the radiation field, has led to a dispersion theory which reproduces the essential features of the observed dispersion, not only in the region of normal dispersion but also in the region of anomalous dispersion near absorption lines with frequencies equal to the eigenfrequencies of the electrons. The theory requires, in accordance with observations, definite maxima in the intensity of the scattered radiation just at those frequencies (resonance radiation).

Nevertheless, it is well known that an attempt to give a definite explanation of the dispersive phenomena on the basis of the classical theory meets with difficulties that are closely connected with the difficulties met when attempting to explain the spectra of the elements using that theory; these latter difficulties, however, were solved by the quantum theory of line spectra. We therefore have the task of describing scattering and dispersive actions of the atom using the quantum theoretical picture of atomic structure. The appearance of a spectral line is, according to this picture, not connected with the presence of elastically oscillating electrons but with transitions from one stationary state to another. Bohr's correspondence principle, though, gives a valuable indication that it might be possible to describe the reaction of atoms to a radiation field using classical concepts. Bohr, Kramers, and Slater[1] have sketched in a recent paper how such a description might be carried out relatively easily. Especially characteristic for this theory is the assumption that the reaction of the atom to the radiation field should be considered primarily to be a reaction of the atom in a definite stationary state; the transitions between two stationary states will take a very short time and the detailed features of these transitions should not play a role when describing optical phenomena. The first step in the description of what happens when an atom in a stationary state is irradiated by monochronatic light will in this scenario be the following. Plane monochromatic waves are incident upon the atom; let the electric vector $\mathfrak{E}(t)$ of these waves at the position of the atom be given by the real part of a vector:

$$\mathfrak{E}(t) = \mathrm{Re}\left(\mathfrak{E}\,e^{2\pi i \nu t}\right), \tag{1}$$

where the components of the time-independent vector $\mathfrak{E}$ are, in general, complex quantities and where ν here and everywhere else in this paper is a positive quantity. The atom will now emit spherical waves into the surrounding space under the action of these waves. The moment $\mathfrak{P}(t)$ of the oscillating dipole which can be considered to be the source of these spherical waves is also given by the real part of some expression:

$$\mathfrak{P}(t) = \mathrm{Re}\left(\mathfrak{P}\,e^{2\pi i \nu t}\right), \tag{2}$$

where $\mathfrak{P}$ is, in general, a complex quantity which depends for a given stationary state on ν and $\mathfrak{E}$; its direction depends on the direction of $\mathfrak{E}$ and its

---

[1] Zs. Phys. **24**, 69 (1924); Phil. Mag. **47**, 785 (1924).

magnitude is, at least in the weak radiation limit, proportional to the absolute magnitude of $\mathfrak{E}$, that is, $\mathfrak{P}$ is a linear vector function of $\mathfrak{E}$. Equation (2) is valid as long as the atom is in the given stationary state.

The assumption made here enables us to a very large extent to explain the dispersion, absorption, and scattering effects of light in gases such as were mentioned at the start of this paper. In particular, as regards the total intensity of the scattered light, one is tempted to put in general the energy S scattered per unit time equal to

$$ S = \frac{(2\pi\nu)^4}{3c^3} \left(\mathfrak{P} \cdot \overline{\mathfrak{P}}\right), \tag{3} $$

where $\overline{\mathfrak{P}}$ is the vector which is the complex conjugate of $\mathfrak{P}$. However, one should expect that this expression is no longer correct in the cases where the atom has made one or more spontaneous transitions to stationary states with a lower energy. In fact, the atom acts according to the point of view of the above-mentioned paper by Bohr, Kramers, and Slater, even when there is no external radiation, as a source for spherical waves with frequencies ν_q which correspond to each transition according to Bohr's frequency condition (spontaneous radiation). The simplest assumption for the description of this radiation is that the atom acts as a classical dipole with a moment which is the real part of the expression

$$ \sum_q \mathfrak{A}_q \, e^{2\pi i \nu_q t}, \tag{4} $$

where the amplitude $\mathfrak{A}$ is a vector connected with Einstein's probability coefficient a_q through the relation

$$ a_q h \nu_q = \frac{(2\pi\nu_q)^4}{3c^3} \left(\mathfrak{A}_q \cdot \overline{\mathfrak{A}}_q\right). \tag{5} $$

In the case of irradiation by monochromatic light the radiation (4) will, in general, give rise to interference with the radiation (2), which will lead to the appearance of more terms in the expression for the scattered radiation because of the finite lifetime of the atom; under certain conditions these terms cannot be neglected as compared to the term (3) (cf. the end of §4). Without considering the limits of the validity of the assumption (2) one of us[2] has recently considered the problem how $\mathfrak{P}$ might depend on ν. The two guiding principles in this investigation were, on the one hand, the fact that experience has shown that the formulæ for $\mathfrak{P}$ from classical dispersion theory are applicable if one thinks that there are classical oscillators in the atom with eigenfrequencies equal to the frequencies of the absorption lines and, on the other hand, the correspondence principle. According to the latter there is a close connection between the actual behaviour of an atomic system and the

---

[2] H. A. Kramers, *Nature* **113**, 673 (1924); **114**, 310 (1924).

action of the system that one would expect according to classical electron theory on the basis of its structure. In particular, the correspondence principle requires that in the region of large quantum numbers one can asymptotically describe the actual properties of the atom using classical electrodynamical laws. Using this requirement it was possible, by comparing the classical dispersion formulæ with the classical response of a multiply periodic system to incident radiation, to construct a dispersion formula adapted to the quantum theory. In the case when the atom is in its ground state this formula is the same as one suggested earlier by Ladenburg[3] on the basis of different considerations.

The aim of the present paper is to show that a further use of the correspondence principle leads to the surprising result that the assumption (2) for the reaction of an atom to incident radiation is too narrowly conceived and that it must, in general, be supplemented by a series of terms:

$$\mathfrak{P}(t) = \text{Re} \left\{ \mathfrak{P} e^{2\pi i \nu t} + \sum_k \mathfrak{P}_k e^{2\pi i (\nu + \nu_k) t} + \sum_l \mathfrak{P}_l e^{2\pi i (\nu - \nu_l) t} \right\}, \quad (6)$$

where $h\nu_k$ or $h\nu_l$ denote the energy differences of the atom in two stationary states, one of which is always the state considered; the vectors $\mathfrak{P}_k$ and $\mathfrak{P}_l$ again depend on $\mathfrak{E}$ and ν, and again in the form of a linear vector function in the first case. Put in words this result means: *An atom will under the action of irradiation by monochromatic light emit not only coherent spherical waves with the same frequency as the incident light, but also sets of incoherent spherical waves with frequencies which are combinations of the same frequency with other frequencies which correspond to all possible transitions to other stationary states.* These extra sets of spherical waves will clearly occur as scattered light; however, they cannot contribute to the dispersion and absorption of the incident light.

Some time ago Smekal[4] reached, on the basis of a consideration connected with the concept of light quanta, the same result, that is, that scattered radiation from an atom may appear with a frequency $\nu + \nu_k$ or $\nu - \nu_l$. We can represent Smekal's arguments more or less as follows. The absorption or emission of light by an atom can be described as a process in which a light quantum of frequency ν is accepted or donated by an atom; as a result the atom makes a transition to a higher or a lower stationary state and changes its energy by an amount equal to $h\nu$ and its momentum by an amount $h\nu/c$. The normal scattering of light by an atom can, on the other hand, be described as the acceptance of a light quantum of frequency ν and the donation of a light quantum of frequency ν'. In this case the atom does not change its stationary state but, in general, will change its velocity. In an arbitrarily chosen frame

---

[3] R. Ladenburg, *Zs. Phys.* **4**, 451 (1921). See also R. Ladenburg and F. Reiche, *Naturwiss.* **11**, 584 (1923).
[4] A. Smekal, *Naturwiss.* **11**, 873 (1923).

of reference the two frequencies, ν and ν', will, in general, be different. In particular, this will be the case for a system of reference in which the atom is at rest (Compton effect). Generalising, Smekal now expresses the conjecture that there must also be processes in the atom in which simultaneously a light quantum is accepted and donated but in which, in contrast to the just mentioned scattering processes, not only the velocity of the atom changes, but the atom also makes a transition to another stationary state. If we neglect the small change in velocity of the atom during the transition and denote the change in energy of the atom in the transition by $h\nu_k$ or $h\nu_l$, depending on whether the transition is in the negative or the positive direction, we find that the frequency of the light quantum emitted in this process is clearly given by $\nu + \nu_k$ or $\nu - \nu_l$ where ν is the frequency of the incident light quantum. One can thus describe this result by saying that if the atom is irradiated by light of frequency ν, light with frequencies $\nu + \nu_k$ or $\nu - \nu_l$ will be emitted by the atom while simultaneously the atoms have a probability of decreasing their energy by $h\nu_k$ or increasing it by $h\nu_l$.

Using light quanta has a special importance because it enables us easily and instructively to connect the macroscopic energy and momentum conservation laws with the ideas of quantum theory. However, by their very nature these considerations do not enable us to reach any conclusions about the corpuscular nature of light as we must require always that the results thus obtained must be brought to agree without contradictions with the wave description of the optical phenomena. One sees immediately that this requirement is met in our case and that the wave theoretical assumption (6) corresponds, in fact, to Smekal's result. We must, though, mention that our later considerations will show that the processes indicated by Smekal are not the only ones that can be connected with the scattering action of atoms. To complete the picture we must also take into account processes in which, to retain the light quantum language, the atom is induced by the radiation to emit two light quanta; one has the frequency ν of the incident light and the other one has a frequency ν' such that the loss of energy, $h(\nu+\nu')$ corresponds to the transition of the atom to a lower state.

In this connection it is of interest to stress that even when we do not neglect the change in the momentum of the atom during the transitions, the necessary requirement of a wave description of the phenomena is satisfied. In fact, we find that if we introduce a frame of reference in which the magnitude of the atomic momentum in its stationary state is $h\nu/c$ and its direction the opposite of that of the incident light (for the processes neglected by Smekal it should be the same as that of the incident light), according to the light quantum calculation the frequency of the scattered light, corresponding to a given transition, will be the same in all directions of space, in agreement with the wave concept of a train of monochromatic waves with its source inside a very small spatial region. We shall not consider further the peculiar fact that the centre of these spherical waves moves with respect to the excited atom. We only mention that assumptions such as (2), (4), and (6) cannot be exactly correct and that they should be suitably modified. However, these modifications would not affect what follows in any essential way.

The idea that one should through use of the correspondence principle connect the scattering action, indicated by Smekal, of atoms irradiated by external radiation through use of the correspondence principle with the scattering action of the atomic system to be expected according to the classical theory was first conceived by Kramers in connection with his work on dispersion theory. The working out of the ideas to be found in the present paper followed from discussions between the two authors.

2. The Effect of External Radiation on a Multiply Periodic System According to Classical Theory

We consider a non-degenerate periodic system the motion of which can be described by the canonical variables $J_1, \ldots, J_s, w_1, \ldots, w_s$. The electrical moment of the system as function of these variables can be represented by the following multiple Fourier series:

$$\mathfrak{M}(t) = \sum_{\tau_1 \ldots \tau_s} \tfrac{1}{2} \mathfrak{C}_{\tau_1 \ldots \tau_s} e^{2\pi i (\tau_1 w_1 + \cdots \tau_s w_s)}. \tag{7}$$

The summation is over all possible positive and negative values of the integers $\tau_1 \ldots \tau_s$. The coefficients $\mathfrak{C}$ are complex vectors depending solely on $J_1 \ldots J_s$. If we again indicate complex conjugate quantities by a bar we have

$$\mathfrak{C}_{\tau_1 \ldots \tau_s} = \overline{\mathfrak{C}}_{-\tau_1 \ldots -\tau_s}. \tag{8}$$

The energy H of the system also depends only on the J. We denote the normal frequencies by

$$\omega_k = \frac{\partial H}{\partial J_k}, \qquad k = 1 \ldots s. \tag{9}$$

We introduce the following notation for a differential operator which occurs repeatedly:

$$\frac{\partial}{\partial J} = \tau_1 \frac{\partial}{\partial J_1} + \cdots \tau_s \frac{\partial}{\partial J_s}, \tag{10}$$

and we also introduce the following abbreviation for the frequencies of the harmonic components which occur in the motion:

$$\omega = \tau_1 \omega_1 + \cdots \tau_s \omega_s = \frac{\partial H}{\partial J}. \tag{11}$$

Our task is to find the electrical moment as function of the time for the case when the atom is subject to a plane monochromatic train of light waves with a wavelength large compared to the extension of the system.

A. On the Scattering of Radiation by Atoms   127

Let the light vector of the incident monochromatic light again be given by Eq. (1). Let $J_1^*, \ldots, J_s^*, w_1^*, \ldots w_s^*$ be a new system of canonical coordinates obtained from the old variables through an infinitesimal contact transformation,

$$J_k^* - J_k = \frac{\partial K}{\partial w_k^*}, \qquad w_k^* - w_k = -\frac{\partial K}{\partial J_k^*}, \qquad k = 1, \ldots s. \tag{12}$$

It is now possible to choose the function $K(J_1^* \ldots J_s^*, w_1^* \ldots w_s^*, t)$ in such a way that the J_k^* become time-independent to a first approximation while the w_k^* increase linearly with time, in such a way that $dw_k^*/dt = \omega_k$. One finds that the function K can be written as follows as the real part of a complex expression:

$$K = \text{Re}\left\{\sum_{\tau_1 \ldots \tau_s} -\frac{1}{2}\frac{(\mathfrak{E} \cdot \mathfrak{C}_{\tau_1 \ldots \tau_s})}{2\pi i(\omega + \nu)} e^{2\pi i(\tau_1 w_1^* + \cdots \tau_s w_s^* + \nu t)}\right\}, \tag{13}$$

where now $\mathfrak{C}$ and ω are the same functions of the J^* as before of the J. If we substitute in (7) the new canonical variables defined through (12) and (13) and at the same time replace the w_k^* by $\omega_k t$ we finally get for the electrical moment of the atom as function of the time the expression:

$$\mathfrak{M}(t) = \mathfrak{M}_0(t) + \mathfrak{M}_1(t), \text{ with } \mathfrak{M}_0 = \sum_{\tau_1 \ldots \tau_s} \tfrac{1}{2} \mathfrak{C}_{\tau_1 \ldots \tau_s} e^{2\pi i \omega t}. \tag{14}$$

Here and in what follows we again again the asterisks at the canonical variables; $\mathfrak{M}_0$ corresponds to the unperturbed motion of the atom. We can write $\mathfrak{M}_1$ as follows as the real part of a $2s$-fold sum:

$$\mathfrak{M}_1(t) = \text{Re}\left\{\sum_{\tau_1 \ldots \tau_s} \sum_{\tau_1' \ldots \tau_s'} \frac{1}{4}\left[\frac{\partial \mathfrak{C}}{\partial J'} e^{2\pi i \omega t} \cdot \frac{(\mathfrak{E} \cdot \mathfrak{C}')}{\omega' + \nu} e^{2\pi i(\omega' + \nu)t}\right.\right.$$

$$\left.\left. - \mathfrak{C} e^{2\pi i \omega t} \frac{\partial}{\partial J}\left(\frac{(\mathfrak{E} \cdot \mathfrak{C}')}{\omega' + \nu}\right) e^{2\pi i(\omega' + \nu)t}\right]\right\}. \tag{15}$$

The summation is over all pairs of combinations of integral values of $\tau_1, \ldots, \tau_s$, $\tau_1', \ldots, \tau_s'$; $\mathfrak{C}$ and $\mathfrak{C}'$ are abbreviations of $\mathfrak{C}_{\tau_1 \ldots \tau_s}$ and $\mathfrak{C}_{\tau_1' \ldots \tau_s'}$; $\partial/\partial J'$ is, by analogy with (10), an abbreviation of $\tau_1' \partial/\partial J_1 + \cdots \tau_s' \partial/\partial J_s$; and ω' an abbreviation of $\tau_1' \omega_1 + \cdots \tau_s' \omega_s$. We shall rewrite expression (15) first by combining those terms for which the sums

$$\tau_1 + \tau_1' = \tau_1^0, \quad \ldots, \quad \tau_s + \tau_s' = \tau_s^0 \tag{16}$$

have the same value. If we then introduce the abbreviation

$$\tau_1^0 \omega_1 + \cdots \tau_s^0 \omega_s = \omega^0, \tag{17}$$

we get

$$\mathfrak{M}_1(t) = \mathrm{Re} \left\{ \sum_{\tau_1^0 \ldots \tau_s^0} \sum_{\tau_1 \ldots \tau_s} \frac{1}{4} \left[\frac{\partial \mathfrak{C}}{\partial J'} \cdot \frac{(\mathfrak{E} \cdot \mathfrak{C}')}{\omega' + \nu} \right. \right.$$
$$\left. \left. - \mathfrak{C} \frac{\partial}{\partial J} \left(\frac{(\mathfrak{E} \cdot \mathfrak{C}')}{\omega' + \nu} \right) \right] e^{2\pi i (\omega^0 + \nu) t} \right\}. \tag{18}$$

When we sum over $\tau_1 \ldots \tau_s$ we must for $\tau_1' \ldots \tau_s'$ substitute the values following from Eqs. (16). We also draw attention to the fact that ω' and ω^0 can take on negative as well as positive values since the summation is over all positive and negative values of the τ_k and τ_k^0. It is clearly necessary that the frequency ν of the incident light not be the same as any of the frequencies ω of the unperturbed motion in order for one to apply this formula.

Equation (18) expresses that the system will under the action of the incident light emit scattered radiation with an intensity proportional to the intensity of the incident light; decomposed into harmonic components it contains both the frequency ν of the incident light as well as frequencies which are the sum or the difference of ν and a frequency ω^0 which can be written in the form (17). It is not necessary that the frequency ω^0 be a frequency of the unperturbed motion. One rather sees from (15) that ω^0 is always of the form $\pm |\omega| \pm |\omega'|$ where $|\omega|$ and $|\omega'|$ are two frequencies which, in fact, occur in the unperturbed motion.

3. Quantum Theory and Coherent Scattered Radiation

If we now proceed on the basis of the quantum theory of multiply periodic systems, we are dealing with a discrete set of stationary states which are given by the quantisation rules

$$J_k = n_k h. \tag{19}$$

The radiation emitted by the unperturbed system in a given stationary state corresponds to transitions to stationary states with a lower energy. Nevertheless we can consider this, according to the correspondence principle, as the appropriate analogue of the radiation to be expected according to the classical theory. Although one can derive this from Eq. (14) for the oscillating electrical moment $\mathfrak{M}_0$ of the unperturbed atom, we should get the quantum theoretical radiation from an oscillating moment given by a formula such as Eq. (4). Each frequency ν_q in this equation corresponds to a classical frequency $\tau_1 \omega_1 + \ldots \tau_s \omega_s$ in such a way that

$$\tau_k = n_k^{(1)} - n_k^{(2)}, \tag{20}$$

where $n_k^{(1)}$ and $n_k^{(2)}$ are the values of the quantum numbers in the initial and the final state, respectively. The value of the classical frequency,

$$\omega = \left(\tau_1 \frac{\partial}{\partial J_1} + \cdots \tau_s \frac{\partial}{\partial J_s}\right) H = \frac{\partial H}{\partial J}, \tag{21}$$

is not the same as the value of the corresponding quantum theoretical frequency ν_q since the latter is given by

$$\nu_q = \frac{1}{h}\left(H^{(1)} - H^{(2)}\right). \tag{22}$$

However, in the limit of large quantum numbers this expression can approximately be written as

$$\nu_q = \frac{\Delta H}{h} = \left(\frac{\Delta J_1}{h}\frac{\partial}{\partial J_1} + \cdots + \frac{\Delta J_s}{h}\frac{\partial}{\partial J_s}\right) H; \tag{23}$$

from this and (19) and (20) it follows directly that in this limit the quantum theoretical frequencies and the classical frequencies are asymptotically the same. In the region of low quantum numbers ν_q is the simple average of the corresponding ω.

In the limit of large quantum numbers the amplitudes $\mathfrak{A}_q$ of the harmonic components of the radiation must asymptotically be the same as the amplitudes $\mathfrak{C}$ of the classical oscillations, while in the region of low quantum numbers one can consider $\mathfrak{A}$ symbolically as a kind of average of $\mathfrak{C}$. We may call the complex vector $\mathfrak{A}_q$ the *characteristic amplitude* for the transition considered. Taking into account that the phase is arbitrary, it is, apart from a complex factor of modulus 1, fully determined and contains five constants. It is clearly related through Eq. (5) to the Einstein coefficient a_q for the probability of a spontaneous transition:

$$a_q h \nu_q = \frac{(2\pi\nu_q)^4}{3c^3}\left(\mathfrak{A}_q \cdot \overline{\mathfrak{A}}_q\right). \tag{24}$$

Bohr[5] has recently emphasised that the polarisation of the emitted radiation is not uniquely determined by the initial state of the transition considered in the case when the state of the system is degenerate. However, in the present paper we restrict ourselves solely to discussing non-degenerate systems where the nature of the emitted radiation presumably is always uniquely determined by the state of the atom.

---

[5] N. Bohr, *Naturwiss.* **12**, 1115 (1924).

Our task now is to find a quantum theoretical expression which is the analogue of the classical formula (18) for the scattering by a system under the influence of external radiation in such a way that the scattering is the same as the classical one in the limit of large quantum numbers. We shall show that this can be achieved by proceeding in a similar way to what Bohr did in the case of frequencies, and interpreting the derivatives occurring in (18) as differences of two quantities in such a way that one naturally obtains formulæ in which only the frequencies and amplitudes characteristic for the transitions occur, while all symbols referring to the mathematical theory of multiply periodic systems have disappeared.

We shall start with that part of the scattered light which has the same frequency ν as the incident light. In the classical case this corresponds to those terms in (18) for which $\tau_1^0 = \cdots \tau_s^0 = 0$, $\tau_k' = -\tau_k$ and which correspond to the following scattering moment:

$$\mathfrak{M}_{kl}(t) = \mathrm{Re} \sum_{\tau_k} \frac{1}{4} \left\{ \frac{\partial \mathfrak{C}}{\partial J} \frac{(\mathfrak{E} \cdot \overline{\mathfrak{C}})}{\omega - \nu} + \mathfrak{C} \frac{\partial}{\partial J} \left(\frac{(\mathfrak{E} \cdot \overline{\mathfrak{C}})}{\omega - \nu} \right) \right\} e^{2\pi i \nu t}. \qquad (25)$$

The expression inside the curly brackets can clearly be written as the derivative of a single expression. If we now combine all pairs of terms for which the quantities τ_k take on numerically the same values but with opposite signs we can rewrite (25) as follows:

$$\mathfrak{M}_{kl}(\nu) = \mathrm{Re} {\sum_{\tau_k}}' \frac{1}{4} \frac{\partial}{\partial J} \left\{ \frac{\mathfrak{C}(\mathfrak{E} \cdot \overline{\mathfrak{C}})}{\omega - \nu} + \frac{\overline{\mathfrak{C}}(\mathfrak{E} \cdot \mathfrak{C})}{\omega + \nu} \right\} e^{2\pi i \nu t}, \qquad (26)$$

where the prime on the summation sign indicates that the summation is over only those τ-combinations for which ω turns out to be positive. One can now obtain a quantum theoretical expression for the scattering moment with frequency ν operating in a given stationary state, which in the large quantum number limit asymptotically is the same as (26) and at the same time agrees with experiments, by writing

$$\mathfrak{M}_{qu}(\nu) = \mathrm{Re} \left[\sum_a \frac{1}{4h} \left(\frac{\mathfrak{A}_a(\mathfrak{E} \cdot \overline{\mathfrak{A}}_a)}{\nu_a - \nu} + \frac{\overline{\mathfrak{A}}_a(\mathfrak{E} \cdot \mathfrak{A}_a)}{\nu_a + \nu} \right) \right.$$
$$\left. - \sum_e \frac{1}{4h} \left(\frac{\mathfrak{A}_e(\mathfrak{E} \cdot \overline{\mathfrak{A}}_e)}{\nu_e - \nu} + \frac{\overline{\mathfrak{A}}_e(\mathfrak{E} \cdot \mathfrak{A}_e)}{\nu_e + \nu} \right) \right] e^{2\pi i \nu t}, \qquad (27)$$

where the first summation is over all frequencies ν_a at which the system shows selective absorption, and the second summation is over all frequencies ν_e appearing in the spontaneous emission. The quantities $\mathfrak{A}_a$ and $\mathfrak{A}_e$ are the amplitudes characteristic for the transitions showing absorption and emission. Equation (27) follows from (26) if one interprets each derivative occurring in (26) as the difference between two quantities referring to two states of motion in which the quantities $J_1 \ldots J_s$ differ by $\tau_1 h \ldots \tau_s h$. Here it does not make sense to consider especially two stationary states; on the one hand, because one would not obtain a quantity that would naturally relate to the reaction of the atom in a given stationary state and, on the other hand, since the values of the amplitudes $\mathfrak{C}$ themselves do not have a special meaning in a stationary state. Rather one should give a meaning to a symbolic average of $\mathfrak{C}$ taken over the range between two stationary states; this would be the meaning of the characteristic amplitude $\mathfrak{A}$ of the corresponding transition. In this way one concludes that the derivatives occurring in (26) must be interpreted as the difference, divided by h, of two quantities corresponding to two transitions characterised by $\tau_1 \ldots \tau_s$ where the stationary state considered is the final state of one transition and the initial state of the other.

Figure A.1 illustrates this procedure. Let the system have two degrees of freedom and let the states of motion be represented by points in a J_1, J_2-plane, as shown in the figure. The stationary states form a point lattice; let P be the stationary state the reaction of which we are studying, and let Q and R represent two stationary states with J-values that are larger, respectively smaller, than those of P by $h\tau_1$ and $h\tau_2$. The derivative occurring in (26) must then be interpreted as the difference between one quantity corresponding to the transition a and one corresponding to the transition e.

We can write Eq. (27) in a slightly different form in the simple case where the vector $\mathfrak{E}$ and all the vectors $\mathfrak{A}_a$ and $\mathfrak{A}_e$ are real and parallel to one another, that is, where the incident light is plane polarised and the electric vector in the radiation corresponding to all the a and e transitions is parallel to the light vector. We introduce the damping time τ_ν of an electron, oscillating classically with frequency ν:

$$\tau_\nu = \frac{3mc^3}{8\pi^2 e^2 \nu^2}, \tag{28}$$

and define the "strength" of a transition by the relation

$$f = a\tau_\nu,$$

where a is the Einstein probability coefficient occurring in (24). We then can write (27) in the form:

$$\mathfrak{M}_{qu}(t) = \mathfrak{E}\frac{e^2}{4\pi^2 m}\left[\sum_a \frac{f_a}{\nu_a^2 - \nu^2} - \sum_e \frac{f_e}{\nu_e^2 - \nu^2}\right]\cos 2\pi\nu t. \tag{29}$$

Kramers[6] gave this formula, which clearly shows the similarity with the classical formulæ, in his first paper on the quantum theoretical dispersion

---
[6] *Nature* **113**, 673 (1924).

theory, while in his second paper[7] he sketched the above derivation. The terms corresponding to the absorption lines correspond to the formula given by Ladenburg.

The direction of $\mathfrak{M}$ is no longer the same as that of $\mathfrak{E}$ in the general case where the vectors $\mathfrak{E}$ and $\mathfrak{A}$ are not real and not parallel to one another, and it is, in general, impossible to write down as simple a formula as (29).

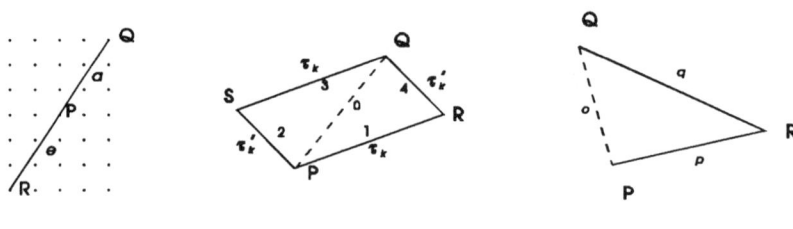

Fig. A.1.   **Fig. A.2.**   **Fig. A.3.**

The moment increases strongly when the frequency of the incident light approaches the frequency ν_a of an absorption line or the frequency ν_e of an emission line, but even if Eq. (27) may be valid otherwise, it must lose its validity in the immediate vicinity of such frequencies, just as the classical formula ceases to be valid when ν approaches ω. At any rate we find that when there is close coincidence with an absorption line, light with the frequency of this line will be strongly scattered. Apparently the resonance radiation observed by Wood and others in the case of metal vapours should be attributed to these strong coherent scattered waves; the presence of these waves is also shown by the fact of the absorption itself as well as by metallic mirror action at high pressures. The resonance radiation will also partly be due to excited states reached by some atoms as a result of the radiation. We do not wish to discuss resonance radiation further; it was merely mentioned to emphasise how suitable the form of expression (27) is to show formally the nature of absorption lines as singularities in the scattering.

On the other hand, although it is true that if ν is close to an emission frequency ν_e expression (27) becomes very large, we cannot conclude that the radiation will enhance the spherical waves of frequency ν_e because of the presence of spontaneous radiation at the frequency ν_e and our lack of knowledge about the phase of the scattered light; there are certain arguments which we do not want to discuss here that indicate, in fact, that the opposite occurs.

From the fact that in each term both the amplitude $\mathfrak{A}$ and its complex conjugate $\overline{\mathfrak{A}}$ occur, so that the uncertainty in the phase of $\mathfrak{A}$ itself drops out, one sees immediately that the scattered radiation given by (27) is coherent

---

[7] *Nature* **114**, 310 (1924). See also J. H. Van Vleck, *Phys. Rev.* **24**, 344 (1924).

with the incident radiation. This coherence is the cause of the dispersion and one sees easily that the incident ray, in general, splits into two polarised parts corresponding to two different refraction indexes. The terms in the second sum in (27) or (29) correspond to negative dispersion corresponding to Einstein's "negative absorption" for $\nu = \nu_e$ just as the normal or positive dispersion corresponds to the normal absorption line for $\nu = \nu_a$.[8]

4. The Incoherent Scattered Radiation

We now proceed to use the correspondence principle to give a quantum theoretical meaning to the other terms in (18) for which ω^0 is non-vanishing. To do this we consider Fig. A.2 which again refers to a system with two degrees of freedom. Let the points P, Q, R, and S in the J_1, J_2-plane correspond to four stationary states such that when the state P corresponds to the quantum numbers $n_k = (n_k)_P$, the states Q, R, and S are characterised by the following quantum numbers:

$$\begin{aligned} (n_k)_Q &= (n_k)_P + \tau_k + \tau_k' = (n_k)_P + \tau_k^0, \\ (n_k)_R &= (n_k)_P + \tau_k, \quad (n_k)_S = (n_k)_P + \tau_k', \end{aligned} \quad (30)$$

where the sets of integers τ_k, τ_k', and τ_k^0 have the same meaning as in Eqs. (16) and (17). The various transitions are denoted by the numbers 0, 1, 2, 3, and

---

[8] Wentzel (*Zs. Phys.* **29**, 306 (1924)) has recently tried to give a quantum theoretical treatment of dispersion which has very little in common with the one given here. Wentzel supports the idea that the quantum theoretical dispersion formula can, in general, not be written in the simple Helmholtz-Ketteler form but must be considered as a kind of distorted classical dispersion formula. This possibility deserves attention since we cannot expect a rigorous derivation of our formula for the induced scattering moment when we use the correspondence principle. Nevertheless, we feel that there are so far no experimental grounds to express doubts about simple formulæ such as (27) or (29). It is true that Wentzel gives as an example dispersion in helium since here the "effective" absorption frequency lies at the short-wavelength side of the limit of the absorption series. (It is nearly 5 per cent larger than the limit frequency.) This situation, however, in no way contradicts the classical formulæ since in the dispersion there are not only contributions from the absorption lines of helium but also from the continuum absorption which extends beyond the series limit on the short-wavelength side. A simple calculation based on the magnitude of this continuum absorption (cf. H. A. Kramers, *Phil. Mag.* **46**, 836 (1923)) or on an extrapolation of the empirical formulæ for the absorption of X-rays to the case of helium shows, in fact, that if one applies the classical formulæ the effect of this continuum absorption on the dispersion in the optical range is the same as that of an absorption line with a frequency about 1.2 times the frequency of the series limit and with an "intensity" of several units, that is, corresponding to several dispersion electrons. Hence, one cannot say that experiments in the case of helium show a failure of the clasical formulæ. (Cf. also K. F. Herzfeld and K. L. Wolf, *Ann. Phys.* **76**, 71 (1925)).

4. Let the transitions 1 and 3, as well as 2 and 4, correspond to spontaneous transitions, that is, let their characteristic amplitudes be different from zero. On the other hand, PQ does not have to correspond to a possible transition, that is, $\mathfrak{A}_0$ may be equal to zero. We denote the frequencies which quantum theoretically correspond to these transitions by ν_0, ν_1, ν_2, ν_3, and ν_4.

First of all, we now must interpret the exponential in (18) as follows:

$$e^{2\pi i(\omega^0+\nu)t} \sim e^{2\pi i(\nu^0+\nu)t}. \tag{31}$$

The quantum theoretical interpretation of the expression within the curly brackets in (18) is obtained by replacing the derivatives by suitably chosen differences, divided by h, the frequencies ω' by the quantum theoretical frequencies, and the amplitudes $\mathfrak{C}$ of the motion by the characteristic amplitudes $\mathfrak{A}$ of the quantum theoretical transitions, and it is, apparently uniquely, determined by the following relations:

$$\left.\begin{array}{c} \dfrac{\partial \mathfrak{C}}{\partial J'} \cdot \dfrac{(\mathfrak{C} \cdot \mathfrak{C}')}{\omega'+\nu} \sim \dfrac{\mathfrak{A}_3 - \mathfrak{A}_1}{h} \cdot \dfrac{1}{2}\left[\dfrac{(\mathfrak{C} \cdot \mathfrak{A}_4)}{\nu_4+\nu} + \dfrac{(\mathfrak{C} \cdot \mathfrak{A}_2)}{\nu_2+\nu}\right], \\[2ex] \mathfrak{C}\,\dfrac{\partial}{\partial J}\left(\dfrac{(\mathfrak{C} \cdot \mathfrak{C}')}{\omega'+\nu}\right) \sim \dfrac{\mathfrak{A}_3 + \mathfrak{A}_1}{h} \cdot \dfrac{1}{2}\left[\dfrac{(\mathfrak{C} \cdot \mathfrak{A}_4)}{\nu_4+\nu} - \dfrac{(\mathfrak{C} \cdot \mathfrak{A}_2)}{\nu_2+\nu}\right]. \end{array}\right\} \tag{32}$$

A number of terms drop out in the subtraction and we find as the interpretation of the expression in braces in (18):

$$\{\tau_k, \tau_k'\} \sim \dfrac{1}{h}\left\{-\dfrac{\mathfrak{A}_1(\mathfrak{C} \cdot \mathfrak{A}_4)}{\nu_4+\nu} + \dfrac{\mathfrak{A}_3(\mathfrak{C} \cdot \mathfrak{A}_2)}{\nu_2+\nu}\right\}. \tag{33}$$

We have tacitly assumed in this transformation that all frequencies ω, ω', and ω^0 were positive and, by quantum theoretical analogy, that the frequencies ν_0, ν_1, ν_2, ν_3, and ν_4, defined by the formulæ

$$h\nu_0 = H(Q) - H(P), \quad h\nu_1 = H(R) - H(P), \quad h\nu_2 = H(S) - H(P),$$
$$h\nu_3 = H(Q) - H(S), \quad h\nu_4 = H(Q) - H(R), \tag{34}$$

all turn out to be positive. Whenever one of these frequencies turns out to be negative, we must in Eq. (33) replace the corresponding characteristic amplitude $\mathfrak{A}$ by the complex conjugate vector $\overline{\mathfrak{A}}$.

As one must take a sum according to (18), it is important to note that the term in (18) obtained by exchanging the values of τ_k and τ_k' corresponds again to the same quadruple of stationary states P, Q, R, S. In this exchange we obtain for the quantity within the braces an expression that one can obtain from (33) by exchanging 1 and 2, and 3 and 4, so that we can write

$$\{T'_k, T_k\} \sim \frac{1}{h}\left\{-\frac{\mathfrak{A}_2(\mathfrak{E}\cdot\mathfrak{A}_3)}{\nu_3+\nu}+\frac{\mathfrak{A}_4(\mathfrak{E}\cdot\mathfrak{A}_1)}{\nu_1+\nu}\right\}. \tag{35}$$

For the sum of the two terms in (18) which correspond according to the quantum theoretical interpretation to the quadrangle PQRS we thus obtain:

$$\{T_k, T'_k\} + \{T'_k, T_k\} \sim \frac{1}{h}\left\{-\frac{\mathfrak{A}_1(\mathfrak{E}\cdot\mathfrak{A}_4)}{\nu_4+\nu}+\frac{\mathfrak{A}_3(\mathfrak{E}\cdot\mathfrak{A}_2)}{\nu_2+\nu}\right.$$
$$\left.-\frac{\mathfrak{A}_2(\mathfrak{E}\cdot\mathfrak{A}_3)}{\nu_3+\nu}+\frac{\mathfrak{A}_4(\mathfrak{E}\cdot\mathfrak{A}_1)}{\nu_1+\nu}\right\}. \tag{35}$$

We now define a complex vector $\mathfrak{M}(P, Q : R)$ which can be assigned to any triple of stationary states P, Q, R:

$$\mathfrak{M}(P, Q : R) = \frac{1}{4h}\left\{\frac{\mathfrak{A}_q(\mathfrak{E}\cdot\mathfrak{A}_p)}{\nu_p+\nu}-\frac{\mathfrak{A}_p(\mathfrak{E}\cdot\mathfrak{A}_q)}{\nu_q+\nu}\right\}e^{2\pi i(\nu_o+\nu)t}. \tag{37}$$

The lower case letters p, q, o refer to the transitions RP, QR, and QP (see Fig. A.3). The frequencies ν_p, ν_q, and ν_o are defined by the relations

$$h\nu_p = H(R) - H(P), \quad h\nu_q = H(Q) - H(R), \quad h\nu_o = H(Q) - H(P), \tag{38}$$

and they can take on negative values. Whenever ν_p (or ν_q) is negative one must replace the characteristic amplitude $\mathfrak{A}_p$ (or $\mathfrak{A}_q$) of the corresponding transition by the complex conjugate vector $\overline{\mathfrak{A}}_p$ (or $\overline{\mathfrak{A}}_q$).

The expression on the right-hand side of (36), multiplied by $\frac{1}{4}e^{2\pi i\nu t}$, can clearly be written as the sum of $\mathfrak{M}(P, Q : R)$ and $\mathfrak{M}(P, Q : S)$, and we can therefore expect from (18) that the scattering moment can be represented as a sum of terms, each of which is of the form Re $\mathfrak{M}(P, Q : R)$. A difficulty arises in that we do not know whether this expression relates to the reaction of the atom in the state P or in the state Q — the state R does not enter this discussion. One can only solve this problem by using considerations that do not directly involve the contents of the correspondence principle. One meets with something similar when discussing the problem of the spontaneous radiation from an unperturbed multiply periodic system. This consists of harmonic components, each of which is connected with a combination of two stationary states and the frequencies (22) and amplitudes $\mathfrak{A}$ of which can be interpreted using the correspondence principle; however, this principle does not allow us to decide in which of these two states the atom is emitting this spontaneous radiation; only considerations connected with the energy conservation law and the nature of the emission can lead to a decision;[9]

---
[9] See N. Bohr, Zs Phys. **13**, 164 (1923).

it is known that this decision is that the emission always takes place from the state with the higher energy, in accordance with Bohr's postulate about radiation. We have to make a similar decision here and we shall assume that the scattering moment (37) corresponds to the state Q if $\nu_0 + \nu$ is positive and to the state P if $\nu_0 + \nu$ is negative. This statement can be justified as follows. Scattering radiation of frequency $\nu_0 + \nu$ means, according to the fundamental laws governing the energy exchange between the radiation field and atoms, that the atom in the state considered has a probability that it loses an amount $h(|\nu_0 + \nu|)$ when it changes its state. On the other hand, irradiation with a frequency ν always gives rise to scattered radiation of frequency ν and the cooperation of the two leads to a reaction in the radiation field such that the atom acquires a probability either to gain or to lose an amount of energy $h\nu$ when it changes its state. An actual change of state of the atom always consists of a transition to another stationary state; in order that the scattered radiation of frequency $|\nu_0 + \nu|$ will lead to the occurrence of such a transition it must therefore be accompanied by action at the frequency ν. If $\nu_0 + \nu$ is positive, this is only possible if the atom simultaneously gains an energy $h\nu$ and loses $h(\nu + \nu_0)$; therefore, in the transition the atom loses altogether an energy $h\nu_0$, that is, it must before the transition be in the state Q and after the transition in the state P. (This is true whether ν_0 is positive or negative.) If, on the other hand, $\nu_0 + \nu$ is negative (here ν_0 must always be negative) the transition can only take place by the atom simultaneously losing both an energy $h\nu$ and an energy $h(-[\nu + \nu_0])$, that is, altogether an energy $-h\nu_0$. Therefore the atom must be in the state P before the transition and in the state Q after it. Transitions of the first kind are mentioned in the above-mentioned paper by Smekal. However, transitions of the second kind, which occur just as naturally from the point of view of light quanta, were not considered.

Using Eqs. (36) and (37) and the just-mentioned decision about the state of the atom which refers to (37) we are now able to indicate a completely general quantum theoretical interpretation of Eq. (18); in fact, we obtain for the scattering moment induced by external radiation in the stationary state P of an atom the expression:

$$\mathfrak{M}(t) = \mathrm{Re}\left\{\sum_Q \sum_R \mathfrak{M}(P,Q:R) + \sum_Q \sum_R \mathfrak{M}(Q,P:R)\right\}, \qquad (39)$$

where in both sums we must sum over all different stationary states R which differ from P and Q. In the first sum we must also sum over all stationary states Q of the atom for which $H(Q) < H(P) + h\nu$; in the second sum over all Q states for which $H(Q) < H(P) - h\nu$. As before, $H(P)$ and $H(Q)$ denote the energies of the atom in the states P and Q.

For the sake of clarity we shall write the general formulæ (37) and (39) in a slightly more special form, namely, in such a way that we indicate the contribution $\mathfrak{M}(|\nu \pm \nu^*|)$, where ν^* is always understood to be a positive quantity, to the total scattering moment of the atom in the state P which corresponds, due to the presence of a state Q, to a definite frequency $|\nu \pm \nu^*|$ in the scattered light. We must distinguish different cases.

Case I. $H(Q) > H(P), \quad H(Q) - H(P) = h\nu^*$

If $\nu > \nu^*$ the state Q contributes to the scattered radiation when $\nu > \nu^*$ and the frequency of this scattered radiation is equal to $\nu - \nu^*$. The states R that are different from P and Q fall into three groups, R_b, R_a, and R_c, depending on whether $H(R)$ is larger than $H(Q)$, smaller than $H(Q)$ but larger than $H(P)$, or smaller than $H(P)$. The absolute magnitudes of the frequencies corresponding to the transitions between P or Q, on the one hand, and R, on the other hand, are denoted by ν_1, ν_2, and so on, corresponding to the numbers in Fig. A.4. We now get terms corresponding to the first sum in (39) and for the scattering moment $\mathfrak{M}(\nu - \nu^*)$ we find

$$\mathfrak{M}(\nu - \nu^*) = \operatorname{Re} \frac{1}{4h} \left\{ \sum_{R_a} \left[\frac{\mathfrak{A}_2(\mathfrak{E} \cdot \overline{\mathfrak{A}}_1)}{\nu_1 - \nu} + \frac{\overline{\mathfrak{A}}_1(\mathfrak{E} \cdot \mathfrak{A}_2)}{\nu_2 + \nu} \right] + \sum_{R_b} \left[\frac{\overline{\mathfrak{A}}_4(\mathfrak{E} \cdot \overline{\mathfrak{A}}_3)}{\nu_3 - \nu} \right. \right.$$

$$\left. \left. - \frac{\overline{\mathfrak{A}}_3(\mathfrak{E} \cdot \overline{\mathfrak{A}}_4)}{\nu_4 - \nu} \right] + \sum_{R_c} \left[-\frac{\overline{\mathfrak{A}}_6(\mathfrak{E} \cdot \mathfrak{A}_5)}{\nu_5 + \nu} - \frac{\mathfrak{A}_5(\mathfrak{E} \cdot \overline{\mathfrak{A}}_6)}{\nu_6 - \nu} \right] \right\} e^{2\pi i (\nu - \nu^*) t}. \quad (40)$$

In the ground state of the atom only this kind of scattered radiation occurs.

Case II. $H(Q) < H(P), \quad H(P) - H(Q) = h\nu^*$

a. Scattered radiation with a frequency $\nu + \nu*$ occurs for all values of the frequency ν. The corresponding terms in (39) all belong to the first sum. The states R, different from P and Q, split into states R_a for which $H(R_a) > H(P)$, states R_b for which $H(P) > H(R_b) > H(Q)$, and states R_c for which $H(Q) > H(R_c)$. The corresponding frequencies are denoted as indicated in Fig. A.5 and are all taken to be positive. For the scattering moment $\mathfrak{M}(\nu + \nu^*)$ we get the expression:

$$\mathfrak{M}(\nu + \nu^*) = \operatorname{Re} \frac{1}{4h} \left\{ \sum_{R_a} \left[\frac{\mathfrak{A}_2(\mathfrak{E} \cdot \overline{\mathfrak{A}}_1)}{\nu_1 - \nu} + \frac{\overline{\mathfrak{A}}_1(\mathfrak{E} \cdot \mathfrak{A}_2)}{\nu_2 + \nu} \right] + \sum_{R_b} \left[-\frac{\mathfrak{A}_4(\mathfrak{E} \cdot \mathfrak{A}_3)}{\nu_3 + \nu} \right. \right.$$

$$\left. \left. + \frac{\mathfrak{A}_3(\mathfrak{E} \cdot \overline{\mathfrak{A}}_4)}{\nu_4 + \nu} \right] + \sum_{R_c} \left[-\frac{\overline{\mathfrak{A}}_6(\mathfrak{E} \cdot \mathfrak{A}_5)}{\nu_5 + \nu} - \frac{\mathfrak{A}_5(\mathfrak{E} \cdot \overline{\mathfrak{A}}_6)}{\nu_6 - \nu} \right] \right\} e^{2\pi i (\nu + \nu^*) t}. \quad (41)$$

b. Scattered radiation with a frequency $\nu^* - \nu$ occurs for all radiation frequencies ν that are smaller than ν^*. The corresponding terms in (39) all belong to the second sum. Using the same notation as in case IIa, we get for the scattering moment $\mathfrak{M}(\nu - \nu^*)$ the expression:

$$\mathfrak{M}(\nu - \nu^*) = \operatorname{Re} \frac{1}{4h} \left\{ \sum_{R_a} \left[\frac{\overline{\mathfrak{A}}_2(\mathfrak{E} \cdot \mathfrak{A}_1)}{\nu_1 + \nu} + \frac{\mathfrak{A}_1(\mathfrak{E} \cdot \overline{\mathfrak{A}}_2)}{\nu_2 - \nu} \right] + \sum_{R_b} \left[-\frac{\overline{\mathfrak{A}}_4(\mathfrak{E} \cdot \overline{\mathfrak{A}}_3)}{\nu_3 - \nu} \right. \right.$$
$$\left. \left. + \frac{\overline{\mathfrak{A}}_3(\mathfrak{E} \cdot \overline{\mathfrak{A}}_4)}{\nu_4 - \nu} \right] + \sum_{R_c} \left[-\frac{\mathfrak{A}_6(\mathfrak{E} \cdot \overline{\mathfrak{A}}_5)}{\nu_5 - \nu} - \frac{\overline{\mathfrak{A}}_5(\mathfrak{E} \cdot \mathfrak{A}_6)}{\nu_6 + \nu} \right] \right\} e^{2\pi i(\nu - \nu^*)t}. \quad (42)$$

Case III. $H(\mathrm{Q}) = H(\mathrm{P})$.

The scattered radiation has the same frequency ν as the incident light.

a. The state Q is not the same as the state P (see Fig. A.6). This case follows from both case I and case IIa when ν^* is equal to zero. Of course, the summation over R_b drops out. The frequencies ν_1 and ν_2 are the same as the frequencies ν_5 and ν_6; we denote them by ν_{12} and ν_{56}, respectively.

$$\mathfrak{M}(\nu) = \operatorname{Re} \frac{1}{4h} \left\{ \sum_{R_a} \left[\frac{\overline{\mathfrak{A}}_2(\mathfrak{E} \cdot \mathfrak{A}_1)}{\nu_{12} - \nu} + \frac{\overline{\mathfrak{A}}_1(\mathfrak{E} \cdot \mathfrak{A}_2)}{\nu_{12} + \nu} \right] \right.$$
$$\left. + \sum_{R_c} \left[-\frac{\overline{\mathfrak{A}}_6(\mathfrak{E} \cdot \mathfrak{A}_5)}{\nu_{56} + \nu} - \frac{\mathfrak{A}_5(\mathfrak{E} \cdot \overline{\mathfrak{A}}_6)}{\nu_{56} - \nu} \right] \right\} e^{2\pi i \nu t}. \quad (43)$$

If we bear in mind the meaning of the vectors $\mathfrak{A}$ we see that the scattered light has no phase relation with the incident light because the phase of the $\mathfrak{A}$ is undetermined.

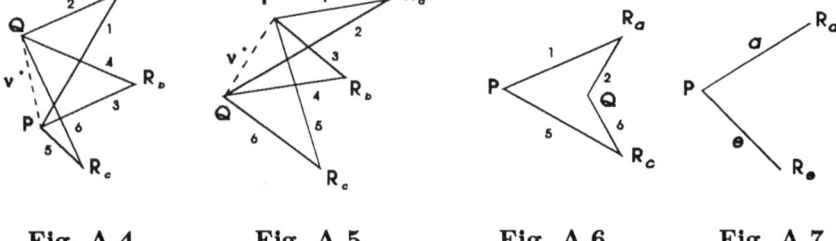

Fig. A.4.   **Fig. A.5.**   **Fig. A.6.**   **Fig. A.7.**

b. The state Q is the same as the state P (see Fig. A.7). If we denote the frequencies of the absorption and the emission lines in the state P by ν_a and ν_e we obtain:

$$\mathfrak{M}(\nu) = \operatorname{Re} \frac{1}{4h} \left\{ \sum_{R_a} \left[\frac{\mathfrak{A}_a(\mathfrak{E}\cdot\overline{\mathfrak{A}}_a)}{\nu_a - \nu} + \frac{\overline{\mathfrak{A}}_a(\mathfrak{E}\cdot\mathfrak{A}_a)}{\nu_a + \nu} \right] \right.$$
$$\left. - \sum_{R_e} \left[\frac{\overline{\mathfrak{A}}_e(\mathfrak{E}\cdot\mathfrak{A}_e)}{\nu_e + \nu} + \frac{\mathfrak{A}_e(\mathfrak{E}\cdot\overline{\mathfrak{A}}_e)}{\nu_e - \nu} \right] \right\} e^{2\pi i \nu t}. \qquad (44)$$

Clearly the fact that the phase of $\mathfrak{A}$ is undetermined does not play a role for this expression, and the scattered radiation is therefore coherent with the incident radiation. Equation (44) expresses exactly the same as Eq. (27).

It is especially interesting to study the singular behaviour of expressions (40), (41), and (42) in the vicinity of some critical values of the frequencies. For instance, we see that in cases I and II the scattering moment approaches very large values when the incident frequency approaches the absorption frequency ν_1 and when the scattered frequency approaches the value ν_2. Of course, the scattering moment cannot become infinite, as would follow from the formulæ; by analogy with the classical scattering formula (18), which is valid only for values of ν that differ from the eigenfrequency ω of the system, we expect that Eq. (40) is valid only as long as $\nu - \nu_1$ is large as compared to the line width of the absorption line. Nevertheless we may assume that Eq. (40) is correct in that it is, indeed, the frequency ν_1 which occupies a critical place for the scattering, in that when the atom is irradiated by light which contains the frequency ν_1 the scattered radiation will have a pronounced maximum at the frequency ν_2. Parallel with this is the fact that in the case of irradiation with the frequency ν_1 the atom acquires a probability to make a transition to the state R_a. Once it has reached this state it may make a spontaneous transition to the state Q and the possibility for this transition corresponds to the emission of spherical waves of frequency ν_2 in the state R_a. The observation of scattered radiation with a frequency ν_2 can thus be due to two different kinds of processes; the scattered radiation is partly emitted by all atoms in the state P which we may, for instance, consider to correspond to the ground state, and partly it comes from the few atoms which through the radiation have reached the state R_a and now spontaneously emit, among other frequencies, the frequency ν_2. This case is clearly similar to the situation mentioned earlier, in that the scattered radiation with the same frequency as that of the incident light comes partly from atoms in the ground state P and partly from atoms in the excited state Q. It is possible that the two mentioned contributions to the observable scattered radiation may be of the same order of magnitude. We shall not further discuss the question of what their relative magnitude will be under various circumstances; this is closely connected with the structure of the absorption line.

However, we want to draw attention to two important points. The first one concerns the relation between the intensities with which the frequencies ν_1 and ν_2 will be scattered in the state P when there is incident radiation which contains the frequency ν_1. Of course, one cannot use Eqs. (40) and (44) to calculate this relation, since these formulæ are invalid for very small values of $\nu - \nu_1$ which are of the same order of magnitude as the line width. We shall, however, consider the case where the difference between the frequency ν of the incident light and ν_1 is such that $\nu - \nu_1$ is large as compared to the line width, but at the same time $\nu - \nu_1 = -\delta$ is so small that the relevant terms in Eqs. (40) and (44) are large compared with all other terms. The scattering moment at the frequency $\nu_1 - \delta$ will then be approximately equal to $\text{Re}\big(\mathfrak{A}_1(\mathfrak{E} \cdot \overline{\mathfrak{A}}_1)/4h\delta\big)e^{2\pi i(\nu_1-\delta)t}$ and the scattering moment at the frequency $\nu_2 - \delta$ to $\text{Re}\big(\mathfrak{A}_2(\mathfrak{E} \cdot \overline{\mathfrak{A}}_1)/4h\delta\big)e^{2\pi i(\nu_2-\delta)t}$. The ratio of the intensities at these two frequencies will thus be independent of δ and equal to $\nu_1^4(\mathfrak{A}_1 \cdot \overline{\mathfrak{A}}_1)/\nu_2^4(\mathfrak{A}_2 \cdot \overline{\mathfrak{A}}_2)$. This is, however, exactly the ratio of the intensities emitted by the atom in the state R_a at the frequencies ν_1 and ν_2. *We may therefore expect quite generally that the intensity relations of the scattered radiation which appears when an atom is irradiated by light with frequencies which contain an absorption frequency will be exactly the same as the intensity relations of the spontaneous radiation at the corresponding frequencies which appear in the excited state.* This expectation or assumption plays an essential role in the discussion of the polarisation in the case of resonance radiation of metallic vapours.[10] One sees that the above considerations give strong support, based on the correspondence principle, for this assumption, independently of a detailed investigation, such as the one partially attempted in the present paper, of how the scattered radiation is produced.

The second point concerns the possibility for certain transitions between stationary states with which the presence of scattered radiation is closely connected. From the earlier discussion it follows immediately that in all special cases considered, the appearance of scattered radiation is connected with the possibility of a jump of the atom from the state P to the state Q in such a way that the probability for the jump per unit time is exactly equal to the energy, divided by $h(|\nu \pm \nu^*|)$, which is emitted per unit time in the form of the scattered radiation of frequency $|\nu \pm \nu^*|$. The probability for a direct jump to the state Q is therefore very large, if in case I the incident radiation contains the frequency ν_1. According to the above we may expect that this probability stands to the probability for a jump to another state Q' which can also be reached from R_a in the same ratio as the probabilities for the atom in the state R_a to jump either to Q or to Q'.

So far we have only talked about the critical position as regards the scattered radiation occupied in case I by the frequency ν_1 of an absorption line. However, we can see from Eq. (40) that the expression for the scattering moment also increases strongly when ν approaches ν_6. This looks strange at first sight, as one might think that irradiation by light containing the frequency ν_6 would lead to strong scattered radiation and that at the same time, because of the energy balance, the atom should show an absorption line with the fre-

---

[10]See N. Bohr, *Naturwiss.* **12**, 1115 (1924) and especially W. Heisenberg, *Zs. Phys.* **31**, 617 (1925), where the correspondence principle was used in an attempt at a quantitave theory of the observations.

quency ν_6. Such a conclusion, however, contradicts Eq. (44), which expresses the fact that for the reaction of the atom to monochromatic irradiation only those frequencies are critical which correspond to a transition to another stationary state. However, one can show easily that Eq. (40) can represent the situation correctly without the coincidence $\nu = \nu_6$ being reflected in some way or other by the absorption or dispersion of the incident light as is required by (44). When ν approaches ν_6 the frequency $\nu - \nu^*$ comes ever closer to the frequency ν_5. This frequency, however, has always been present in the spontaneous radiation by the atom; that part of the electric moment of the atom corresponding to this frequency is given by the expression $\operatorname{Re}\mathfrak{A}_5 e^{2\pi i \nu_5 t}$. For small values of the difference $\nu - \nu_6 = -\delta$ the induced moment that is according to (40) equal to $\operatorname{Re}\{-(\mathfrak{A}_5(\mathfrak{E} \cdot \overline{\mathfrak{A}_6})/4h\delta)e^{2\pi i (\nu_5 - \delta)t}\}$ will become more and more able to interfere with the spontaneous radiation in such a way that in the expression for the scattered intensity terms appear containing the light vector $\mathfrak{E}$ linearly. The fact that these terms do not become infinitesimal, notwithstanding the fact that there is a difference δ in frequency is naturally explained by the fact that the atom has a finite lifetime—just because of its spontaneous emission. The actual intensity of the scattered radiation induced in the atom by the incident radiation depends on the phase relation between the induced scattering moment and the moment, already present, of the spontaneous emission, and calculations show that for an appropriate phase relation the intensity of the induced scattered radiation, when there is irradiation present with a frequency close to $\nu = \nu_q$, will not show peculiar behaviour at the same time that Eq. (40) holds. A simple calculation will show this.

We assume for the sake of simplicity that the characteristic amplitude of the spontaneous scattered radiation at a frequency ν_0 in a stationary state P is real and its absolute magnitude is p. Let the mean lifetime of P be given by $1/a$. Let the frequency of the induced scattering moment be $\nu_0 - \delta$ where δ is small compared to ν_0 but large compared to a. Let the absolute magnitude of its amplitude be q. Let $t = 0$ denote the moment the atom reached the state P and let the phase difference between the two scattered moments at that time be φ. The energy emitted during the lifetime T is then proportional to the expression

$$s(T, \varphi) = \int_0^T \{p \cos 2\pi \nu_0 t + q \cos[2\pi(\nu_0 - \delta)t + \varphi]\}^2 \, dt. \tag{45}$$

The average energy emitted during the lifetime of the state P is thus proportional to the expression

$$S(\varphi) = a \int_0^\infty s(T, \varphi) e^{-aT} \, dT. \tag{46}$$

Neglecting small quantities of order δ/ν_0 we find

$$S(\varphi) = \frac{p^2}{2a} + \frac{pq}{a^2 + (2\pi\delta)^2}(a\cos\varphi + 2\pi\delta\sin\varphi) + \frac{q^2}{2a}, \qquad (47)$$

or

$$S(\varphi) = \frac{p^2}{2a} + \frac{pq}{\sqrt{a^2 + (2\pi\delta)^2}}\sin(\varphi + \alpha) + \frac{q^2}{2a}, \text{ with } \tan\alpha = \frac{a}{2\pi\delta}. \qquad (48)$$

Let now, in accordance with (40), the moment q be given by the expression

$$q = -\frac{Epp'}{4h\delta}, \qquad (49)$$

where E denotes the amplitude of the electric vector in the incident waves while p' is a quantity of the same order of magnitude as p. If we further neglect terms of relative order δ^2/a^2 (48) takes on the following form:

$$S(\varphi) = \frac{p^2}{2a} - \frac{p^2 p' E}{8\pi h} \cdot \frac{\sin(\varphi + \alpha)}{\delta^2} + \frac{p^2 p'^2 E^2}{32 h^2 a}\frac{1}{\delta^2}. \qquad (50)$$

It is clear from this expression that the second term can always be of the same order of magnitude as the third one, if the incident radiation is sufficiently weak (E very small), and that for fixed $\varphi + \alpha$ their dependence on δ is the same. The third term would be exactly cancelled by the second if $\sin(\varphi + \alpha)$ on average were equal to $\pi p' E/4ha$. In order that our considerations hold this quantity must be small compared to 1. Comparison with (49) and the fact that for very small values of $2\pi\delta$ this quantity must in (49) be replaced by an expression of the form $\sqrt{a^2 + (2\pi\delta)^2}$ show us therefore that our considerations can only be valid if the induced scattering moments are also smaller than the moments of the spontaneous scattering moments in the atom in those cases where the induced moment reaches a maximum for certain critical frequencies. This restriction would seem to be a priori a natural one.

Our equations (40) and (44), based upon the correspondence principle, have thus led to assuming a phase relation of the kind considered. The fact that the phase of the induced scattering moment depends not only on $\mathfrak{A}_5$ but also on $\overline{\mathfrak{A}_6}$ and that this vector in the stationary state P has, of course, a completely undetermined phase, shows that this assumption can be considered to be a natural one.

So far we have considered only the critical values of the frequency ν for case I when the scattered radiation has the frequency $\nu - \nu^*$ and is connected with a transition to a state with higher energy. The discussion is essentially the same for the cases IIa and IIb. In case IIa there are two kinds of critical frequencies. One is again given by the absorption frequency ν_1. Irradiation with light containing the frequency ν_1 gives rise to scattered radiation that reaches a maximum of intensity at the frequency ν_2 (see Fig. A.5). Parallel with that is the fact that the atoms through the irradiation can be lifted into the state R_a and that ν_2 is spontaneously emitted from this state. Second, ν_6 is a critical frequency which, however, should not lead to an intensity maximum of scattered radiation at the frequency ν_5 which is already present in the form of spontaneous radiation. In case IIb one needs consider only the frequencies ν_3 and ν_4 (ν_2 and ν_5 are always larger than ν^* and ν must always be smaller than ν^*). Irradiation by light containing ν_3 gives rise to scattered light with a maximum at ν_4. This corresponds to the fact that the atom acquires through the radiation a larger probability to make a transition to the state R and in that state ν_4 is emitted spontaneously. On the other hand, irradiation with ν_4 leads to a strong scattering moment of the atom with frequency ν_3. However, the intensity of the scattered radiation may not reach an intensity maximum at the frequency ν_3 which is already present from the beginning, since we must assume that the scattering moment with the frequency ν_4 of the incident radiation does not have a maximum.

Summarising, we can say that an atom when irradiated with light containing an absorption or emission frequency of the atom will imitate the radiation properties of that state which together with the given state characterises the relevant absorption or emission frequency.

5. Concluding Remarks

One should require that the types of scattered radiation considered in the present paper and the jumps between stationary states to which they give rise leave the distribution in the black-body radiation as well as the statistical equilibrium distribution of the atoms unchanged. The assumptions needed to have this come true must clearly show a great similarity with the assumptions that enabled Pauli[11] to describe the equilibrium between free electrons and black-body radiation. For instance, in case I the probability for a direct jump from P to Q must be represented by a series of terms which among others contains a term proportional to the radiation density $\varrho(\nu_1)$ at the frequency ν_1 and a term proportional to the product $\varrho(\nu_1)\varrho(\nu_2)$. On the other hand, it follows from the above that there is no term proportional to $\varrho(\nu_6)$.

The considerations of the present paper show that it should hardly be possible to satisfy the requirement from the correspondence principle that the scattered radiation induced by external radiation in an atomic system must in the limit of large quantum numbers be the same as the scattered radiation following from the classical theory by anything but formulæ of the kind of (39) where $\mathfrak{M}$ is of the nature characterised by (37). Even if the description of the actual scattering moment would require a less simple expression for $\mathfrak{M}$ one can hardly avoid the consequences that the moment of the scattered radiation will take on an especially large value whenever a condition of the form $\nu + \nu_p = 0$ or $\nu + \nu_q$ is satisfied. It also seems to be impossible to arrange the

---

[11]W. Pauli Jr., *Zs. Phys.* **18**, 272 (1923).

formulæ in such a way that the correspondence principle is satisfied without the critical values for the frequencies corresponding to transitions for which neither the initial nor the final state is the same as the state of the atom which is considered. We have shown that the difficulty which seems to arise about the question of the exchange of energy between the atom and the radiation field can be solved naturally by assuming that during the lifetime of the state considered the atom can emit spontaneous radiation. One can conclude from this that the idea that the scattering action of an atom under external irradiation must be considered to be the action of the atom in a definite stationary state is necessarily accompanied by the idea that one must also describe the spontaneous radiation of an atom as radiation in a definite stationary state and not as the action shown by the atom only when it makes a transition between stationary states. This is just the hypothesis introduced by Slater which gave rise to the ideas further developed by Bohr, Kramers, and Slater. We mentioned in §1 that the general basic elements of these ideas can be formulated by saying that the role played by the atom in optical phenomena can always be related to the interactions between the radiation field and the atom in a stationary state; the peculiar processes which we can describe as "transitions from one stationary state to another" can be assumed to last for a very short time and, as far as they are known and analysed, optical phenomena do not give any information about them.

B. Some Remarks on Heisenberg's Quantum Mechanics

Abstract. In Heisenberg's quantum mechanics Born and Jordan's so-called "quantum conditions" mean that an electron which is bound in an atom behaves in high-frequency oscilating electric fields exactly like a bound electron in the classical theory.

Heisenberg[1] describes the motion of an electron in an atomic system symbolically by means of a number of expressions

$$A_{m,n} e^{2\pi i \nu_{m,n} t}, \tag{1}$$

which can be assigned to various possible transitions from a stationary state m to a state n; m and n are just indices and need not be quantum numbers. The quantity

$$\nu_{m,n} = \frac{W_m - W_n}{h} \tag{2}$$

is the frequency of the light which, according to Bohr, can be emitted in such a transition; it must be counted negative if the energy W_m in the state m is smaller than the energy n in the state n:

$$\nu_{m,n} = -\nu_{n,m}. \tag{3}$$

The quantity $A_{m,n}$ is a vector the components of which are, in general, complex; it represents the "characteristic amplitude"[2] which one can assign to the transition $m \to n$, that is, the light emitted by an atom in the state m with a frequency $\nu_{m,n}$ can be ascribed in the framework of Maxwell's radiation theory to an oscillating dipole, the electric moment of which is given by the real part of the above expression. If $W_m < W_n$ we must write

---

[1] W. Heisenberg, Zs. Phys. **33**, 879 (1925). I express my gratitude to Mr. Heisenberg, who kindly communicated to me the most recent developments of his theory, made by Born, Jordan, and himself.
[2] H. A. Kramers and W. Heisenberg, Zs. Phys. **31**, 681 (1925).

$$A_{m,n} = \overline{A}_{n,m}, \tag{4}$$

where the bar indicates the conjugate complex value. We introduce Cartesian coordinates $q^{(1)}$, $q^{(2)}$, and $q^{(3)}$ and denote the component of $A_{m,n}$ along the $q^{(k)}$-axis by $A_{m,n}^k$. We can express the symbolic motion of the electron through the formula

$$q^k = \{q_{m,n}^k\} = \left\{\frac{1}{2e} A_{m,n}^k e^{2\pi i \nu_{m,n} t}\right\}. \tag{5}$$

The quantities within the braces are elements of an infinite Hermitean matrix. It is now important that in a natural manner one can define for these matrices not only addition but also multiplication in such a way that both operations are associative, that the addition is commutative, and that the multiplication is distributive. If $\{a_{m,n}\}$, $\{b_{m,n}\}$, .. are matrices, we have

$$\begin{aligned}\{a_{m,n}\} + \{b_{m,n}\} &= \{a_{m,n} + b_{m,n}\}, \\ \{a_{m,n}\} \times \{b_{m,n}\} &= \left\{\sum a_{m,j} b_{j,n}\right\}.\end{aligned} \tag{6}$$

The summation is here over all rows of a and all columns of b. The multiplication is, in general, not commutative. As to the time-dependence of the matrix elements, we see that if this dependence is in the form indicated by (5), this form is retained in the multiplication, since from Bohr's frequency condition (2) it directly follows that $\nu_{m,j} + \nu_{j,n} = \nu_{m,n}$. The components $p^{(1)}$, $p^{(2)}$, and $p^{(3)}$ of the electron's momentum which in classical (non-relativistic) mechanics are given by

$$p^k = m\frac{dq^k}{dt}, \tag{7}$$

can in non-relativistic quantum mechanics be described by the matrices

$$p^k = \{p_{m,n}^k\} = \left\{\frac{\pi i m \nu_{m,n}}{e} A_{m,n}^k e^{2\pi i \nu_{m,n} t}\right\}. \tag{8}$$

In classical mechanics there exists, if we consider the motion of an electron in a conservative field of force, a Hamiltonian $H(q^k, p^k)$ with a constant value during the motion equal to what we call the energy of the system. In quantum mechanics this corresponds to a certain function $H(q^k, p^k)$ of the matrices q^k, p^k such that in the matrix which represents[3] H all elements which depend on the time $(n \neq m)$ will vanish, that is, that H is a "diagonal matrix." The

---

[3] It is immediately clear that if H can be expanded in integral positive and negative powers of p^k and q^k H itself will also again be a matrix. One thus arrives at the conclusion that it is possible to define a very general class of "functions of matrices."

element $H_{m,m}$ of this matrix will be equal to what we call the energy W_m in the stationary state m.

The p's and q's must satisfy not only the energy equation

$$H(q^k, p^k) = \text{constant} \quad \text{(that is, a diagonal matrix)}, \tag{9}$$

but also other conditions which are closely connected with the quantum conditions in the theory of periodic systems and which, with Heisenberg, we shall call the "quantum conditions." Born and Jordan, to whom we are indebted for the systematic development of Heisenberg's quantum mechanics, have shown that these conditions can be written in the following form:

$$q^k q^l - q^l q^k = 0, \tag{10}$$
$$p^k q^l - q^l p^k = 0, \quad (k \neq l) \tag{11a}$$
$$= \frac{h}{2\pi i}, \quad (k = l) \tag{11b}$$
$$p^k p^l - p^l p^k = 0. \tag{12}$$

Each of these conditions expresses an infinite number of relations between the quantities $A_{m,n}$ and $\nu_{m,n}$: (10), (11a), and (12) mean that all elements of certain matrices vanish, while (11b) means that the matrix $p^k q^l - q^l p^k$ is a diagonal matrix the elements of which are equal to $h/2\pi i$. The quantum mechanical problem is defined by Eqs. (9) to (12).

The aim of the present paper is to draw attention to a certain interpretation of the quantum conditions (10) to (12) which is closely connected with the interpretation given by Heisenberg to the quantum condition, and which he uses in his paper cited earlier.[4] To do this we start from the expressions that Kramers and Heisenberg, using the correspondence principle, derived for the "scattering dipole moment" excited in an atom by a monochromatic light ray. The derivation was based on the principle that the relations to be found should contain exclusively quantities which in principle could be observed, viz., the frequencies and the characteristic amplitudes. In fact, Heisenberg's whole theory is an ingenious attempt to apply this principle as consistently as possible to the problem of determining the energy of the stationary states and the transition probabilities for transitions between them. He then came to the conclusion that this appeared, indeed, to be possible and that even in simple cases one obtained results that contradicted the way in which one had previously determined the energy of stationary states, which was based on an

---

[4] This condition refers to a system with one degree of freedom and expresses that the diagonal elements of the matrix $pq - qp$ are equal to $h/2\pi i$. It is therefore incomplete compared to Born and Jordan's conditions. Heisenberg draws attention to the fact that it is identical with the contents of the "sum rule" recently proposed by Kuhn (*Zs. Phys.* **33**, 408 (1925)) and Thomas (*Naturwiss.* **13**, 627 (1925)).

analysis of the solution of the classical equations of motion. Although the existence of this contradiction in the paper by Kramers and Heisenberg was not suspected or accepted, the results of that paper are in excellent agreement with quantum mechanics in the form developed by Heisenberg, Born, and Jordan, and can be derived directly from it. In matrix terms the expression of the scattering moment of an atom under the influence of an oscillating electric field with an electric force given by the real part of

$$Ee^{2\pi i \nu t}, \tag{13}$$

(E is a complex vector) takes the following simple form:[5]

$$e\Delta q^k = \frac{e^2}{h} \text{Re} \sum_l E^l \left[\{q^k_{m,n}\} \left\{\frac{q^l_{m,n}}{\nu_{m,n}+\nu}\right\} - \left\{\frac{q^l_{m,n}}{\nu_{m,n}+\nu}\right\} \{q^k_{m,n}\} \right] e^{2\pi i \nu t}. \tag{14}$$

Here $e\Delta q^k$ is a matrix the element of which in the column characteriseed by m and the row characterised by n represents the component along the q^k-axis of the scattering moment of the atom assigned to the $m \to n$ transition with frequency $|\nu_{m,n}+\nu|$. To determine whether this scattering moment is excited in the atom when it is in the state m or rather in the state n needs a special consideration based upon the principle of the conservation of energy, in this case upon a calculation involving light quanta. For our aim it is superfluous to dwell further on such a consideration but it is important to mention in passing that, on the one hand, it is absolutely necessary in order to arrive at a physical interpretation of Eq. (14), as well as any formula in the new theory, while, on the other hand, it is foreign to the formalism of that theory. The relation is here of exactly the same nature as that between the theory of light quanta and the wave theory of light; both are necessary for a description of phenomena, and nevertheless it has not been possible, at any rate so far, to combine them into an organic unity.

At first sight expression (14) differs so much from the formulæ of the classical electron theory that one might ask whether it can make sense to say that (14) refers to a classically defined "electron" bound in a field of force. However, one can give an affirmative answer to this question. In fact, we can require that for very large values of the frequency ν of the incident light the nature of the forces that bind the electron must have very little effect on the scattering moment, so that Eq. (14) asymptotically in the region of large ν will reproduce the values of the scattering moment of a free electron according to the classical theory, at least with the same approximation which one can use when the binding forces are classically taken into

---

[5] Compare Eq. (37) of the paper by Kramers and Heisenberg. Here and henceforth Re stands for the "real part of."

account. Consider an electron which describes according to the laws of classical electrodynamics a periodic motion in a three-dimensional field of force with potential $U(q^1, q^2, q^3)$. We shall denote the Cartesian coordinates q^k and the corresponding momenta p^k as functions of time in the unperturbed motion by q_0^k and p_0^k. If the motion is perturbed by an oscillating field of the form (13) we can write

$$q^k = q_0^k + \Delta q^k, \qquad p^k = p_0^k + \Delta p^k. \tag{15}$$

A simple mechanical consideration now shows us that for large values of ν the expressions for Δq^k and Δp^k, expanded in negative powers of ν, will look as follows:

$$\Delta q^k = -\frac{1}{\nu^2} \frac{e}{4\pi^2 m} \mathrm{Re} E^k e^{2\pi i \nu t} + \frac{C_1}{\nu^3} + \cdots, \tag{16}$$

$$\Delta p^k = -\frac{1}{\nu} \frac{e}{2\pi} \mathrm{Re}\, iE^k e^{2\pi i \nu t} + \frac{C_2}{\nu^3} + \cdots, \tag{17}$$

where the coefficients C_1 and C_2, which depend on the nature of the binding field of force, represent sums of periodic terms with frequencies of the form $\nu \pm \omega$, where ω is a frequency occurring in the unperturbed motion. If we now expand expression (14) in negative powers of ν we get

$$\Delta q^k = \frac{1}{\nu} \frac{e}{h} \mathrm{Re} \sum_l E^l \left[\{q_{m,n}^k\} \{q_{m,n}^l\} - \{q_{m,n}^l\} \{q_{m,n}^k\} \right] e^{2\pi i \nu t}$$

$$- \frac{1}{\nu^2} \frac{e}{h} \mathrm{Re} \sum_l E^l \left[\{q_{m,n}^k\} \{\nu_{m,n} q_{m,n}^l\} - \{\nu_{m,n} q_{m,n}^l\} \{q_{m,n}^k\} \right] e^{2\pi i \nu t}$$

$$+ \frac{C_3}{\nu^3} + \cdots . \tag{18}$$

If we now use (8) and introduce the same simplified notation as applied earlier in (10) to (12) we can write

$$\Delta q^k = \frac{1}{\nu} \frac{e}{h} \mathrm{Re} \sum_l E^l \left(q^k q^l - q^l q^k \right) e^{2\pi i \nu t}$$

$$- \frac{1}{\nu^2} \frac{e}{2\pi h m} \mathrm{Re}\, i \sum_l E^l \left(p^l q^k - q^k p^l \right) e^{2\pi i \nu t} + \frac{C_3}{\nu^3} + \cdots . \tag{19}$$

We can now satisfy the requirement that the atom in the limit of large ν values behaves like a free electron by demanding that (19) have the same form as (16). This gives

$$q^k q^l - q^l q^k = 0, \tag{20}$$

$$p^k q^l - q^l p^k = 0 \ (k \neq l); \quad = \frac{h}{2\pi i} \ (k = l). \tag{21}$$

These conditions are identical with (10) and (11). The conditions (12) are also necessary in order that the atom behave in the required approximation as a free electron. To see this, one needs only investigate the quantum mechanical expression for Δp^k which corresponds to (14). It is true that such an expression is not given in the paper by Kramers and Heisenberg, but it can be derived directly by the methods used there and it is

$$\Delta p^k = \frac{e}{h}\text{Re}\sum_l E^l \left[\{p^k_{m,n}\}\left\{\frac{q^l_{m,n}}{\nu_{m,n}+\nu}\right\} - \left\{\frac{q^l_{m,n}}{\nu_{m,n}+\nu}\right\}\{p^k_{m,n}\}\right] e^{2\pi i \nu t}. \tag{22}$$

Expansion in powers of $1/\nu$ gives

$$\Delta p^k = \frac{1}{\nu}\frac{e}{h}\text{Re}\sum_l E^l \left(p^k q^l - q^l p^k\right) e^{2\pi i \nu t}$$

$$+ \frac{1}{\nu^2}\frac{e}{2\pi h}\text{Re}\,i\sum_l E^l \left(p^l q^k - q^k p^l\right) e^{2\pi i \nu t} + \frac{C_4}{\nu^3} + \cdots. \tag{23}$$

Equating the terms with $1/\nu$ in (23) and (17) again gives (21), while equating the terms with $1/\nu^2$ gives the required relations

$$p^k p^l - p^l p^k = 0. \tag{24}$$

We have thus reached the following conclusion: *Born and Jordan's quantum conditions mean that an electron bound in an atom will in a high-frequency oscillating electric field asymptotically behave just like a bound electron in the classical theory.*

This formulation of the quantum conditions may have some importance for the study of atomic systems with several electrons.

Before concluding we may draw attention to an interpretation of the conditions (10) to (12) which is directly connected with the correspondence principle. In the limiting region of large quantum numbers these conditions go over into the following relations of classical mechanics:

$$(q^k q^l) = 0, \quad (p^k q^l) = 0 \ (k \neq l); \ = 1 \ (k = l), \quad (p^k p^l) = 0. \tag{25}$$

The expression (ab) is here the well known Poisson bracket:

$$(ab) = \sum_k \left(\frac{\partial a}{\partial I_k}\frac{\partial b}{\partial w_k} - \frac{\partial a}{\partial w_k}\frac{\partial b}{\partial I_k}\right), \tag{26}$$

where the canonical variables I_k, w_k are nothing but the uniformisation variables (the w's are Stäckel's "angle variables") which play such an important role in the quantum theory of periodic systems. In that theory the quantities I_k were put equal to integral multiples of Planck's constant. However, from the point of view of Heisenberg's theory these "quantum conditions" must be considered to be obsolete. The methods developed in the paper by Kramers and Heisenberg, when applied to (26), immediately give the result that Eqs. (10) to (12) are the proper quantum theoretical "translation" of the relations (26). Pauli has also already pointed to this interpretation of (10) to (12).

C. Wave Mechanics and Half-Odd-Integral Quantisation

Abstract. In this paper we consider a method to give an approximate solution of the Schrödinger eigenvalue and eigenfunction problem for an arbitrary one-dimensional system. We show in §1 that the so-called half-odd-integral quantisation is a natural first approximation. In §2 we give approximation formulæ for a graphical determination of the eigenfunctions. We discuss in §3 the relation between this method and the systematic approximations considered by Brillouin and Wentzel. In §4 we consider central motion and give approximate formulæ for spectral problems.

1. The Approximate Solution of the Wave Equation

Consider the problem of the quantisation of a system of one degree of freedom with a motion which is oscillatory in nature. According to Schrödinger one finds the stationary states by looking for the eigenvalues E_n for which the differential equation

$$\varphi'' + \frac{y}{K^2}\varphi = 0, \tag{1}$$

with

$$y = 2m(E_n - V(x)) \tag{2}$$

($K = h/2\pi$, m is the mass, and $V(x)$ is the potential energy), has real solutions φ_n which are everywhere finite. Let the quantum number n denote the number of zeros of φ_n between the two values x_1 and x_2 where y vanishes; these corresponds to the turning points of the classical motion. In the case when n is a large number we can use an elementary treatment to construct a function ψ which is an approximate solution of (1) in the range $x_1 < x < x_2$. Because of the wave nature of φ in this range, and on the basis of the consideration that y for large n changes little over a wavelength, we are led to the following Ansatz for ψ:

$$\psi = g(x) \cos f(x), \tag{3}$$

where $g(x)$ is a "smooth" function of the same kind as $y(x)$ while $f(x_2)-f(x_1)$ is of the order of magnitude of $n\pi$.

We obtain another expression for $f(x)$ by bearing in mind that for constant y the wavelength must be equal to $2\pi K y^{-1/2}$. This gives for $f(x)$ the approximate condition

$$f(x + 2\pi K y^{1/2}) - f(x) = 2\pi,$$

or

$$2\pi K y^{-1/2} f'(x) = 2\pi, \qquad f(x) = \frac{1}{K}\int^x y^{1/2}\,dx. \qquad (4)$$

We get an expression for the function $g(x)$ which takes into account the x-dependence of the wave amplitude by looking at the differential equation

$$\varphi'' + \frac{y(x_0) + y'(x_0)(x - x_0)}{K^2}\varphi = 0, \qquad (5)$$

which is practically the same as (1) over the range of the order of a wavelength. Neglecting small quantities proportional to the second or higher powers of y' we get the following solution of (5):

$$\varphi = \cos\left(\frac{\sqrt{y}}{K}(x - x_0)\right) - \frac{y'}{4y}\left[(x - x_0)\cos\left(\frac{\sqrt{y}}{K}(x - x_0)\right)\right.$$
$$\left. + \frac{\sqrt{y}(x - x_0)^2}{K}\sin\left(\frac{\sqrt{y}}{K}(x - x_0)\right)\right].$$

Hence we find that the amplitude of the oscillating function φ to first approximation over a range of the order of a wavelength can be written as

$$1 - \frac{y'(x - x_0)}{4y}.$$

Thus we get the following equation for the function $g(x)$:

$$\frac{g'}{g} = -\frac{y'}{4y}, \qquad \text{or} \qquad g = y^{-1/4}.$$

Our approximation for the eigenfunction thus has the form

$$\psi = y^{-1/4}\cos\left[\frac{1}{K}\int^x \sqrt{y}\,dx\right]. \qquad (6)$$

We next ask how we can determine the eigenvalue E_n and the integration constant in (4) so that ψ is, indeed, an eigenfunction of the problem. To answer this we cannot just consider the properties of ψ, since ψ becomes infinite at x_1 and at x_2 and complex for $x < x_1$ and $x > x_2$. Let us therefore consider the solution of (1) in the neighbourhood of x_1. If y' has the value α

C. Wave Mechanics and Half-Odd-Integral Quantisation

at x_1 and if we denote $x - x_1$ by ξ we have the following form for (1) in the vicinity of x_1:

$$\varphi'' + \frac{\alpha}{K^2}\xi\varphi = 0. \tag{7}$$

The solution of this equation[1] can be written in the form

$$\varphi = \xi^{1/2} Z_{1/3}\left(\frac{2\sqrt{\alpha}}{K}\xi^{3/2}\right), \tag{8}$$

where Z is a solution of the Bessel equation. However, the discussion is easier if we write the solution in the form of a definite integral:

$$\varphi = C \int \exp\left[\left(\frac{\alpha}{K^2}\right)^{1/3}\xi t + \tfrac{1}{3}t^3\right] dt, \tag{9}$$

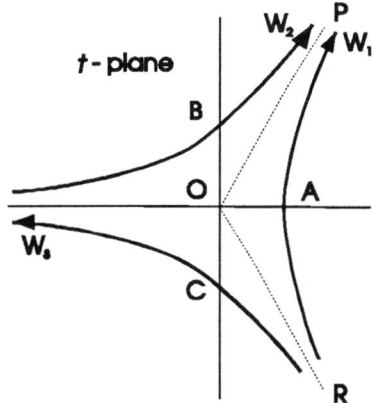

Fig. C.1.

which also shows more clearly than expression (8) that φ is an entire function. The integral (9) is a solution for all ξ if the integration contour asymptotically follows the directions with argument $\pm\pi/3$ and π (see Fig. C.1). Further one sees easily that the integral along W_1 for negative real ξ tends to zero as ξ increases and corresponds to the particular solution of (1) corresponding to a wavefunction of a stationary state. If we take the integral from R to O and then from O to P we get, expanding the integrand in powers of ξ and putting $C = -i$:

$$\varphi = 3^{-1/6}\Gamma(\tfrac{1}{3}) + 3^{1/6}\Gamma(\tfrac{2}{3})\left(\frac{\alpha}{K^2}\right)^{1/3}\xi + \cdots \frac{\alpha}{K^2}\xi^3 + \cdots. \tag{10}$$

---

[1] See Watson, *Theory of Bessel Functions*, Cambridge (1922), pp. 188ff for the literature about this equation and its solutions.

The asymptotic expansions for large value of the argument $(\alpha/K^2)^{1/3}\xi$ are important. One easily gets those using the saddle point method.[2] For negative real ξ we need the saddle point at A $(OA = \sqrt{(\alpha/K^2)^{1/3}|\xi|})$ and the line of steepest descent along which we must take the integral becomes the hyperbola W_1 with asymptotes OP and OR. One easily finds

$$\varphi = \sqrt{\pi}\left[\left(\frac{\alpha}{K^2}\right)^{1/3}|\xi|\right]^{-1/4}\exp\left[-\tfrac{2}{3}\left(\frac{\alpha}{K^2}\right)^{1/2}|\xi|^{3/2}\right]. \tag{11}$$

In the case of positive real ξ we use the saddle points B and C,

$$OB = OC = \sqrt{\left(\frac{\alpha}{K^2}\right)^{1/3}\xi}.$$

The lines of steepest descent are the third-degree curves W_2 and W_3, and the integral has to be taken over both W_2 and W_3. We then get an asymptotitc expression, differing from (11),[3]

$$\varphi = 2\sqrt{\pi}\left[\left(\frac{\alpha}{K^2}\right)^{1/3}\xi\right]^{-1/4}\cos\left[\tfrac{2}{3}\left(\frac{\alpha}{K^2}\right)^{1/2}\xi^{3/2} - \frac{\pi}{4}\right]. \tag{12}$$

We can compare this expression directly with the one we get from (6) for values of α in the vicinity of x_1. We put $y = \alpha\xi/K^2$ and find:

$$\psi = \left(\frac{\alpha}{K^2}\right)^{-1/4}\xi^{-1/4}\cos\left[\tfrac{2}{3}\left(\frac{\alpha}{K^2}\right)^{1/2}\xi^{3/2} - \beta\right], \tag{13}$$

where $-\beta$ is the integration constant in the integral (4) which still has to be determined. Comparing (12) and (13) we see that the function given by (6) approximates the particular solution of (1) required by wave mechanics only when $\beta = \pi/4$, that is, it must be possible to write ψ in the form

$$\psi = y^{-1/4}\cos\left[\frac{1}{K}\int_{x_1}^{x}y^{1/2}\,dx - \frac{\pi}{4}\right]. \tag{14}$$

We have introduced here the condition that $\varphi(x_1)$ is positive; this can be done.[4]

---

[2] See, e.g., Courant-Hilbert, *Mathematische Physik*, p. 435.
[3] This is the so-called Stokes phenomenon; see, e.g., Watson, *Bessel Functions*, p. 201.
[4] The fact that, indeed, we can consider for the behaviour of φ in the vicinity of x_1 the simple equation (7) follows as soon as one can assume that $y'(x_1)(x_2 - x_1)$, $y''(x_1)(x_2 - x_1)^2$, and so on, are of the same order of magnitude as the values of y between x_1 and x_2. In fact, a more exact consideration shows that if we assume that ξ is, say, of the order of magnitude $(x_2-x_1)/n^{2/5}$, that is, when (13) and (6) differ only by amounts of relative order $n^{-2/5}$, the differences between expressions (11) and (12) and the function (10) will also be of the relative order $n^{-2/5}$ and, moreover, that for negative ξ the function φ of (11) is small of order $\exp(-n^{2/5})$.

C. Wave Mechanics and Half-Odd-Integral Quantisation

We can proceed completely similarly to consider the other turning point and one finds easily that it must be possible to write expression (6) in the form

$$\psi = y^{-1/4} \cos\left[\frac{1}{K}\int_{x_2}^{x} y^{1/2}\,dx - \frac{\pi}{4} \pm \frac{\pi}{2}\right], \tag{15}$$

in order that ψ is an approximation of an eigenfunction. The plus- or minus-signs occur depending on whether $\varphi(x_2)$ is positive or negative.

The condition that (14) and (15) describe the same function is

$$\frac{1}{K}\int_{x_1}^{x} y^{1/2}\,dx - \frac{\pi}{4} = \frac{1}{K}\int_{x_2}^{x} y^{1/2}\,dx - \frac{\pi}{4} \pm \frac{\pi}{2} + 2\pi m, \tag{16}$$

where m is an integer. Hence follows immediately that

$$\int_{x_1}^{x_2} y^{1/2}\,dx = 2\pi K(2m \pm \tfrac{1}{2}),$$

or, if we multiply by 2, introduce the classical phase integral, and substitute for K its value $h/2\pi$,

$$2\int_{x_1}^{x_2} y^{1/2}\,dx = \oint p\,dx = (2m + \tfrac{1}{2})h. \tag{17}$$

We find thus that our approximate representation (6) of the Schrödinger eigenfunctions corresponds just to those energy values which would in the old quantum theory correspond to the half-odd-integral quantisation. The relative accuracy of these energy values will be better than $1/n$ (in general, of order $1/n^2$) whereas the energy values calculated by the integral quantisation differ, in general, from the Schrödinger eigenvalues by amounts of the order $1/n$.[5] We have thus gained a certain understanding why the half-odd-integral quantisation gives a better approximation to the results of quantum mechanics than the integral quantisation — a fact which is well known from many examples.

---

[5] In fact, the distance between two successive zeroes of the function (6) will differ from the distance between the corresponding zeroes of the Schrödinger eigenfunction only by small amounts of the order $(x_2 - x_1)/n^3$, if we assume that $y'(x)(x_2 - x_1)$ is of the same order as y in the interval $x_1 < x < x_2$. The error in the distance between the smallest and the largest zero will therefore, if we apply (6), be of the order $(x_2 - x_1)/n^2$, that is, small as compared to the distance between two successive zeroes. A relative change of the energy values of the order of magnitude $1/n$ would produce a distortion of the position of the zeroes that would be incompatibale with the properties of the eigenfunctions in the vicinity of the two turning points.

2. Actual Calculation of the Eigenfunctions

Our considerations provide us with a simple approximate determination of the eigenfunctions which may be of practical use in cases where they cannot be expressed in terms of elementary functions. If we use in the interval $x_1 < x < x_2$ the function

$$\varphi = y^{-1/4} \cos\left(\int_{x_1}^{x} \sqrt{y}\, dx - \frac{\pi}{4}\right), \qquad x_1 < x < x_2, \tag{18a}$$

which can be plotted graphically once y is known, we can in the immediate vicinity of the turning points write (cf. (10), (12), and (13)):

$$\varphi = \frac{1}{2\sqrt{\pi}}\left(\frac{\alpha_1}{K^2}\right)^{-1/6}\left[3^{-1/6}\Gamma(\tfrac{1}{3})\right.$$
$$\left. + 3^{1/6}\Gamma(\tfrac{2}{3})\left(\frac{\alpha_1}{K^2}\right)^{1/3}(x - x_1) + \cdots\right], \qquad x \text{ close to } x_1, \tag{18b}$$

$$\varphi = \frac{(-1)^n}{2\sqrt{\pi}}\left(\frac{\alpha_2}{K^2}\right)^{-1/6}\left[3^{-1/6}\Gamma(\tfrac{1}{3})\right.$$
$$\left. - 3^{1/6}\Gamma(\tfrac{2}{3})\left(\frac{\alpha_2}{K^2}\right)^{1/3}(x - x_2) + \cdots\right], \qquad x \text{ close to } x_2. \tag{18c}$$

Here $\alpha_1 = y'(x_1)$, $\alpha_2 = y'(x_2)$, and $n = $ the number of zeroes between x_1 and x_2. These formulæ determine the tangents to the φ-curve in the points of inflexion x_1 and x_2.

The exponential decrease of φ for $x < x_1$ and $x > x_2$ is given by the formulæ

$$\varphi = \frac{1}{2}\left(\frac{\alpha_1}{K^2}\right)^{-1/4}(x_1 - x)^{-1/4} \exp\left[-\frac{2}{3}\left(\frac{\alpha_1}{K^2}\right)^{1/2}(x_1 - x)^{3/2}\right],$$
$$x < x_1 \tag{18d}$$

$$\varphi = \frac{(-1)^n}{2}\left(\frac{\alpha_2}{K^2}\right)^{-1/4}(x - x_2)^{-1/4} \exp\left[-\frac{2}{3}\left(\frac{\alpha_2}{K^2}\right)^{1/2}(x - x_2)^{3/2}\right],$$
$$x > x_2 \tag{18e}$$

It is, however, simpler, instead of explicitly using Eqs. (18b to e), to evaluate once for all the values of the integral

$$\omega(\xi) = \frac{-i}{2\sqrt{\pi}}\int_{W_1} e^{\xi t + \frac{1}{3}t^3}\, dt \tag{18f}$$

from $\xi = -\infty$ to say, the first zero of $\omega(\xi)$ (see the Table in the Appendix) and to use in the vicinity of the turning points the following expressions for φ:

C. Wave Mechanics and Half-Odd-Integral Quantisation

$$\varphi = \left(\frac{\alpha_1}{K^2}\right)^{-1/6} \omega\left[\left(\frac{\alpha_1}{K^2}\right)^{1/3}(x-x_1)\right], \tag{18g}$$

from $x = -\infty$ to the first zero, and

$$\varphi = (-1)^n \left(\frac{\alpha_2}{K^2}\right)^{-1/6} \omega\left[\left(\frac{\alpha_2}{K^2}\right)^{1/3}(x_2-x)\right], \tag{18h}$$

from the nth zero to $x = \infty$.

I have applied the approximate formulæ (18) to the case of the harmonic oscillator and found that even for $n = 1$ and $n = 0$ they are able to construct the eigenfunctions with good accuracy.

3. Comparison with Systematic Approximation Methods

The function (6) derived using elementary considerations is closely connected with the procedure used by Brillouin[6] and Wentzel[7] for successive calculations of the eigenvalues and the eigenfunctions where the solution of Hamilton's differential equaton is the first step. These authors use the following Ansatz for the solution of (1):

$$\varphi = \exp\left[\frac{i}{K}\left(S_0 + KS_1 + K^2 S_2 + \cdots\right)\right], \tag{19}$$

and one finds easily

$$S_0' = \pm\sqrt{y}, \qquad S_0 = \pm\int \sqrt{y}\,dx,$$

$$S_1' = \frac{i}{2}\frac{S_0''}{S_0'}, \qquad S_1 = \tfrac{1}{2}i\ln S_0' = \tfrac{1}{4}i\ln y + \text{const.}$$

If one stops at the second approximation one thus obtains

$$\varphi = y^{-1/4} \exp\left(\pm\frac{i}{K}\int \sqrt{y}\,dx\right). \tag{20}$$

The plus- and minus-signs correspond to two particular solutions and the function represented by (6) corresponds just to half the sum of these two solutions.[8] The interesting fact is now that (6) is not at all a unique function

---

[6] L. Brillouin, *C.R.* **183**, 24 (1926).
[7] G. Wentzel, *Zs. Phys.* **38**, 518 (1926).
[8] To obtain (6) directly from the Riccati equation $(h/2\pi i)\,y' = p^2 - y^2$ employed by Wentzel (l.c. p. 518) one must not take $y_0 = \pm p$ as the first approximation, but write something like

$$y_0 = ip\tan\left(\frac{2\pi}{h}\int p\,dx\right).$$

like the eigenfunction which we are trying to approximate. It can represent the latter only in the interval $x_1 < x < x_2$, and then only if one determines the integration constant in the manner described in §1: the turning points themselves are singular points and the function does not return to its old value when the variable goes around them. One also sees from (18) that the eigenfunction approximated by (6) in the regions outside the turning points is approximated by other approximate solutions of (1), for instance, for $x < x_1$ by

$$\varphi = \frac{1}{2}(-y)^{-1/4} \exp\left(\frac{-1}{K}\int_x^{x_1} \sqrt{-y}\, dx\right). \tag{21}$$

We must also note here that there is no practical advantage to prefer the simpler Eq. (18d) to Eq. (21) for the calculation of the eigenfunction. It is true that it is a closer solution of the differential equation (1) but the difference is negligible because of the fast exponential drop.

The nature of the many-valuedness of the functions which one obtains through the approximation (19) becomes clear if one writes down the differential equations corresponding to the successive approximations. For instance, the functions (6) and (20) satisfy the differential equation

$$\varphi'' + \left(\frac{y}{K^2} - \frac{5y'^2 - 4yy''}{16y^2}\right)y = 0. \tag{22}$$

Equation (22) has singularities in the turning points. In a region far from these the solutions of (22) are the same as those of (1) up to quantities of the order K^2, but in their vicinity the solutions are completely different. It is known from mathematics how careful one must be when one investigates the solution of a differential equation by looking at a solution of an "approximated" differential equation.[9] In our case the solution of (22) always represents, away from the turning points, an asymptotic expression of a solution of (1); the same particular solution of (1), however, is approximated in different parts of the x-plane by different particular solutions of (22). It appears therefore that the beautiful method used by Wentzel to calculate the eigenvalues by considering the complex integral $\oint (\varphi'/\varphi)\, dx$ around the turning points needs a further justification. Nevertheless, the examples given indicate that it is blameless. If one applies it to the approximate solution (20) one immediately gets the half-odd-integral quantisation.

4. Application to Central Motion

In the case of a central potential one can, according to Schrödinger, reduce the problem to determining the eigenvalues and eigenfunctions of the differential equation

---

[9] See, e.g., Schlesinger, *Differentialgleichingen*, p. 199ff.

C. Wave Mechanics and Half-Odd-Integral Quantisation   159

$$\varphi'' + \frac{y}{K^2}\varphi = 0, \qquad y = 2m(E - V(r)) - \frac{K^2 k(k+1)}{r^2}, \tag{23}$$

where the azimuthal quantum number k can take the integral values 0, 1, 2, ... and where $V(r)$ is the potential of the central force.

If there are two turning points r_1 and r_2 with $r \neq 0$ — when $k \neq 0$ this is the case for all existing problems — the problem is similar to the one treated in §§ 1 and 2 since the fact that the variable r here goes from 0 to ∞ does not necessitate any changes to the approximation considered. However, in the problems of atomic theory where the series terms or the X-ray levels can be described by a central field one often encounters the case where the assumptions on which the justification of the approximation described in §§ 1 and 2 is based are no longer satisfied, and at first sight the justification of the half-odd-integral quantisation seems to be questionable. I am thinking here of the cases where k is equal to 0, 1, or 2, which correspond for non-hydrogenlike terms to the so-called penetrating orbits. The function y has in these cases a very steep maximum very close to the smallest turning point x_1 — in the case of $k = 0$ where r_1 has become equal to zero it will even become infinite for $r = 0$ — and the application of the differential equation (7) to discuss the eigenfunction in the neighbourhood of this point becomes illusory. However, in these cases one can give a simple approximate description of the behaviour of the eigenfunction in the vicinity of $r = 0$, which will suffice for most practical purposes and which at the same time shows in what sense one can retain the half-odd-integral quantisation. To find that description we note that in the vicinity of the smallest turning point r_1 and of the maximum of y (this means in the vicinity of $r = 0$ in the $k = 0$ case) the field of force is to a very good accuracy a Coulomb field, and can thus be described by the potential

$$V(r) = -\frac{Ne^2}{R} + a. \tag{24}$$

Here N describes an effective nuclear charge while the constant a is a measure of the so-called external screening. It is usually small compared with the maximum of $y/2m$. For hydrogen-unlike orbits the eigenvalue E is small compared with its maximum value and we can therefore state that one can approximate the function y of (23) for values of r of the order of r_1 or smaller by the formula

$$\frac{y}{K^2} = \frac{2mNe^2}{K^2}\frac{1}{r} - \frac{k(k+1)}{r^2}. \tag{25}$$

If we use as unit of length the radius K^2/me^2 of the first "hydrogen orbit" the differential equation (23) takes the form

$$\varphi'' + \left(\frac{2N}{r} - \frac{k(k+1)}{r^2}\right)\varphi = 0. \tag{26}$$

It corresponds clearly to the parabolic orbits in a Coulomb field. The particular solution of (26) which does not become infinite at $r = 0$ can be expressed as follows in terms of a Bessel function of order $2k+1$:

$$\varphi = \sqrt{r} J_{2k+1}\left(\sqrt{8Nr}\right). \tag{27}$$

If we use the formulae given, for instance, by Jahnke and Emde and of their asymptotic series $P_p(x)$ and $Q_p(x)$, only taking the first term into account, we find for (27) the asymptotic representation

$$\varphi = \frac{1}{\sqrt{\pi}} \left(\frac{r}{2N}\right)^{1/4} \cos\left[\sqrt{8Nr} + \frac{2k(k+1) + \frac{3}{8}}{\sqrt{8Nr}} - \pi(k + \tfrac{1}{2}) - \frac{\pi}{4}\right]. \tag{28}$$

In particular, for $k = 0$ and $k = 1$ this formula also holds reasonably accurately for those values of r for which we can still use the representation (25).

We now consider the values which the approximate solution (6) of the differential equation (1) can take on in the same range of r, that is, we introduce the step similar to Eq. (13) in §1. To do this we put

$$\frac{y}{K^2} = \frac{2N}{r} - \frac{l^2}{r^2}, \tag{29}$$

where l is kept undetermined for the time being, and we evaluate the integral $K^{-1} \int_{r_1}^{r} \sqrt{y}\, dr$ where we put for the turning point $r_1 = l^2/2N$. We find

$$\frac{1}{K}\int_{r_1}^{r} \sqrt{y}\, dr = 2\sqrt{2Nr - l^2} - 2l \arctan\sqrt{\frac{2Nr}{l^2} - 1}$$

$$= \sqrt{8Nr} + \frac{2l^2}{\sqrt{8Nr}} - \pi l,$$

where in the last expression we have neglected terms of relative order $1/r$. With the same accuracy the function (6) now takes the form

$$\psi = \left(\frac{r}{2N}\right)^{1/4} \cos\left[\sqrt{8Nx} + \frac{2l^2}{\sqrt{8Nx}} - \pi l - \beta\right], \tag{30}$$

where $-\beta$ is again the undetermined integration constant of the integral in (4). A comparison of (28) and (30) shows that (6) approximates, indeed, the eigenfunction provided we put, as in §1,

$$\beta = \frac{\pi}{4}, \tag{31}$$

and, moreover, assume that

$$l = k + \tfrac{1}{2}. \tag{32}$$

C. Wave Mechanics and Half-Odd-Integral Quantisation

To find an approximate solution of (23) we are thus led to consider the mechanical problem for which the radial momentum p_r is given by

$$p_r^2 = y = 2m\left(E - V(r) - \frac{K^2}{2m}\frac{(k+\frac{1}{2})^2}{r^2}\right). \tag{33}$$

It is well known that (33) corresponds just to the classical equation for the radial momentum in a central field if we use the half-odd-integral quantisation for the angular momentum of the particle. Since the difference between the terms containing k in (23) and (25) is hardly noticeable as soon as one considers values of r some distance from the maximum of y, we shall once again assume that Eq. (18e) is the approximate solution of (23) where we must, however, use function (33) for y.

We can now again apply the considerations of §1 to the largest turning point r_2 and we are again led to Eqs. (16) and (17), because of (15). The half-odd-integral quantisation thus turns out to be, also for for the smallest value of the azimuthal quantum number, the natural method for an approximate calculation of the eigenvalues.[10]

For the actual construction of the eigenfunctions of (23) one must now proceed as follows. For the smallest values of r one uses

$$\varphi = \sqrt{\pi r}\, J_{2k+1}\left(\sqrt{8Nr}\right). \tag{34a}$$

One joins this function to the function

$$\varphi = y^{-1/4}\cos\left(\int_{r_1}^{r}\sqrt{y}\,dr - \frac{\pi}{4}\right), \tag{34b}$$

where y is given by (33). In the second turning point and beyond one can then again use the old Eqs. (18c) and (18e) or (18h).

The considerations of this section may be of some help for the problem of the calculation of the transition probabilities in series and X-ray spectra. Using the method applied by Fuess and Hartree one can construct a central field such that the observed terms correspond to a half-odd-integral quantisation of the azimuthal and radial momenta. Using that field one can to a certain approximation construct the corresponding eigenfunctions and use those to evaluate the corresponding characteristic oscillation amplitudes pertaining to the transitions.

---

[10] If, for the application of Eq. (6), one wants to retain expression (23) for y, that is, if one puts $l^2 = k(k+1)$ in (29) one must change the quantisation rule so that one puts $(1/h)\oint \sqrt{y}\,dr + \sqrt{k(k+1)} - (k+\frac{1}{2}) = n + \frac{1}{2}$ (n an integer).

Appendix

Mr. M.van der Held has kindly performed a numerical calculation of the function $\omega(\xi)$ defined by Eq. (18f). He finds the following values:

ξ	ω	ξ	ω	ξ	ω	ξ	ω	ξ	ω
-9.0	$4.4 \cdot 10^{-9}$	-1.9	0.071	-0.8	0.299	0.3	0.761	1.4	0.871
-5.0	0.00019	-1.8	0.082	-0.7	0.333	0.4	0.803	1.5	0.823
-4.0	0.0017	-1.7	0.096	-0.6	0.369	0.5	0.843	1.6	0.761
-3.5	0.0045	-1.6	0.110	-0.5	0.408	0.6	0.877	1.7	0.688
-3.0	0.0121	-1.5	0.127	-0.4	0.451	0.7	0.904	1.8	0.603
-2.5	0.028	-1.4	0.144	-0.3	0.493	0.8	0.931	1.9	0.507
-2.4	0.034	-1.3	0.164	-0.2	0.538	0.9	0.944	2.0	0.403
-2.3	0.039	-1.2	0.186	-0.1	0.584	1.0	0.950	2.1	0.290
-2.2	0.045	-1.1	0.211	0	0.629	1.1	0.947	2.2	0.172
-2.1	0.054	-1.0	0.239	+0.1	0.674	1.2	0.933	2.3	0.048
-2.0	0.062	-0.9	0.268	+0.2	0.719	1.3	0.907	2.4	-0.079

The following integral is of interest for the normalisation of the eigenfunction:

$$\int_{-\infty}^{2.3} \omega^2 \, d\xi = 1.54.$$

D. The Scattering of Light by Atoms

§ 1. In this paper we want to make a few theoretical remarks about dispersion theory, that is, the formula which shows how the refractive index of a homogeneous and isotropic medium depends on the wavelength. It is useful first to summarise some results of the classical theory of this topic.

Maxwell's theory of electromagnetic phenomena is able to describe in a rather simple way most effects connected with the propagation of light waves in matter. In fact, if we start from the well-known Maxwell equations we need only assume that the magnetic induction B is the same as the magnetic field strength H, whereas the electric displacement D is a linear vector function of the electric field strength E:

$$B_k = H_k, \qquad D_k = \sum_l \varepsilon_{kl} E_l, \qquad k = 1, 2, 3. \tag{1}$$

In an inhomogeneous medium the ε_{kl} coefficients will be functions of space, and the light rays will, in general, be curved lines. In a homogeneous medium, these coefficients will be independent of position, and light will propagate along straight lines. Assuming the ε_{kl} to be real one can obtain the laws for double refraction in crystals. In order to describe as well absorption or the magnetic rotation of the plane of polarisation we must assume these coefficients to be complex. For an isotropic medium we can put

$$\varepsilon_{kl} = \varepsilon \delta_{kl}, \tag{2}$$

where δ_{kl} is the Weierstrass symbol. There is thus only a single dielectric constant ε which must be assumed to be complex if we want to account for possible absorption. If we write

$$\sqrt{\varepsilon} = n(1 - i\varkappa), \qquad \lambda = \frac{2\pi c}{\omega n}, \tag{3}$$

where n and $\varkappa$ are real quantities, the propagation of light will be represented by the factor:

$$e^{-2\pi \varkappa x/\lambda} e^{i(\omega t - 2\pi x/\lambda)}.$$

The wave propagation speed thus depends on n, whereas the absorption also depends on $\varkappa$.

If one wants to describe the natural rotation of the polarisation plane, Eqs. (1) will no longer suffice; one must add to the expressions for the D_k terms containing partial derivatives of E_k with respect to the coordinates. We shall be concerned solely with the simple case where Eqs.(1) suffice.

To take dispersion effects into account, we must assume that the ε_{kl} coefficients are also functions of the frequency ω of the light waves. Electron theory where one considers phenomena from an atomistic point of view explains dispersion by examining induced oscillations of the electric particles in the atoms due to the forces of the electromagnetic field. As a result of these oscillations the atoms emit secondary spherical waves in all directions in space; these themselves satisfy the Maxwell equations in empty space. In the medium the macroscopic waves, which are described by Maxwell's phenomenological theory, are the result of a superposition of all these microscopic waves. Lorentz, Nathanson, Ewald, and Oseen have shown that the laws for dispersion (like those for absorption) can be derived from the laws describing scattering of light by a single atom. The macroscopic scattering of light in a medium at any rate plays a practically negligible role; this rather paradoxical effect can be explained by considering the phase relations between the secondary waves of the various atoms in the medium.

Let us consider an atom or molecule with a size which is very small as compared to the wavelength, and let us represent the components of the electric force at the position of the atom by the real part of a complex quantity, which contains the time exponentially:

$$E_k = \text{Re}\left(A_k e^{i\omega t}\right). \tag{4}$$

The constants A_k are the, in general complex, components of the vector amplitude of the electric force. Under the action of the force (4) the particles in the atom will execute forced oscillations which have a harmonic component with the same frequency ω as the external force. As a result the atom acts like an electric dipole with a moment, which is a vector, with a harmonic factor which can be written as follows:

$$P_k = \text{Re}\left(B_k e^{i\omega t}\right). \tag{5}$$

To a first approximation which will be realised when the external force can be considered to be weak, the vector B will be a linear vector function of the vector A:

$$B_k = \sum \zeta_{kl} A_l. \tag{6}$$

The quantities B, ζ, and A will, in general, be complex. The ζ coefficients may be called the *polarisation coefficients* of the atom.

Assuming that the oscillating dipoles emit spherical waves according to the classical radiation laws we can express the ε_{kl} coefficients as functions of the ζ_{kl} coefficients once the distribution of the atoms in space is known. If we are dealing with a gas with a small refractivity and if all atoms are the same, we have

$$\varepsilon_{kl} = \delta_{kl} + 4\pi N \zeta_{kl}, \tag{7}$$

where N is the number of atoms per unit volume. If the density of the medium is higher, the relation between the ε and the ζ will be more complicated. As an example we remind ourselves of the Lorenz-Lorentz formula, which is valid for isotropic media.

§ 2. We shall restrict ourselves in what follows to the case of an isotropic atom in which case we have only a single ζ coefficient which, in general. will takle on complex values:

$$B_k = \zeta A_k, \qquad \zeta = \xi + i\eta. \tag{8}$$

The simplest classical model of such an atom consists of one electron which can execute free harmonic oscillations of frequency ω about an equilibrium position. A simple, well-known calculation gives:

$$\zeta = \frac{e^2}{m} \frac{1}{\omega_1^2 - \omega^2 + i\delta}, \qquad \delta = \frac{2e^2\omega^3}{3mc^3}. \tag{9}$$

Here e and m are the electron charge and mass, while c is the velocity of light. Usually δ/ω^2 is very small, and when the difference $\omega_1 - \omega$ is large compared to δ/ω we can write:

$$\zeta = \frac{e^2}{m} \frac{1}{\omega_1^2 - \omega^2}. \tag{10}$$

If we consider the case where the atom contains several electrons which are independently connected with different equilibrium positions in the atom, we find a more general kind of formula (Sellmeier formula):

$$\zeta = \sum_k \frac{e^2}{m} \frac{f_k}{\omega_k^2 - \omega^2}, \tag{11}$$

which is valid for those regions of the spectrum where $\omega_k - \omega$ is not too small. This formula corresponds to a model where the atom contains a large number of electrons with f_k of them having the frequency ω_k. It describes the experiments very exactly, but one knows that one must attribute to the f_k coefficients values which, far from being integers, are more often much smaller than unity. Luckily, recent investigations[1] using Bohr's atomic theory, which rejects the above-mentioned simple model of classical electron theory, have led exactly to Eq. (11), which made it possible to express the amplitude of the induced dipole as function of the amplitude of the external field and of

---

[1] Ladenburg, *Zs. Phys.* **4**, 451 (1921); H. A. Kramers, *Nature* **113**, 673 (1924); H. A. Kramers and W. Heisenberg, *Zs. Phys.* **31**, 681 (1925).

the wave frequency. As in the classical model, the frequencies ω_k indicate the position of the atomic absorption lines, but the quantities f_k can take on any value whatever. According to the theory, the value of f_k is directly connected with the intensity of the corresponding absorption line. This intensity can be described as follows.

We introduce the atomic absorption coefficient α, which will be a function of the frequency ω and which is equal to the atomic absorption cross-section (that is, the atom if hit by a a system of waves of frequency ω absorbs an energy equivalent to the energy traversing an area α prependicular to the direction of the waves). A simple calculation shows that α depends on the imaginary part η of the polarisation coefficient ζ:

$$\alpha = -\frac{4\pi\omega}{c}\eta. \tag{12}$$

In the case under consideration η is practically equal to zero in the whole spectral region except in the absorption lines. Hence we can define as the intensity of the absorption line corresponding to the frequency ω_k the integral:

$$a_k = \int \alpha(\omega)\,d\omega, \tag{13}$$

taken over a small frequency range, containing ω_k. The quantities a_k differ by only a factor from the well known Einstein B coefficients which describe the rate of Bohr absorption transitions. In fact, we have

$$B = \frac{ca}{\hbar\omega}. \tag{14}$$

For the classical model with a single electron bound to an equilibrium position we can use (9) to find the value of this integral:

$$a_1 = \frac{2\pi^2 e^2}{mc}. \tag{15}$$

Comparing (10) and (11) we see that in this case the value of f_1 is equal to unity. In the general case we have:

$$f_k = \frac{mca_k}{2\pi^2 e^2}. \tag{16}$$

We may ask the question how one should generalise Eq. (11) when the absorption spectrum contains regions where there is continuous absorption. We can find such a generalisation directly by noting that if we use (13) and (16) we can write Eq. (11) in the form:

$$\zeta = \frac{c}{2\pi^2}\int_0^\infty \frac{\alpha(\omega')\,d\omega'}{\omega'^2 - \omega^2}. \tag{17}$$

In the case where we have only absorption lines, this integral reduces to a sum of terms each corresponding to a single line. If there is at the same time a region of continuous absorption, the integral will still have an exact meaning provided ω lies in a region where $\alpha(\omega')$ vanishes. This clearly corresponds to the fact that Eq. (11) is only an approximation valid in regions without absorption lines and that one could think that Eq. (17) will always be an approximate formula.

Now we shall, however, show that Eq. (17) is exact and that one can give it an exact meaning even in the case when $\alpha(\omega') \neq 0$. To do this it suffices to accept two facts: first, that in that case one must take the Cauchy principal value of the integral[2] and, second, that the integral, which is always real, represents the real part ξ of the polarisation coefficient ζ, rather than the coefficient itself:

$$\xi = \frac{c}{2\pi} \int_0^\infty \frac{\alpha(\omega')\, d\omega'}{\omega'^2 - \omega^2}. \tag{18}$$

I have given this formula two years ago in an unpublished communication to the Royal Danish Academy.[3] Recently, Kronig[4] and Kallmann and Mark[5] have given relations equivalent to (18). The latter authors have made an interesting application to the problem of the refractibility of X-rays in matter. We shall return to that question in what follows.

§ 3. Equation (18) has a very simple mathematical meaning. In fact, let us use (12) to introduce the quantity η instead of α. Equation (18) then takes the form

$$\xi = -\frac{2}{\pi} \int_0^\infty \frac{\omega'\eta(\omega')\, d\omega'}{\omega'^2 - \omega^2}. \tag{19}$$

The functions ξ and η have been defined only for positive values of ω. Let us now allow that negative values are also possible by assuming that ξ is an even and η an odd function:

$$\xi(-\omega) = \xi(\omega), \qquad \eta(-\omega) = -\eta(\omega), \qquad \zeta = \xi + i\eta. \tag{20}$$

We can then write (19) in the form

$$\xi(\omega) = \frac{1}{\pi} \int_{-\infty}^\infty \frac{\eta(\omega')\, d\omega'}{\omega - \omega'}. \tag{21}$$

---

[2] The principal value integral is defined by the equation

$$\int_0^\infty d\omega' \ldots = \lim_{\varepsilon \to 0} \left(\int_0^{\omega-\varepsilon} d\omega' \ldots + \int_{\omega+\varepsilon}^\infty d\omega' \ldots \right).$$

[3] Cf *Nature* **117**, 775 (1925).
[4] R. de L. Kronig, *J. Opt. Soc. Am.* **12**, 547 (1926).
[5] Kallmann and Mark, *Ann. Physik* **82**, 585 (1927).

This is a well-known and often studied mathematical relation which enables one to find the real part of an analytic holomorphic function in the half-plane below the real axis if the imaginary part takes the values η on that axis. *The dispersion formula, put in the form (21) thus shows us that the polarisation coefficient ζ behaves as an analytical function of the frequency, which is holomorphic for negative values of the imaginary parts of the frequency.* We shall not discuss here the mathematical investigations connected with Eq. (21) which are concerned with, among other points, the way the functions ξ and η vanish at infinity. We note merely that $\eta(\omega)$ can be considered as the real part of an analytical function, the imaginary part of which is equal to $-i\xi(\omega)$; this leads to the inversion formula:

$$\eta(\omega) = -\frac{1}{\pi}\int_{-\infty}^{\infty}\frac{\xi(\omega')\,d\omega'}{\omega-\omega'}. \tag{22}$$

The existence of the analytical function ζ with the above mentioned properties allows a physical interpretation if we look for the significance which one can attach to complex values of the frequency of the incident waves. Let us assume that there is a system of waves incident upon the atom which has an electric vector represented by the real part of the expression

$$f(t)e^{i(a-ib)t}, \tag{23}$$

where the constant b is positive and where $f(t) = 1$ for $t < T$ and $f(t) = 0$ for $t > T$ (with positive T). Expanding expression (23) in a Fourier integral and applying Eqs. (5) and (8), we find that the polarisation of the atoms is given by the real part of the expression:

$$F(a,b,t,T)e^{i(a-ib)t}, \tag{24}$$

where F is given by:

$$F(a,b,t,T) = \frac{i}{2\pi}\int_{-\infty}^{+\infty}\frac{\xi+i\eta}{\omega-a+ib}e^{i(\omega-a+ib)(t-T)}\,d\omega. \tag{25}$$

Now, if b could take on any positive value, this expression would not converge, as $T \to \infty$, to a definite limit unless ξ and η represented the real and imaginary parts of an analytic function ζ which is holomorphic for all values of $a - ib$ with positive b; the function $F(a,b,t,T)$ then reduces for $t < T$ to $\zeta(a - ib)$. We have already seen that the dispersion formula (21) shows us just that this condition is satisfied. In other words, *the dispersion formula makes certain that one can define a finite refractive index for complex values of the frequency such that the amplitude of the corresponding waves growth with time.*

§ 4. Let us as an example of an application of (21) and (22) consider the case of X-ray dispersion which was studied by Kallmann and Mark (loc. cit.).

D. The Scattering of Light by Atoms

Neglecting absorption, which occurs on the side of increasing wavelengths starting from the limit ω_k of the K absorption, we can write approximately:

$$\eta(\omega) = 0, \quad 0 < \omega < \omega_k; \qquad \eta(\omega) = -C/\omega^4, \quad \omega_k < \omega < \infty, \qquad (26)$$

since the continuous absorption is roughly proportional to the cube of the wavelength. Using (21) we find:[6]

$$\xi(\omega) = \frac{C}{\pi}\left(-\frac{1}{\omega^2 \omega_k^2} + \frac{1}{\omega^4}\log\frac{\omega_k^2}{|\omega^2 - \omega_k^2|}\right). \qquad (27)$$

We see immediately that (26) and (27) can be represented by an analytic function which is holomorphic below the real axis and which vanishes at infinity:

$$\zeta(\omega) = \xi(\omega) + i\eta(\omega) = \frac{C}{\pi}\left(-\frac{1}{\omega^2\omega_k^2} + \frac{1}{\omega^4}\log\frac{\omega_k^2}{\omega_k^2 - \omega^2}\right). \qquad (28)$$

This function has two singularities on the real axis, namely at $\omega = \pm\omega_k$. This is due to the discontinuities of η that we have introduced at those values. In reality there is no such a sharp discontinuity and ζ remains finite on the real axis and holomorphic in the lower half-plane.

Equations (21) and (22) do not directly apply to the harmonic oscillator of the classical electron theory. In fact, Eq. (9) gives an analytical expression for ζ which has two poles in the upper half-plane and a far removed pole in the lower half-plane. Nevertheless, by adding to (9) a function of ω, the value of which is negligible when δ/ω^2 is small, one can remove this pole.

In the case of the harmonic oscillator the energy of the light which is absorbed goes completely into the secondary scattered waves. According to the classical laws the energy scattered per unit time in the shape of spherical waves of frequency ω will be equal to

$$\frac{\omega^4}{3c^3}|B|^2 = \frac{\omega^4}{3c^3}|\zeta|^2|A|^2 = \frac{\omega^4}{3c^3}\left(\xi^2 + \eta^2\right)|A|^2.$$

On the other hand, the energy absorbed per unit time will be equal to $-\frac{1}{2}\omega\eta|A|^2$ (see (12)). Hence, the condition that the absorbed energy can be found in the scattered wave of frequency ω is given by:

$$\frac{\omega^4}{3c^3}\left(\xi^2 + \eta^2\right) = -\tfrac{1}{2}\omega\eta,$$

or

$$\frac{-\eta}{\xi^2 + \eta^2} = \frac{2\omega^3}{3c^3}. \qquad (29)$$

---

[6] For an experimental verification of this relation see: J. A. Prins, Zs. Phys. **47**, 479 (1928).

As $-i\eta/(\xi^2 + \eta^2)$ is the imaginary part of the analytical function $1/(\xi + i\eta)$ we have:

$$\zeta = \xi + i\eta = \frac{1}{F(\omega) + 2i\omega^3/c^3}, \tag{30}$$

where $F(\omega)$ is an even analytical function which takes on real values on the real axis and which must be such that ζ is holomorphic in the lower half-plane. Equation (9) has the same form as (30); it gives for F:

$$F(\omega) = \frac{m}{e^2}\left(\omega_1^2 - \omega^2\right), \tag{31}$$

but in the case of this function ζ is not holomorphic everywhere in the lower half-plane. It would be interesting to know whether there exist functions F for which this holomorphy condition on ζ is satisfied.

In general, only part of the absorbed energy can be found in the scattered waves of frequency ω. Even if the absorbed energy is not transformed into heat, atomic theory requires, in general, that the irradiated atoms scatter not only waves with a frequency ω which are coherent with the incident light, but also waves with frequencies different from ω.[7]

§ 5. So far we have treated the scattering of light by an atom in a semi-classical, semiquantal way. In fact, the description of this scattering using the concept of an oscillating dipole which is excited by the incident waves is completely classical and we needed to introduce Bohr's atomic theory and quantum mechanics only when deriving Eq. (11). In that derivation the incident waves were treated in a formalistic manner as constituting an oscillating perturbing force with a potential which could be derived directly from the classical electron theory, that is, the electromagnetic field acting on the atom turns out to be treated classically. This derivation was completely insufficient when we were dealing with the case where the frequency of the incident light was the same as an absorption frequency. The principal reason for this difficulty was the impossibility of introducing the analogue of the radiation friction in quantum mechanics; this friction is in the classical theory represented by the imaginary term in Eq. (9). We can also not claim to have shown the validity of Eq. (18) using quantum mechanics; we have solely given formal arguments for the plausibility of this formula. Luckily, Dirac's recent studies[8] on the rigorously quantal treatment of the radiation field have yielded results which appear to confirm the validity of our formulæ and which therefore give a strong argument in favour of the validity of the classical description of dispersion phenomena using complex polarisation coefficients.

The result obtained in § 3 for the refractive index of an isotropic medium applies to the case where the refractibility is small (Eq. (7)). As the formulation of this result also has a physical meaning for a denser transparent

---

[7] H. A. Kramers and W. Heisenberg, Zs. Phys. **31**, 681 (1925).
[8] P.A.M. Dirac, Proc. Roy. Soc. **114**, 710 (1927); Zs. Phys. **44**, 585 (1927).

medium for which this condition is not satisfied, it would be interesting to know if for such a medium the real and imaginary parts of $\varepsilon-1$ (ε = dielectric constant) are still connected in the same way as ξ and η. Van der Plaats's experimental results[9] seem to speak in favour of this hypothesis.

I am grateful to Mr. Levi Civita and Mr. de Laer Kronig for several important remarks about the subject of this paper and I express to them my sincere thanks.

---

[9] B. J. van der Plaats, *Ann. Physik* **47**, 429 (1915).

E. General Theory of Paramagnetic Rotation in Crystals

A year ago I published a theoretical paper[1] discussing the magneto-optic effects studied by J. Becquerel and W. J. de Haas. In a second paper[2] Becquerel and the present author applied the results obtained. As the theory was unable to explain satisfactorily a number of problems and as new experiments have been performed since then, it seems useful to me to reconsider the question of paramagnetic rotation in crystals in a more general way. In the present paper I give the results of this study without considering applications. Several results are of a general theoretical interest apart from their possible application to Becquerel and de Haas's work.

In the first section we give a general study of the optical properties of a birefringent crystal with magnetic rotatory power. We show that if the deviation from isotropy is small, the introduction of a *rotation vector* makes it possible the describe rather simply the properties of light beams in any directions.

In the second section we prove a general theorem connected with the properties of an atomic system in a purely electric external field. We prove that the energy levels are necessarily doubly degenerate if the number of electrons in the system is odd.

In the third section we examine the magnetisation produced by a magnetic field of arbitrary direction in an atom in such a degenerate state.

In the fourth section we discuss the paramagnetic rotation in a crystal containing atoms of the kind considered in § 3.

In the fifth section we examine the effect of the magnetic interaction between the atoms on the paramagnetic rotation in a crystal.

---

[1] H. A. Kramers, *Proc. Amsterdam Acad.* **32**, 1176 (1929).
[2] H. A. Kramers and J. Becquerel, *Proc. Amsterdam Acad.* **32**, 1190 (1929).

1. Phenomenological Theory of the Optical Properties of a Crystal Having Magnetic Rotatory Power

We can represent the electric force vector in a crystal in which plane monochromatic light waves propagate by

$$\text{Re } \mathbf{E} e^{2\pi i(\sigma_1 x + \sigma_2 y + \sigma_3 z - \nu t)}, \tag{1}$$

where Re indicates the real part and where $\mathbf{E}$ is a vector with, in general, complex components. Similarly, the electric displacement vector is represented by

$$\text{Re } \mathbf{D} e^{2\pi i(\sigma_1 x + \sigma_2 y + \sigma_3 z - \nu t)}. \tag{2}$$

We assume that the components of $\mathbf{E}$ and $\mathbf{D}$ are related to one another by

$$D_k = \sum_{l=1}^{3} \varepsilon_{kl} E_l, \tag{3}$$

where the ε are the *dielectric coefficients* which can also be complex. We shall assume that for the frequencies of visible light the magnetic force and induction vectors will be equal to one another. They are represented by substituting a complex vector $\mathbf{B}$ for $\mathbf{E}$ in (1).

If we neglect absorption, the most general case will be the one where the matrix of the ε_{kl} coefficients will be Hermitean,[3] that is,

$$\varepsilon_{kl} = \varepsilon_{lk}^{*},$$

where the $*$ indicates a complex conjugate value.

In that case one finds that ν in (1) and (2) is always real for real σ.

Without loss of generality we can write:

$$\sigma_1 = \sigma, \quad \sigma_2 = \sigma_3 = 0.$$

Using the Maxwell equations to eliminate $\mathbf{B}$ we are led to the following equations:

$$\left. \begin{array}{l} 0 = \sum_l \varepsilon_{1l} E_l, \\[4pt] \left(\dfrac{\sigma c}{\nu}\right)^2 E_2 = \sum_l \varepsilon_{2l} E_l, \\[4pt] \left(\dfrac{\sigma c}{\nu}\right)^2 E_3 = \sum_l \varepsilon_{3l} E_l. \end{array} \right\} \tag{4}$$

The solution of (4) will be rather complicated for arbitrary values of ε_{kl}, and $\mathbf{E}$ will not be at right angles to the wave direction. If the deviation from

---

[3] Compare: C. G. Darwin, *Trans. Camb. Phil. Soc.* **23**, 160 (1924).

isotropy is small everything becomes simpler. In that case, which is often realised in nature, we put

$$\begin{aligned}
&\varepsilon_{11} = \varepsilon + \delta_{11}, \quad &&\varepsilon_{12} = \delta_{12} + i\omega_3, \quad &&\varepsilon_{13} = \delta_{13} - i\omega_2, \\
&\varepsilon_{21} = \delta_{21} - i\omega_3, \quad &&\varepsilon_{22} = \varepsilon + \delta_{22}, \quad &&\varepsilon_{23} = \delta_{23} + i\omega_1, \\
&\varepsilon_{31} = \delta_{31} + i\omega_2, \quad &&\varepsilon_{32} = \delta_{32} - i\omega_1, \quad &&\varepsilon_{13} = \varepsilon + \delta_{33},
\end{aligned} \qquad (5)$$

where the δ form a real symmetric tensor while the ω are the components of an axial vector. We assume that the δ and the ω are small of first order in ε. Substituting (5) into (4) we find a secular equation with a solution which has the following form if we neglect second order terms:

$$\left(\frac{\sigma c}{\nu}\right)^2 = \varepsilon' \pm r,$$

$$\varepsilon' = \varepsilon + \tfrac{1}{2}(\delta_{22} + \delta_{33}), \quad r = \sqrt{\tfrac{1}{4}(\delta_{22} - \delta_{33})^2 + \delta_{23}^2 + \omega_1^2}. \qquad (6)$$

For the ratios of the components of **E** we find

$$E_1 : E_2 : E_3 = 0 : -(\delta_{23} + i\omega_1) : -\tfrac{1}{2}(\delta_{22} - \delta_{33}) \pm r. \qquad (7)$$

The electric vector is thus at right angles to the direction of propagation.

If we put

$$2\,\frac{\delta_{23} + i\omega_1}{\delta_{22} - \delta_{33}} = -\tan\vartheta\, e^{i\varphi}, \qquad 0 \leqslant \vartheta \leqslant \pi, \qquad (8)$$

which gives

$$r = \tfrac{1}{2}(\delta_{22} - \delta_{33}) \sec\vartheta,$$

Eq. (7) takes the form

$$E_2 : E_3 = \cos\tfrac{1}{2}\vartheta\, e^{i\varphi/2} : \sin\tfrac{1}{2}\vartheta\, e^{-i\varphi/2},$$

or

$$E_2 : E_3 = -\sin\tfrac{1}{2}\vartheta\, e^{i\varphi/2} : \cos\tfrac{1}{2}\vartheta\, e^{-i\varphi/2}, \qquad (9)$$

depending on whether one chooses the upper or the lower sign in (6).

Equations (9) and (6) define two elliptically polarised waves which travel without change in the x-direction. One finds the difference in their wavelengths from (6) in the case when they have the same frequency.

If we represent an arbitrary elliptic vibration with its plane at right angles to the x-axis and for which $E_2 : E_3 = \cot\tfrac{1}{2}\alpha\, e^{i\gamma}$ by a point on a sphere (α is the polar distance and γ the azimuth) we recover the results of Gouy and Poincaré[4] on the effect of superimposing birefringence and rotation of the polarisation plane. In fact, an arbitrary monochromatic wave propagating in the x-direction is represented by an electric vector which, at every point

---

[4] See: J. Becquerel, *J. de Phys.* **9**, 343 (1928).

in space, describes an ellipse. While moving uniformly in the propagation direction the ellipse changes its shape. A simple calculation shows that this change can be described as a uniform rotation of the represenative point on the sphere about an axis with the polar coordinates ϑ and φ. A displacement dx corresponds to a rotation over an angle

$$\frac{2\pi\nu}{c\sqrt{\varepsilon'}}\sqrt{\tfrac{1}{4}(\delta_{22}-\delta_{33})^2+\delta_{23}^2+\omega_1^2}\,dx.$$

This rotation can be considered to be the superposition of a rotation about the axis $\alpha=\vartheta$, $\gamma=0$, over an angle

$$\frac{2\pi\nu}{c\sqrt{\varepsilon'}}\sqrt{\tfrac{1}{4}(\delta_{22}-\delta_{33})^2+\delta_{23}^2}\,dx$$

(pure birefringence) and a rotation about the axis $\alpha=\pi/2$, $\gamma=\pi/2$ over an angle $(2\pi\nu/c\sqrt{\varepsilon'})\omega_1\,dx$ (pure rotation).

This is Poincaré's result. If we introduce another system of polar coordinates, β and δ, on the sphere such that β is the complement of the angular distance between the representative point and the point $\alpha=\gamma=\pi/2$ and δ is the azimuth reckoned from the point $\alpha=\pi$, the ratio of the minor to the major axis of the ellipse will be $\tan\tfrac{1}{2}\beta$, while $\tfrac{1}{2}\delta$ gives the direction of the major axis with respect to the y- axis.

We see that ω_1 alone determines the pure rotation for a propagation parallel to the x-axis. Hence, for an arbitrary direction of the light propagation it will always be the component of the vector $\omega_1,\omega_2,\omega_3$ along that direction which determines the rotation. On this rotation is superimposed the birefringence which one finds from the tensor in exactly the same way as if there were no rotation. We shall call the vector $\omega_1,\omega_2,\omega_3$ the *rotation vector*.

If the anisotropy is no longer small, everything becomes more complicated. One can no longer distinguish simply between the effects of the real and the imaginary parts of the polarisation coefficients ε_{kl}.

2. General Theorem on the Properties of an Atomic System in an Electric Field

Consider a system of atomic nuclei and electrons in an external force field. Darwin[5] has calculated the classical expression for the energy as function of the coordinates and the momenta, taking into account the relativistic corrections up to terms containing $1/c^2$. This expression contains two parts: H_1+H_2; H_2 is the part due to the external forces whereas H_1 contains those terms which remain when the forces vanish. The terms H_1 are even in the momenta; as an operator in the Schrödinger equation H_1 is thus a real expression. On the other hand, H_2 contains terms linear in the momenta which

---
[5] C. G. Darwin, *Phil Mag.* **39**, 537 (1920).

E. General Theory of Paramagnetic Rotation in Crystals

lead to imaginary terms in the corresponding operator. However, these terms disappear when the field is purely electric in origin. Therefore. in that case the Schrödinger equation, which determines the wavefunctions in the stationary states, is real. Hence follows the well-known theorem that if φ is a solution, φ^* will also be a solution for the same energy value.

As the condition that the state considered is non-degenerate we have

$$\varphi^* = a\varphi = aa^*\varphi^*, \tag{10}$$

where a is a constant, or

$$|a|^2 = 1.$$

If condition (10) is satisfied we can thus write the wavefunction in the real form $a^{1/2}\varphi$.

If we take the electron spin into account the above theorem can be generalised in an interesting way. Consider first the expression for the energy when the spins are no longer neglected. This expression was given by Heisenberg[6] in his paper on the helium spectrum. Recently, Breit[7] has devoted to this question a very important paper in which he studies the interaction between particles using Dirac's spin theory. Breit's and Heisenberg's expressions are not completely identical; nonetheless they have in common that the part H_1 of the energy which is independent of the external forces, considered as a function of the momentum components and of the spins, only contains even terms with real coefficients. The contribution H_2 of the forces contains terms of degree 0, 1, and 2. If the field is purely electric the coefficients of the terms of degree 0 and 2 are real whereas those of degree 1 are imaginary.

Let n be the number of electrons. Neglecting possible nuclear spins which have, in general, no effect on the energy, a solution of the Schrödinger equation will be given by 2^n wavefunctions which we denote by

$$\varphi_{s_1...s_k...s_n}, \tag{11}$$

where s_k is the angular momentum (divided by $\hbar$) of the spin of the k-th electron in a specified direction ($s_k = +\frac{1}{2}$ or $-\frac{1}{2}$).

Given the special properties of the energy operator in the absence of external magnetic forces which we shall indicate presently, one can state the following theorem: *when Eq. (11) satisfies the Schrödinger equation, an other solution, corresponding to the same energy eigenvalue, will be given by*

$$\varphi'_{s_1...s_n} = (-1)^{\sum_k s_k - \frac{1}{2}n} \varphi^*_{-s_1...-s_n}. \tag{12}$$

We prove this statement through a direct calculation. Denote the components of the spin of the k-th eectron, divided by $\frac{1}{2}\hbar$, by s^k_x, s^k_y, s^k_z. The way they operate on a function (11) is described by the Pauli matrices:

---

[6] W. Heisenberg, *Zs. Phys.* **39**, 514 (1926).
[7] G. Breit, *Phys. Rev.* **34**, 553 (1929); see especially his Eq. (48).

$$s_x^k \to \begin{vmatrix} 1 & 0 \\ 0 & -1 \end{vmatrix}, \quad s_y^k \to \begin{vmatrix} 0 & 1 \\ 1 & 0 \end{vmatrix}, \quad s_z^k \to \begin{vmatrix} 0 & -i \\ i & 0 \end{vmatrix}, \tag{13}$$

where x is the direction specified when writing down (11). We have thus:

$$s_x^k \varphi_{s_k} = (-1)^{s_k - \frac{1}{2}} \varphi_{s_k}, \quad s_y^k \varphi_{s_k} = -\varphi_{-s_k}, \quad s_z^k \varphi_{s_k} = -i(-1)^{s_k - \frac{1}{2}} \varphi_{-s_k}. \tag{14}$$

The omitted indices $s_{k'}$ ($k' \neq k$) are unchanged.

We write the Schrödinger equation in the form

$$H(s_x^k, s_y^k, s_z^k) \varphi_{s_k} = E \varphi_{s_k}. \tag{15}$$

The energy H also depends, apart from the spin components, on the coordinates and the momenta of the electrons and of the nuclei. We now take the complex conjugate of (15), bearing in mind that the operator representing a momentum is imaginary and that the terms of odd degree in the s_k are multiplied by an odd number of momenta:

$$H(-s_x^{*k}, -s_y^{*k}, -s_z^{*k}) \varphi_{s_k}^* = E \varphi_{s_k}^*.$$

Using (13) we find

$$H(-s_x^k, -s_y^k, s_z^k) \varphi_{s_k}^* = E \varphi_{s_k}^*. \tag{16}$$

If we write

$$\varphi_{s_k}'' = \varphi_{-s_k}^*,$$

Eq. (16) becomes

$$H(s_x^k, -s_y^k, -s_z^k) \varphi_{s_k}'' = E \varphi_{s_k}''. \tag{17}$$

We now introduce the function φ' defined by (12):

$$\varphi_{s_k}' = (-1)^{\sum_k s_k - \frac{1}{2}n} \varphi_{s_k}''.$$

Using (14) one sees easily that the s_x, s_y, s_z act upon φ' in the same way as the $s_x, -s_y, -s_z$ act upon φ''. Hence it follows from (17) that

$$H(s_x^k, s_y^k, s_z^k) \varphi_{s_k}' = E \varphi_{s_k}',$$

which proves our theorem.

Let us now examine whether the stationary state corresponding to (11) can be non-degenerate. The condition is that the functions (12) are the same as the functions (11), apart from a common factor which we denote by a:

$$(-1)^{\sum s_k - \frac{1}{2}n} \varphi_{-s_k}^* = a \varphi_{s_k}.$$

Taking the complex conjugate and replacing s_k by $-s_k$ we have:

$$(-1)^{\sum s_k - \frac{1}{2}n} \varphi_{s_k} = a^* \varphi^*_{-s_k} = a^* a (-1)^{-\sum s_k + \frac{1}{2}n} \varphi_{s_k}.$$

Hence:

$$a^* a = (-1)^n.$$

This condition can only be satisfied if n is even; in that case we can put $a = 1$. If n is odd, we have the following theorem: the stationary states of an atomic system are always degenerate if the system contains an odd number of electrons, and the degree of degeneracy is an even number.

If the number of electrons is even and if the stationary state is non-degenerate the energy will to a first approximation not be affected by a magnetic field. In fact, the perturbing energy Ω corresponding to such a field contains in first approximation only linear terms in the momentum of the kind fpf and linear terms in the spins of the kind gs (f and g are functions of the coordinates). It suffices to consider one term $T = fpf = -i\hbar f(\partial/\partial x_k)f$ and one term $T = gs_x^k$. In order to calculate the perturbation energy corresponding to these terms we must know the following integrals:

$$\overline{T} = \sum_k \int \varphi^*_{s_k} T \varphi_{s_k} \, d\tau = \sum_k (-1)^{\sum s_k - \frac{1}{2}n} \int \varphi_{-s_k} T \varphi_{s_k} \, d\tau$$
$$= \sum_k (-1)^{-\sum s_k - \frac{1}{2}n} \int \varphi_{s_k} T \varphi_{-s_k} \, d\tau. \quad (18)$$

For the two kinds of terms that we have in mind we always have

$$\int \varphi_{s_k} T \varphi_{-s_k} = - \int \varphi_{-s_k} T \varphi_{s_k} \, d\tau. \quad (19)$$

If we introduce (19) into the last part of (18) this becomes equal to the negative of the second member (note that $(-1)^{\sum s_k} = (-1)^{-\sum s_k}$ if n is even) so that $\overline{T}$ is equal to zero. The non-degenerate stationary states are thus always non-magnetic.

If the number of electrons is odd the levels will, in general, be magnetic. We shall study this problem in more detail in the next section.

3. Magnetisation of a Doubly Degenerate Stationary State

Consider a stationary state of an atom with an odd number of electrons placed in an electric field. Assume that the degree of degeneracy is the smallest possible, that is, two. Let us study the effect on the energy of a uniform, weak magnetic field. In the energy operator this field gives rise to the following terms:

$$\Omega = (\mathbf{H} \cdot \mathbf{Q}), \quad \mathbf{Q} = \frac{e}{2mc} \sum (\mathbf{P}_k + 2\mathbf{S}_k), \quad (20)$$

where $\mathbf{P}_k$ is the classical angular momentum of the k-th electron. We shall take for the two wavefunctions just the functions $\varphi_1 = \varphi_{s_k}$ and $\varphi_2 = \varphi'_{s_k}$ we discussed in the preceding section. One checks easily that these functions are orthogonal to one another and that the integrals $\int |\varphi_{s_k}|^2 \, d\tau$ and $\int |\varphi'_{s_k}|^2 \, d\tau$ are equal to one another so that we can assume that they are normalised at the same time. The stationary state will in a magnetic field be split into two states with wavefunctions given by

$$\varphi_I = \alpha\varphi_1 + \beta\varphi_2, \qquad \varphi_{II} = -\beta^*\varphi_1 + \alpha^*\varphi_2, \qquad \alpha^*\alpha + \beta^*\beta = 1, \qquad (21)$$

while the term which is linear in H in the magnetisation energy E_H is given by the roots of the equation

$$\begin{vmatrix} (\mathbf{Q}_{11} \cdot \mathbf{H}) - E_H & (\mathbf{Q}_{12} \cdot \mathbf{H}) \\ (\mathbf{Q}_{21} \cdot \mathbf{H}) & (\mathbf{Q}_{22} \cdot \mathbf{H}) - E_H \end{vmatrix} = 0, \quad \mathbf{Q}_{kk'} = \sum \int \varphi_k^* \mathbf{Q} \varphi_{k'} \, d\tau. \quad (22)$$

One easily checks that the sum of the real vectors $\mathbf{Q}_{11}$ and $\mathbf{Q}_{22}$ vanishes. The components of $\mathbf{Q}_{12}$ are complex, in general. We put

$$\mathbf{Q}_{11} = \mathbf{M}_1, \quad \mathbf{Q}_{12} = \mathbf{M}_2 + i\mathbf{M}_3, \quad \mathbf{Q}_{21} = \mathbf{M}_2 - i\mathbf{M}_3, \qquad (23)$$

where the $\mathbf{M}$ are real vectors.

We find

$$E_H = \pm\sqrt{(\mathbf{M}_1 \cdot \mathbf{H})^2 + (\mathbf{M}_2 \cdot \mathbf{H})^2 + (\mathbf{M}_3 \cdot \mathbf{H})^2}. \qquad (24)$$

If the $+$sign corresponds to the state I and the $-$sign to the state II, we have

$$\frac{\alpha}{\beta} = \frac{([\mathbf{M}_2 + i\mathbf{M}_3] \cdot \mathbf{H})}{|E_H| - (\mathbf{M}_1 \cdot \mathbf{H})}. \qquad (25)$$

For the magnetic moment of the atom in the state I we find

$$\mathcal{M}_I = \sum \int \varphi_I^* \mathbf{Q} \varphi_{II} \, d\tau = \frac{(\mathbf{M}_1 \cdot \mathbf{H})\mathbf{M}_1 + (\mathbf{M}_2 \cdot \mathbf{H})\mathbf{M}_2 + (\mathbf{M}_3 \cdot \mathbf{H})\mathbf{M}_3}{|E_H|}. \qquad (26)$$

The magnetic moment in the state II is equal to that in the state I but with the opposite sign:

$$\mathcal{M}_{II} = -\mathcal{M}_I. \qquad (27)$$

The magnetic moment is thus governed by a symmetric tensor of second rank with components given by

$$t_{xx} = \sum_k M_{kx}^2, \qquad t_{xy} = \sum_k M_{kx} M_{ky}, \qquad \ldots,$$

while the expression $t_{xx} H_x^2 + 2t_{xy} H_x H_y + \ldots$ is never negative.

By changing the axes one can always simplify Eqs. (24) and (26) so that they take the form

$$E_H = \pm \sqrt{\mu_x^2 H_x^2 + \mu_y^2 H_y^2 + \mu_z^2 H_z^2}, \tag{28}$$

$$\mathcal{M}_{Ix} = \frac{\mu_x^2 H_x}{|E_H|}, \quad \mathcal{M}_{Iy} = \frac{\mu_y^2 H_y}{|E_H|}, \quad \mathcal{M}_{Iz} = \frac{\mu_z^2 H_z}{|E_H|}. \tag{29}$$

The components of the magnetic moment always satisfy the relation

$$\frac{\mathcal{M}_x^2}{\mu_x^2} + \frac{\mathcal{M}_y^2}{\mu_y^2} + \frac{\mathcal{M}_z^2}{\mu_z^2} = 1. \tag{30}$$

The direction of the magnetic field which produces the moment $\mathcal{M}_x$, $\mathcal{M}_y$, $\mathcal{M}_z$ is along the normal to the tangential plane to the ellipsoid (30) in the point $\mathcal{M}_x$, $\mathcal{M}_y$, $\mathcal{M}_z$.

One sees that the Zeeman effect for the spectral lines of the atom, in general, splits the line into four, which can reduce to two in special cases.

4. Paramagnetic Rotation in a Crystal

Let us consider a crystal with a lattice which contains atoms with an odd number of electrons. Let us assume that the energy levels of each of these atoms are, under the action of the forces due to the other atoms in the lattice, split into levels which have the smallest possible degree of degeneracy, that is, two. We have found in the preceding section that these levels are "magnetic," that is, that each of them can be further split into two by a magnetic field. At sufficiently low temperatures all these magnetic atoms will be in their ground state. If the crystal is put in a uniform magnetic field each of these levels will be split into two states, I and II; the probability that an atom is in the state I or in the state II is given by the Boltzmann formula.

As a result, a light beam passing through the crystal will show paramagnetic rotation of the plane of polarisation. In fact, the contribution of an atom in the state I to the rotatory power will be equal and opposite to that of the same atom in the state II if we neglect the effect of the magnetic splitting of the higher levels, that is, if we neglect the diamagnetic rotation. The observed rotation will thus be due to the fact that the state II has a different probability from the state I; this is paramagnetic rotation, that is, a rotation the cause of which is the same as that of paramagnetism and with a magnitude which depends on the temperature.[8]

We have shown in the first section that the rotation of the plane of polarisation is characterised by a rotation vector in the crystal the components of which are the imaginary parts of the dielectric constants. The electron theory of dispersion shows that those coefficients can be calculated using the quantities which describe the polarisation of a single atom in the electromagnetic field of the incident light.

---

[8] J. Becquerel, *Le Radium* **5**, 12 (1908).

We represent, as in § 1, the electric force in that field by

$$\text{Re } \mathbf{E} e^{-2\pi i \nu t},$$

and we denote the polarisation of the atom by

$$\text{Re } \mathbf{P} e^{-2\pi i \nu t},$$

The relation between $\mathbf{E}$ and $\mathbf{P}$ is given by

$$P_k = \sum_l p_{kl} E_l. \tag{31}$$

For the dielectric coefficients we have

$$\varepsilon_{kl} = 1 + 4\pi \sum_{at} p_{kl}, \tag{32}$$

where the summation is over all atoms in a unit volume. Equation (32) does no longer hold when one cannot neglect the forces between the moments of neighbouring atoms and one must use a Lorenz-Lorentz type formula. Nonetheless it seems to be allowable to use (32) if one is only interested in the imaginary parts of the ε_{kl}, which occur in the theory of the rotatory power and which are always very small as compared to unity.

According to the quantal dispersion theory Eq. (32) takes the form[9]

$$\mathbf{P}_f = \sum_g \left(\frac{\mathbf{P}_{fg}(\mathbf{P}_{gf} \cdot \mathbf{E})}{h(\nu_{gf} - \nu)} - \frac{\mathbf{P}_{gf}(\mathbf{P}_{fg} \cdot \mathbf{E})}{h(\nu_{fg} - \nu)} \right). \tag{33}$$

The polarisation refers to the ground state, f, and the summation is over all states g with a higher energy.

The quantity $\nu_{fg} = -\nu_{gf}$ is an absorption frequency. The vectors $\mathbf{P}_{fg}$ are defined by

$$\mathbf{P}_{fg} = \sum \int \varphi_f^* \mathbf{P} \varphi_g \, d\tau = \mathbf{P}_{gf}^*,$$

where $\mathbf{P}$ is the classical expression for the polarisation of the atom while φ_f and φ_g are the wavefunctions in the states f and g.

Comparison of (33), (31), (32), and (5) leads to the following expression for the rotation vector in the crystal:

$$\omega = \frac{i\nu}{h} \sum_{at} \sum_g \frac{[\mathbf{P}_{fg} \wedge \mathbf{P}_{gf}]}{\nu^2 - \nu_{gf}^2}. \tag{34}$$

---

[9] H. A. Kramers and W. Heisenberg, *Zs Phys.* **31**, 681 (1925); Born and Jordan, *Elementare Quantenmechanik*, p. 259.

E. General Theory of Paramagnetic Rotation in Crystals

To find out how ω depends on the magnetic field we first calculate that part, ω_I, of ω which is the contribution of an atom in the state I, defined by (21). We introduce the following abbreviations:

$$\mathbf{R}_1 = \sum \int \varphi_1^* \mathbf{P} \varphi_g \, d\tau, \quad \mathbf{R}_2 = \sum \int \varphi_2^* \mathbf{P} \varphi_g \, d\tau, \quad \nu_{g1} = \nu_{g2} = \nu_g,$$

where φ_1 and φ_2 are, as in §3, the functions φ_{s_k} and φ'_{s_k} defined in §2. We also write

$$\frac{i\nu}{h} \sum_g \frac{[\mathbf{R}_1 \wedge \mathbf{R}_1^*]}{\nu^2 - \nu_g^2} = \mathbf{A}_1 = -\frac{i\nu}{h} \sum_g \frac{[\mathbf{R}_2 \wedge \mathbf{R}_2^*]}{\nu^2 - \nu_g^2}, \tag{35}$$

$$\frac{i\nu}{h} \sum_g \frac{[\mathbf{R}_1 \wedge \mathbf{R}_2^*]}{\nu^2 - \nu_g^2} = \mathbf{A}_2 + i\mathbf{A}_3. \tag{36}$$

The fact that the first and the last parts of Eq. (35) are equal follows immediately if one takes into account the twofold degeneracy of the levels g which is also expressed by the formulæ of §2. For two states g and g' which are conjugated like φ and φ' in §2 one has, in fact, $[\mathbf{R}_{1g} \wedge \mathbf{R}_{1g}^*] = -[\mathbf{R}_{2g'} \wedge \mathbf{R}_{2g'}^*]$.

The quantities $\mathbf{A}_1, \mathbf{A}_2, \mathbf{A}_3$ are three real vectors. To be exact, they define a second-order tensor, as we shall see in a moment.

Using (34) we find for ω_I

$$\omega_I = (\alpha^*\alpha - \beta^*\beta)\mathbf{A}_1 + \alpha^*\beta(\mathbf{A}_2 + i\mathbf{A}_3) + \alpha\beta^*(\mathbf{A}_2 - i\mathbf{A}_3). \tag{37}$$

If we use (25), which defines α and β, (37) reduces to

$$\omega_I = \frac{(\mathbf{M}_1 \cdot \mathbf{H})\mathbf{A}_1 + (\mathbf{M}_2 \cdot \mathbf{H})\mathbf{A}_2 + (\mathbf{M}_3 \cdot \mathbf{H})\mathbf{A}_3}{|E_H|}. \tag{38}$$

One sees easily that

$$\omega_{II} = -\omega_I. \tag{39}$$

We saw in §3 that by changing the axes we can make the vectors (M_{1x}, M_{2x}, M_{3x}), (M_{1y}, M_{2y}, M_{3y}), and (M_{1z}, M_{2z}, M_{3z}) mutually orthogonal; the magnitude of these vectors are denoted by μ_x, μ_y, μ_z. Using that coordinate system (38) takes the form

$$\omega_I = \frac{\mu_x H_x}{|E_H|}\mathbf{A}_1 + \frac{\mu_y H_y}{|E_H|}\mathbf{A}_2 + \frac{\mu_z H_z}{|E_H|}\mathbf{A}_3, \tag{40}$$

and if we write

$$\frac{\mu_x H_x}{|E_H|} = \alpha_1, \quad \frac{\mu_y H_y}{|E_H|} = \alpha_2, \quad \frac{\mu_z H_z}{|E_H|} = \alpha_3, \quad \alpha_1^2 + \alpha_2^2 + \alpha_3^2 = 1,$$
$$A_{1x} = a_11, \quad A_{1y} = a_{21}, \quad A_{1z} = a_{31}, \quad A_{2x} = a_{12}, \ldots, \tag{41}$$

we have

$$\omega_{Ik} = \sum_l a_{kl}\alpha_l. \tag{42}$$

We must thus calculate the tensor a_{kl} to find the effect of a single atom on the rotation. In simple cases one often finds that the a_{kl} vanish for $k \neq l$, that is, that the principal axes of the surface $\sum a_{kl}x_k x_l = 1$ are the same as the directions along which the magnetisation is parallel to the magnetic field.

For a given temperature the ratio of the probabilities for the states I and II is equal to $e^{-|E_H|/kT} : e^{+|E_H|/kT}$. Hence the average contribution ω_{at} of this atom to the rotation vector of the crystal is equal to

$$\omega_{at} = -\omega_I \tanh \frac{|E_H|}{kT}. \tag{43}$$

For $kT \ll |E_H|$ only the state II remains and ω_{at} has reached its maximum value which is independent of $|H|$ and T:

$$\omega_{at} = -\omega_I, \qquad kT \ll |E_H|. \tag{44}$$

If, on the other hand, $kT \gg |E_H|$ the rotation is linear in H and we have

$$\omega_{at} = -\omega_I \frac{|E_H|}{kT} = -\frac{\mu_x H_x \mathbf{A}_1 + \mu_y H_y \mathbf{A}_2 + \mu_z H_z \mathbf{A}_3}{kT}. \tag{45}$$

The rotation vector of the whole crystal is $\sum_{at} \omega_{at}$ where the summation is over all magnetic atoms in a unit volume. If all atoms are of the same kind and are in completely identical situations one only needs multiply ω_{at} by the number per unit volume. The rotation depends on H and T as

$$\varrho = \varrho_\infty \tanh \frac{\mu H}{kT}, \tag{46}$$

where ϱ_∞ and μ may still vary with change of direction of H or of the light beam. The principal axes of the magnetisation, the quantities μ_x, μ_y, μ_z, and the properties of the a_{kl} tensor are connected with the symmetry of the crystal.

If the various atoms are no longer of the same kind or if they are not all in the same situation, the rotation will be a sum of expressions of the kind (46), where each of them corresponds to a group of atoms in identical situations. The saturation rotation will depend in a rather complicated manner on the direction of the magnetic field. In the case of weak fields the components of ω will depend linearly on the components of $\mathbf{H}$ (see (45)) and will always change with temperature like $1/T$.

5. Effect of the Magnetic Interactions between the Atoms

If the magnetisation produced by the field in the crystal becomes large, as one should expect at very low temperatures, the magnetic field acting on a magnetic atom will no longer be equal to the external field. The magnetisation will, in general, not be uniform. To simplify we shall, however, assume that the magnetisation M is uniform. This condition will be satisfied if the crystal is very flat, or very long, in the direction of the field and — more generally — when it is in the shape of an ellipsoid. Denoting the field acting on an atom by $\mathbf{H}_{at}$ we have

$$\mathbf{H}_{at} = \mathbf{H} + \frac{4\pi}{3}(\beta - \alpha)\mathbf{M}. \tag{47}$$

We have, as a first approximation, assumed that the field due to the neighbouring atoms can be represented by the expression $(4\pi/3)\beta\mathbf{M}$; in most simple cases the factor β is equal to 1. On the other hand, $-(4\pi/3)\alpha\mathbf{M}$ represents the demagnetisation force. For a very long crystal we have $\alpha = 0$, for a sphere $\alpha = 1$, and for a flat crystal $\alpha = 3$.

If the difference between $\mathbf{H}_{at}$ and $\mathbf{H}$ cannot be neglected we must replace $\mathbf{H}$ by $\mathbf{H}_{at}$ in the formulæ for the paramagnetic rotation. To study the effect of such a substitution we take the case where the rotation is exactly given by

$$\varrho = \varrho_\infty \tanh \frac{\mu H_{at}}{kT}. \tag{48}$$

To further simplify we also assume that the directions of the magnetisation and of the field are parallel. We note that if the direction of $\mathbf{H}$ is fixed, the atomic magnetisation is given by

$$\mathbf{M} = \mathbf{M}_\infty \tanh \frac{\mu H_{at}}{kT}. \tag{49}$$

The magnetisation is thus proportional to the rotation and we can write

$$H_{at} = H - a\frac{\varrho}{\varrho_\infty}. \tag{50}$$

In this formula a is a magnetic field equal to the correction field $(4\pi/3)(\alpha - \beta)M_\infty$ which acts on the atoms when they are in the state II.

The equation for the rotation thus becomes

$$\frac{\varrho}{\varrho_\infty} = \tanh \frac{\mu(H - (a\varrho/\varrho_\infty))}{kT}. \tag{51}$$

If a is positive, the difference between (46) and (51) will be of the same kind as the one found by Becquerel and de Haas for the rotation in xenotime between the measurements at 1.4 K and those at 4.2 K. If all metallic atoms in xenotime are Gd^{+++}, each carrying 7 Bohr magnetons, one finds $a = 7900$ G in the case when $\alpha - \beta = 2$ (very flat crystal). A smaller value appears to suffice to explain the experiments. The above differences were in a previous paper[10] hypothetically represented by a formula, differing from (51), which was based on completely different ideas. The explanation which I give here seems to me to be much more probable.

---

[10] H. A. Kramers and J. Becquerel, *Proc. Amsterdam Acad.* **32**, 1190 (1929); J. Becquerel, W. J. de Haas, and H. A. Kramers, *Proc. Amsterdam Acad.* **32**, 1206 (1929).

F. Classical Relativistic Spin-Theory and Its Quantization

I. Introduction

In the course of the development of modern atomic theory the Zeeman effect has repeatedly played a prominent part. Two outstanding instances hereof may be briefly recalled in the following. Immediately after its discovery in 1896 it proved — on the basis of Lorentz' analysis — to lend a most convincing support to the idea, that small, negatively charged particles, identical with those discovered in the cathode rays, were present in the atom and constituted, through their vibrations, the source of electromagnetic disturbances giving rise to spectral lines. The fact, however, that many spectral lines show a Zeeman effect of the so-called anomalous type, remained for a long time a serious difficulty. On the one hand, Lorentz' and Voigt's formal treatments were far from satisfactory from a physical point of view. On the other hand, the development of Bohr's views on the origin of spectral lines during the years 1913–1925 was hardly fit to encourage the optimistic view that the anomalous Zeeman effect might be a simple consequence of the quantum laws governing the behaviour of electrons inside the atom.

In 1925 a way out of this difficulty was offered by Uhlenbeck's and Goudsmit's hypothesis of the electron spin, according to which an electron should possess — besides its mass and charge — an intrinsic rotational moment accompanied by a magnetic moment. Among the experimental facts leading to this hypothesis, the laws of the anomalous Zeeman effect, as formulated by Landé, ranked first.

II. Classical Spin Problem. Aim of This Paper

A point electron (mass m, charge $-e$), moving in a central field of force gives rise to a magnetic moment equal to its rotational moment multiplied by $-e/2mc$. In order to explain the anomalous Zeeman effect Uhlenbeck and Goudsmit had to assume that the corresponding ratio between magnetic and rotational moment of the electron spin is twice as large, viz. $-e/mc$. The analysis of possible classical models of a rotating electron showed that — although a difference between the two said ratios was to be anticipated — arguments along these lines would be insufficient to predict in an unambiguous way the factor 2 required by experiment.

Dirac's ingenious treatment of the relativistic wave equation of the electron (1928), in which the idea of electronic spin was not primarily introduced, seems to have thoroughly changed the aspect of the theoretical problems involved. In fact, the physical content of Dirac's linear equations—when interpreted in the limit of small velocities—reflects exactly all the properties of the electron including those pertaining to the spin, both the factor 2 and the value $h/2$ for the spin moment appearing automatically. Thus the optimistic view, to which we alluded in part I, appears to be justified after all, and one is tempted to adopt Dirac's elegant formalism as a primary basis for our description of the electron's behaviour.

Should therefore any investigation which approaches the spin properties from a purely classical point of view, such as for instance Uhlenbeck's and Goudsmit's original treatment, be rejected as inappropriate? There are several reasons which urge us to be cautious with our answer. The famous difficulty of the negative mass—even though it be mitigated to a considerable extent by the hole theory—shows us that even Dirac's theory cannot be considered as a satisfactory foundation. Furthermore we may recall the anomalous value of the magnetic moment of the proton which was recently discovered by Stern & Frisch. There is no a priori theoretical reason why the Dirac equations should apply to the electron and not to the proton.

In view of this situation, it is perhaps not without interest that even a consideration, in which the idea of electronic spin is introduced in a purely classical way, affords a simple interpretation of the value of the ratio between the electron's magnetic and rotational moment.[1] The argument rests uniquely on the principle that a consistent set of relativistically invariant equations of motion should be established, which—in a system of coordinates moving with the electron—reduces to the well known laws, expresses how only the electric field (in first approximation) governs the acceleration and how only the magnetic field governs the precession of the spin vector. Considerations pertaining to a detailed classical model of the electron do not enter at all.

In this paper we will show that a classical spin theory developed along these lines is intimately connected with Dirac's theory of the electron. In fact, if a process of quantization is applied in which the quantum number of the spin is put equal to $\frac{1}{2}$ and if the classical Hamiltonian is chosen in an appropriate way, the result will be identical with Dirac's formalism.

III. Equations of Motion

In our paper cited above the equations governing the precession of the spin vector were written in the relativistically invariant form

$$\frac{d\mathbf{S}}{d\tau} = \alpha[\mathbf{S} \wedge \mathbf{F}], \tag{1}$$

---

[1] H. A. Kramers, *Physica* **1**, 825 (1934).

where $d\tau$ denotes the element of eigenzeit, whereas **S** and **F** are two complex vectors:

$$\mathbf{S} = \mathbf{A} + i\mathbf{B}, \qquad \mathbf{F} = \mathbf{H} + i\mathbf{E}. \tag{2}$$

A and **B**, which characterize the spin, transform under a Lorentz transformation as **H** and **E** (magnetic and electric field strength). A Lorentz transformation corresponds to a (generally complex) orthogonal transformation of the components of **S** and **F**. The condition that **B** always vanishes in an inertial system moving with the electron leads to the relativistically invariant relation

$$\mathbf{B} = \frac{1}{c}[\mathbf{A} \wedge \mathbf{v}], \tag{3}$$

where **v** is the velocity of the electron.

In a system in which $\mathbf{v} = 0$, the real part of (1) reduces to the unrelativistic classical description of the behaviour of a spinning electron with spin vector **A** (*i.e.* vector of rotational moment) in a magnetic field **H**, α being the ratio between the magnetic and rotational moment:

$$\dot{\mathbf{A}} = \alpha[\mathbf{A} \wedge \mathbf{H}]. \tag{4}$$

The imaginary part of (1) reduces in the same system to

$$\dot{\mathbf{B}} = \alpha[\mathbf{A} \wedge \mathbf{E}]. \tag{5}$$

If the reaction of the spin on the orbital motion may be considered as very small, this motion will — always for $\mathbf{v} = 0$ — obey the law:

$$m\dot{\mathbf{v}} = -e\mathbf{E}, \tag{6}$$

so that (5) takes the form

$$\dot{\mathbf{B}} = -\frac{\alpha m}{e}[\mathbf{A} \wedge \dot{\mathbf{v}}]. \tag{7}$$

If, now, we derive (3) with respect to the time and put $\mathbf{v} = 0$, we obtain a formula which, when comparing with (7), leads immediately to:

$$\alpha = -\frac{e}{mc}. \tag{8}$$

If one wishes to take the reaction of the spin on the orbit into account without abandoning the rigorous validity of (3), equation (1) has to be considered as a first approximation only. The procedure to be followed in order to develop a more complete theory along these lines is not unambiguously prescribed. At present we will leave this question apart; we do not know if its treatment will lead to results of physical interest.[2]

---

[2] The development of the theory to higher approximations seems to require that to the electron, besides an electrical charge (monopole) and a magnetic moment (dipole), should be attributed also an electrical quadrupole, a magnetic octupole, and so on. These poles of higher order disappear automatically if a quantization is applied which gives the electron a spin moment of only $h/2$.

IV. Canonical Form

Before quantizing the equations of motion (6) and (1), it will be necessary first to establish a Hamiltonian equation, from which they both can be simultaneously derived. For this purpose let us first consider the equations (4), in which only real vectors occur. They can be written in canonical form if — in agreement with the ordinary treatment of a dipole in a field — the energy is taken to be:

$$H_A = -\alpha (\mathbf{A} \cdot \mathbf{H}). \tag{9}$$

There is only one degree of freedom, and for the canonical coordinates one may choose:

$$p = A_1, \quad q = \arctan \frac{A_2}{A_3}, \quad (A_2 + iA_3 = \sqrt{A^2 - p^2} \cdot e^{iq}).$$

This corresponds to the following values for the Poisson brackets:

$$\{A_1 A_2\} = \frac{\partial A_1}{\partial p}\frac{\partial A_2}{\partial q} - \frac{\partial A_1}{\partial q}\frac{\partial A_2}{\partial p} = -A_3, \quad \{A_2 A_3\} = -A_1, \quad \{A_3 A_1\} = -A_2.$$

The equations governing the change of $\mathbf{A}$,

$$\dot{A}_k = -\sum_l \{A_k A_l\} \frac{\partial H_A}{\partial A_l},$$

are seen to be identical with (4) if the expression (9) for H_A is adopted.

The equations (1) can be treated in an exactly analogous way:

$$\left.\begin{aligned} H_S &= -\alpha (\mathbf{S} \cdot \mathbf{F}), \\ \{S_1 S_2\} &= -S_3, \quad \text{cycl.,} \\ \frac{dS_k}{d\tau} &= -\sum_l \{S_k S_l\} \frac{\partial H_S}{\partial S_l}. \end{aligned}\right\} \tag{10}$$

Formally we have still to do with a system of one degree of freedom, the canonical coordinates being, for instance:

$$p = S_1, \quad q = \arctan \frac{S_2}{S_3},$$

but these coordinates are complex and so is the Hamiltonian H_S. This circumstance need not alarm us. Separating everywhere real and imaginary parts:

$$p = p' + ip'', \quad q = q' - iq'', \quad H = H'(p'p''q'q'') + iH''(p'p''q'q''),$$

it is easily verified that the complex equations of motion,

F. Classical Relativistic Spin-Theory and Its Quantization 191

$$\frac{dp}{d\tau} = -\frac{\partial H}{\partial q}, \quad \frac{dq}{d\tau} = \frac{\partial H}{\partial p}, \quad (d\tau \text{ real}),$$

correspond to real canonical equations of a system of two degrees of freedom in which either H' or H'' is taken as Hamiltonian:

$$\frac{dp'}{d\tau} = -\frac{\partial H'}{\partial q'}, \quad \frac{dq'}{d\tau} = \frac{\partial H'}{\partial p'}, \quad \frac{dp'}{d\tau} = \frac{\partial H''}{\partial q''}, \quad \frac{dq''}{d\tau} = -\frac{\partial H''}{\partial p'},$$

or (11)

$$\frac{dp''}{d\tau} = -\frac{\partial H'}{\partial q''}, \quad \frac{dq''}{d\tau} = \frac{\partial H'}{\partial p''}, \quad \frac{dp''}{d\tau} = -\frac{\partial H''}{\partial q'}, \quad \frac{dq'}{d\tau} = -\frac{\partial H''}{\partial p''}.$$

This consideration shows that — besides (10) — two alternative ways of deriving the equations from a real Hamiltonian offer themselves. For this purpose we have to introduce besides $\mathbf{S}$ its complex conjugate vector $\mathbf{S}^*$:

$$H'_S = -\alpha \frac{(\mathbf{S} \cdot \mathbf{F}) + (\mathbf{S}^* \cdot \mathbf{F}^*)}{2}, \quad \{S_k S_l^*\} = 0,$$
$$\{S_1 S_2\} = -2S_3, \text{ cycl.}, \quad \{S_1^* S_2^*\} = -2S_3^*, \text{ cycl.};$$
(10a)

$$H''_S = -\alpha \frac{(\mathbf{S} \cdot \mathbf{F}) - (\mathbf{S}^* \cdot \mathbf{F}^*)}{2i}, \quad \{S_k S_l^*\} = 0,$$
$$\{S_1 S_2\} = -2iS_3, \text{ cycl.}, \quad \{S_1^* S_2^*\} = -2iS_3^*, \text{ cycl.};$$
(10b)

$$\frac{dS_k^{(*)}}{d\tau} = -\sum_l \{S_k^{(*)} S_l\} \frac{\partial H_S}{\partial S_l} - \sum_l \{S_k^{(*)} S_l^*\} \frac{\partial H_S}{\partial S_l^*},$$

where $S_k^{(*)}$ means either S_k or S_k^*.

The system is now explicitly treated as one of two degrees of freedom; the expressions for the Poisson brackets are found by taking the four real canonical variables (compare (11)) explicitly into account.

For completeness we might finally mention the alternative of (10) which arises when, instead of $\mathbf{S}$, its complex conjugate $\mathbf{S}^*$ is introduced:

$$\left. \begin{array}{l} H_S^* = -\alpha(\mathbf{S}^* \cdot \mathbf{F}^*), \\ \{S_1^* S_2^*\} = -S_3^*, \text{ cycl.}, \\ \dfrac{dS_k^*}{d\tau} = -\sum_l \{S_k^* S_l^*\} \dfrac{\partial H_S^*}{\partial S_l^*}. \end{array} \right\}$$
(10d)

From (2) we find

$$(\mathbf{S} \cdot \mathbf{F}) = [(\mathbf{A} \cdot \mathbf{H}) - (\mathbf{B} \cdot \mathbf{E})] + i[(\mathbf{A} \cdot \mathbf{E}) + (\mathbf{B} \cdot \mathbf{H})].$$

It would therefore appear most natural to adopy the form H'_S given by (10a):

$$H'_S = -\alpha[(\mathbf{A} \cdot \mathbf{H}) - (\mathbf{B} \cdot \mathbf{E})],$$

since, in an inertial system moving with the electron, it reduces to the familiar real energy expression (9). If we adopt the simpler form H_S or H_S^* given by (10) or (10c) it looks at first sight as if we formally introduced an imaginary electric moment of the electron equal to $-i\alpha \mathbf{A}$ or $+i\alpha \mathbf{A}$ respectively.

The equations of motion (6), in which the spin is neglected, can be derived from the familiar Hamiltonian equation:

$$H_0 \equiv \frac{1}{2m}\left\{-\frac{(\varepsilon + e\Phi)^2}{c^2} + \left(\mathbf{p} + \frac{e}{c}\mathbf{\Psi}\right)^2\right\} = -\tfrac{1}{2}mc^2, \qquad (12)$$

$$\frac{d\varepsilon}{d\tau} = \frac{\partial H_0}{\partial t}, \quad \frac{dt}{d\tau} = -\frac{\partial H_0}{\partial \varepsilon}, \quad \frac{dp_i}{d\tau} = -\frac{\partial H_0}{\partial x_i}, \quad \frac{dx_i}{d\tau} = \frac{\partial H_0}{\partial p_i},$$

where ε is the energy and $\mathbf{p}$ the momentum of the system, Φ is the scalar and $\mathbf{\Psi}$ the vector potential of the external field, while $d\tau$ denotes again the element of eigenzeit.

Within the limits of the validity of our classical analysis (*i.e.* $\mathbf{S}$ so small that the reaction of the spin on the orbital motion is negligible) a Hamiltonian $\overline{H}$ which simultaneously governs the orbital motion and the spin precession will be simply obtained by replacing H_0 in (12) by the sum of H_0 and the Hamiltonian which governs the spin:

$$\overline{H} \equiv \frac{1}{2m}\left\{-\frac{(\varepsilon + e\Phi)^2}{c^2} + \left(\mathbf{p} + \frac{e}{c}\mathbf{\Psi}\right)^2\right\} + \frac{e}{mc}(\mathbf{S}\cdot\mathbf{F}) = -\tfrac{1}{2}mc^2,$$

or, multiplying by $-2m$:

$$-2m\overline{H} \equiv H \equiv \frac{(\varepsilon + e\Phi)^2}{c^2} - \left(\mathbf{p} + \frac{e}{c}\mathbf{\Psi}\right)^2 - \frac{2e}{c}(\mathbf{S}\cdot\mathbf{F}) = m^2c^2. \qquad (13)$$

Here we based ourselves on (10). Using instead (10a), (10b) or (10c) we might replace $(\mathbf{S}\cdot\mathbf{F})$ in (13) by $\tfrac{1}{2}[(\mathbf{S}\cdot\mathbf{F})+(\mathbf{S}^*\cdot\mathbf{F}^*)]$, $-\tfrac{1}{2}i[(\mathbf{S}\cdot\mathbf{F})-(\mathbf{S}^*\cdot\mathbf{F}^*)]$ or $(\mathbf{S}^*\cdot\mathbf{F}^*)$ respectively. In view of the approximation involved ($\mathbf{S}$ very small), either of these four expressions may be chosen, although the choice $\tfrac{1}{2}[(\mathbf{S}\cdot\mathbf{F})+(\mathbf{S}^*\cdot\mathbf{F}^*)]$ might seem to be the most natural one.

It seems difficult, if not impossible, to establish a Hamiltonian equation, in virtue of which the condition (3), which has practically the effect of reducing the two degrees of freedom of the spin to only one, is automatically fulfilled.[3]

---

[3] This question cannot be settled, anyhow, before a classical system of equations of motion has been established, which allows the condition (3) to be rigorously fulfilled, and not only approximately as in the case of the Hamiltonians considered above.

V. Quantization

In order to quantize the motion governed by the Hamiltonian equation (13), or one of its alternatives, H must be considered as an operator H_{op} acting on a wavefunction ψ:

$$H_{op}\psi = m^2c^2\psi. \tag{14}$$

If the spin is to be given a quantum number $\frac{1}{2}$ and if we restrict ourselves to (13), the following familiar expressions will have to be adopted:

$$\varepsilon = i\hbar\frac{\partial}{\partial t}, \qquad p_i = -i\hbar\frac{\partial}{\partial x_i}, \tag{15}$$

$$S_1 = \frac{\hbar}{2}\begin{Vmatrix} 1 & 0 \\ 0 & -1 \end{Vmatrix}, \quad S_2 = \frac{\hbar}{2}\begin{Vmatrix} 0 & 1 \\ 1 & 0 \end{Vmatrix}, \quad S_3 = \frac{\hbar}{2}\begin{Vmatrix} 0 & -i \\ i & 0 \end{Vmatrix}. \tag{16}$$

The commutation properties of these expressions satisfy the necessary conditions corresponding to the properties of the analogous Poisson brackets. The introduction of the Pauli spin matrices for $\mathbf{S}$ means that ψ, besides on x_1, x_2, x_3, t, depends on a spin variable which only can take two values, for instance the eigenvalues $\pm\frac{1}{2}\hbar$ of S_1, so that ψ can be represented as a set of two wave components ψ_+, ψ_-.

The relativistic invariance of this choice of $\mathbf{S}$ is made clear by the investigations of Weyl and van der Waerden; it follows from the fact that to each Lorentz transformation a unimodular transformation of the wave components may be assigned in such a way that ψ_+^2, $\psi_+\psi_-$ and ψ_-^2 transform like $-F_2+iF_3$, F_1 and F_2+iF_3 respectively. With this convention the Pauli matrix components will transform exactly like the components of $\mathbf{F}$ ($=\mathbf{H}+i\mathbf{E}$). It is interesting to note that the components of $\mathbf{S}$ remain Hermitical even when they undergo a complex orthogonal transformation.

Introducing (15) and (16) into (13) in order to construct H_{op}, the equation (14) assumes exactly the form which Dirac obtained by "squaring" his linear equations. The latter may be got back from (14) by observing that the H_{op} thus constructed factorizes in the following way:

$$H_{op} = \left\{\frac{\varepsilon+e\Phi}{c} - \frac{2}{\hbar}\left(\left[\mathbf{p}+\frac{e}{c}\mathbf{\Psi}\right]\cdot\mathbf{S}\right)\right\}\left\{\frac{\varepsilon+e\Phi}{c} + \frac{2}{\hbar}\left(\left[\mathbf{p}+\frac{e}{c}\mathbf{\Psi}\right]\cdot\mathbf{S}\right)\right\}.$$

Consequently, putting

$$\left\{\frac{\varepsilon+e\Phi}{c} + \frac{2}{\hbar}\left(\left[\mathbf{p}+\frac{e}{c}\mathbf{\Psi}\right]\cdot\mathbf{S}\right)\right\}\psi = mc\chi, \tag{17}$$

χ will satisfy the equation

$$\left\{\frac{\varepsilon+e\Phi}{c} - \frac{2}{\hbar}\left(\left[\mathbf{p}+\frac{e}{c}\mathbf{\Psi}\right]\cdot\mathbf{S}\right)\right\}\chi = mc\psi. \tag{18}$$

Since ψ and χ are both two-component wavefunctions, (17) and (18) represent a system of four simultaneous equations. They are equivalent with Dirac's equations.

If, in (13), we had substituted $(\mathbf{S^*} \cdot \mathbf{F^*})$ for $(\mathbf{S} \cdot \mathbf{F})$, we should have had to introduce for the components of $\mathbf{S^*}$ exactly the same matrices (16) as given for those of $\mathbf{S}$, but the two-component wavefuncion χ_+, χ_- to be introduced now would have to be such that χ_+^2, $\chi_+\chi_-$ and χ_-^2 transform like $-F_2^* + iF_3^*$, F_1^* and $F_2^* + iF_3^*$ respectively. It is easily verified that the corresponding Hamiltonian H_{op}^* factorizes as follows:

$$H_{op}^* = \left\{\frac{\varepsilon + e\Phi}{c} + \frac{2}{h}\left(\left[\mathbf{p} + \frac{e}{c}\boldsymbol{\Psi}\right] \cdot \mathbf{S^*}\right)\right\}\left\{\frac{\varepsilon + e\Phi}{c} - \frac{2}{h}\left(\left[\mathbf{p} + \frac{e}{c}\boldsymbol{\Psi}\right] \cdot \mathbf{S^*}\right)\right\},$$

and that we consequently are led precisely to the equations (17) and (18) again. Since χ_+^2, $\chi_+\chi_-$ and χ_-^2 transform like $-\psi_-^{*2}$, $\psi_-^*\psi_+^*$ and $-\psi_+^{*2}$ we see that χ_+ and χ_- transform like ψ_-^* and $-\psi_+^*$ respectively; this result is well known from van der Waerden's analysis.

VI. Concluding Remarks

In the foregoing we have established the intimate connection between Dirac's linear equations and a purely classical analysis. It would lead us too far to trace this connection in further detail. It may be pointed out, however, that it has been brought about by choosing for the spin part of the Hamiltonian the complex expression (10) or (10c). It involves an imaginary electrical moment, but, from a formal point of view, it is much simpler than the real expression H_S' in (10a) and H_S'' in (10b). From a physical point of view it would perhaps be more natural to choose the Hamiltonian (10a). This would correspond to the Hamiltonian equation:

$$H \equiv \frac{(\varepsilon + e\Phi)^2}{c^2} - \left(\mathbf{p} + \frac{e}{c}\boldsymbol{\Psi}\right)^2 - \frac{e}{c}[(\mathbf{S} \cdot \mathbf{F}) + (\mathbf{S^*} \cdot \mathbf{F^*})] = m^2c^2.$$

The straightforward quantization of this equation, however, leads not to the simple Dirac theory, but to a Schrödinger equation of a more complicated type. It would, indeed, lead us to consider Ψ in (4) as depending on two spin variables, both of which take only two values. Thus Ψ would now be a four-component wavefunction and its four components would satisfy four simultaneous differential equations of the second degree.

G. On the Eigenvalue Problem in a One-Dimensional Field of Force

Abstract. We give a simple proof of the existence of an infinite sequence of allowed energy intervals which are separated by forbidden intervals.

The problem mentioned in the title of this paper has often been discussed in the mathematical literature; in particular, Haupt[1] has in connection with his investigations of the oscillation problem solved the problem of the existence and the properties of the eigenvalues of Hill's differential equation. However, it may be of some interest to give in what follows a more direct proof of those properties.

§ 1. We start in this section by giving some important facts connected with the Floquet theorem.

If in the Schrödinger equation

$$\frac{d^2\varphi}{dx^2} + (E-u)\varphi = 0, \tag{1}$$

the potential energy is a real periodic function of period a,

$$U(x+a) = U(x), \tag{2}$$

we are dealing with Hill's differential equation and if $\varphi(x)$ is a solution of (1), $\varphi(x+a)$ will also be a solution.

As an arbitrary solution can always be represented through a base of two linearly independent solutions φ_1 and φ_2, we have the equations

$$\begin{aligned}\varphi_1(x+a) &= \alpha_{11}\varphi_1(x) + \alpha_{12}\varphi_2(x), \\ \varphi_2(x+a) &= \alpha_{21}\varphi_1(x) + \alpha_{22}\varphi_2(x).\end{aligned} \tag{3}$$

Hence it follows that

$$\begin{vmatrix} \varphi_1(x+a) & \varphi_2(x+a) \\ \varphi_1'(x+a) & \varphi_2'(x+a) \end{vmatrix} = \begin{vmatrix} \varphi_1(x) & \varphi_2(x) \\ \varphi_1'(x) & \varphi_2'(x) \end{vmatrix} \cdot \begin{vmatrix} \alpha_{11} & \alpha_{12} \\ \alpha_{21} & \alpha_{22} \end{vmatrix}. \tag{4}$$

---
[1] O. Haupt, *Math. Ann.* **76**, 67 (1914); **79**, 278 (1919). See also G. Hamel, *Math. Ann.* **73**, 371 (1912) and M. J. O. Strutt, *Math. Ann.* **101**, 559 (1929).

One sees easily from (1) that the expression

$$c = \begin{vmatrix} \varphi_1 & \varphi_2 \\ \varphi_1' & \varphi_2' \end{vmatrix} \tag{5}$$

is independent of x and, if φ_1 and φ_2 are linearly independent, is a non-vanishing constant; hence it follows from (4) that the determinant of the matrix of the αs is equal to unity:

$$\alpha_{11}\alpha_{22} - \alpha_{12}\alpha_{21} = 1. \tag{6}$$

If we try to construct a solution φ which is a linear combination of φ_1 and φ_2 satisfying the condition

$$\varphi(x+a) = \lambda \varphi(x), \tag{7}$$

λ must satisfy the secular equation

$$\begin{vmatrix} \alpha_{11} - \lambda & \alpha_{12} \\ \alpha_{21} & \alpha_{22} - \lambda \end{vmatrix} = 0. \tag{8}$$

Using (6) we can also write this equation in the form

$$\lambda^2 - (\alpha_{11} + \alpha_{22})\lambda + 1 = 0. \tag{9}$$

The quantity

$$f = \alpha_{11} + \alpha_{22} \tag{10}$$

is clearly independent of the choice of base; it is always real since one can choose the base to be real which leads to solely real φ.

One must distinguish the following three cases:

1. $\qquad |f| > 2.$

The secular equation has two different, real roots λ_1 and λ_2, the product of which is equal to 1, and there are two different real solutions satisfying condition (7).

2. $\qquad |f| < 2.$

Equation (9) has two conjugate complex solutions with modulus 1, and there are two conjugate solutions satisfying (7):

$$\varphi(x+a) = e^{i\mu}\varphi(x), \qquad \varphi^*(x+a) = e^{-i\mu}\varphi^*(x), \qquad (\mu \text{ real}). \tag{11}$$

3. $\qquad |f| = \pm 2.$

G. On the Eigenvalue Problem in a One-Dimensional Field of Force

The roots of the secular equations are equal to one another and equal to ±1.

In this case there is certainly *one* solution of (1) which satisfies the condition

$$\varphi_{\mathrm{I}}(x+a) = \pm\varphi_{\mathrm{I}}(x), \tag{12}$$

while any other solution satisfies a condition of the form

$$\varphi_{\mathrm{II}}(x+a) = \pm\varphi_{\mathrm{II}}(x) + \alpha\varphi_{\mathrm{I}}(x). \tag{13}$$

In general we have $\alpha \neq 0$. However, it can happen that we have $\alpha = 0$; in that case *any* solution of (1) will satisfy the condition

$$\varphi(x+a) = \pm\varphi(x).$$

If we assume that an eigenvalue E is characterised in that there is a corresponding solution of (1) which remains everywhere bounded, it follows that

In the case 1. there are no eigenvalues;

in the case 2. there is a doubly degenerate eigenvalue;

and in the case 3. there is a non-degenerate or a doubly degenerate eigenvalue, depending on whether we have $\alpha \neq 0$ or $\alpha = 0$.

§ 2. The behaviour of the quantity f introduced through (10) as (analytical) function $f(E)$ of the eigenvalue parameter E determines the eigenvalue spectrum of our problem.

In the next section we prove that the function $f(E)$ always has the following properties.

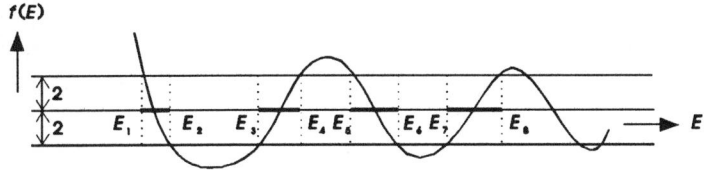

Fig. G.1.

As $E \to -\infty$, $f(E)$ tends to $= \infty$. As E increases from $-\infty$ to $+\infty$, $f(E)$ goes through an infinite number of consecutive minima and maxima. The minima are never larger than -2 and the maxima never smaller than $+2$; however, they tend asymptotically to those values. (These properties have been sketched in Fig. G.1.) They lead to the conclusion that the eigenvalues correspond to a sequence of continuous intervals $E_1 \to E_2$, $E_3 \to E_4$, and so on, which stretch, starting from a smallest value E_1, to infinity.

Inside each range the eigenvalues are twofold degenerate; the limits correspond, in general, to non-degenerate eigenvalues, as we shall show presently.

An exception occurs only when two consecutive ranges have a common limit: in that case, corresponding to a maximum (minimum) value of f of exactly $+2$ (or -2) we are dealing just with the $\alpha = 0$ case whereas otherwise, when $f \neq \pm 2$ the $\alpha \neq 0$ case is always realised.

One sees easily that extrema of f with an absolute value exactly equal to 2 can, in fact, occur. We give an example: a function $U(x)$ of period a can always be considered as a periodic function of period $a' = 2a$. If we do so, we are led to a new set of αs which are connected with the old ones through the relation

$$\alpha'_{kl} = \alpha_{k1}\alpha_{1l} + \alpha_{k2}\alpha_{2l}.$$

In particular, we have

$$f' = \alpha'_{11} + \alpha'_{22} = \alpha_{11}^2 + \alpha_{22}^2 + 2\alpha_{12}\alpha_{21}$$
$$= (\alpha_{11} + \alpha_{22})^2 - 2(\alpha_{11}\alpha_{22} - \alpha_{12}\alpha_{21}) = f^2 - 2.$$

The zeros of f thus correspond to minima of f' which are exactly equal to -2. In this way we have changed every allowed energy range formally into two energy ranges in contact with one another.

The theories of Mathieu and Lamé functions give explicit examples of the situation sketched in the above.

Kronig and Penney[2] have recently discussed an instructive example of a periodic function where one can obtain the eigenfunctions and eigenvalues through elementary calculations.

§ 3. To prove the statements made in § 2 we formulate our problem slightly differently. We consider U in (1) to be an arbitrary real function — subject to restrictions without importance for physics — and we prove a theorem about the properties of the solutions of (1) in an arbitrary interval $\xi \to x$ ($\xi < x$) of the range of the independent variable. The properties of the solutions of Hill's differential equation then follow by assuming simply that the function U is periodic with period $x - \xi = a$.

Let φ_1, φ_2 again be an arbitrary base of the solutions of (1). We define the quantities α_{kl} ($k, l = 1, 2$) through

$$\begin{aligned}\varphi_k(x) &= \alpha_{k1}\varphi_1(\xi) + \alpha_{k2}\varphi_2(\xi), \\ \varphi'_k(x) &= \alpha_{k1}\varphi'_1(\xi) + \alpha_{k2}\varphi'_2(\xi),\end{aligned} \quad (14)$$

The connection between these αs and the ones defined by (3) follows from what we have just said.

Solving (14) we get

---

[2] R. de L. Kronig and W. G. Penney, *Proc. Roy. Soc.* **A130**, 499 (1930).

G. On the Eigenvalue Problem in a One-Dimensional Field of Force

$$\alpha_{11} = \frac{1}{c}\begin{vmatrix}\varphi_1(x) & \varphi_1'(x)\\ \varphi_2(\xi) & \varphi_2'(\xi)\end{vmatrix}, \qquad \alpha_{22} = -\frac{1}{c}\begin{vmatrix}\varphi_2(x) & \varphi_2'(x)\\ \varphi_1(\xi) & \varphi_1'(\xi)\end{vmatrix},$$

$$\alpha_{12} = -\frac{1}{c}\begin{vmatrix}\varphi_1(x) & \varphi_1'(x)\\ \varphi_1(\xi) & \varphi_1'(\xi)\end{vmatrix}, \qquad \alpha_{21} = \frac{1}{c}\begin{vmatrix}\varphi_2(x) & \varphi_2'(x)\\ \varphi_2(\xi) & \varphi_2'(\xi)\end{vmatrix},$$

(15)

where c is again defined by (5). The determinant of the αs is again equal to 1, that is, Eq. (6) of §1 holds; the proof is the same as before.

We can write the sum

$$f(x, \xi; E) = \alpha_{11} + \alpha_{22} \qquad (16)$$

in the form

$$f = \frac{1}{c}\left\{\begin{vmatrix}\varphi_1(x) & \varphi_2(x)\\ \varphi_1'(\xi) & \varphi_2'(\xi)\end{vmatrix} - \begin{vmatrix}\varphi_1'(x) & \varphi_2'(x)\\ \varphi_1(\xi) & \varphi_2(\xi)\end{vmatrix}\right\}; \qquad (17)$$

it is therefore invariant under a change in the base φ_1, φ_2. To study the extrema of f as function of E we shall calculate the quantities

$$f_E = \frac{\partial f}{\partial E} \qquad f_{EE} = \frac{\partial^2 f}{\partial E^2}.$$

To do this we use the functions $\varphi_{kE} = \partial \varphi_k / \partial E$ which, according to (1), satisfy the inhomogeneous differential equation

$$\varphi_{kE}'' + (E - U)\varphi_{kE} = -\varphi_k. \qquad (18)$$

One sees easily that the solution of the inhomogeneous differential equation

$$\Psi'' + (E - U)\Psi = -K \qquad (19)$$

which satisfy the condition

$$\Psi(\xi) = \Psi'(\xi) = 0, \qquad (20)$$

is given by

$$\Psi = \frac{1}{c}\int_\xi^x \begin{vmatrix}\varphi_1(x) & \varphi_2(x)\\ \varphi_1(t) & \varphi_2(t)\end{vmatrix} K(t)\, dt. \qquad (21)$$

Hence

$$\varphi_{kE}(x) = \frac{1}{c}\int_\xi^x \begin{vmatrix}\varphi_1(x) & \varphi_2(x)\\ \varphi_1(t) & \varphi_2(t)\end{vmatrix} \varphi_k(t)\, dt,$$

$$\varphi_{kE}'(x) = \frac{1}{c}\int_\xi^x \begin{vmatrix}\varphi_1'(x) & \varphi_2'(x)\\ \varphi_1(t) & \varphi_2(t)\end{vmatrix} \varphi_k(t)\, dt,$$

(22a)

$$\varphi_{kE}(\xi) = 0, \qquad \varphi_{kE}'(\xi) = 0 \qquad (22b)$$

is a solution of (18). With this choice of $\varphi_{kE}(x)$ we need, when we differentiate (17) with respect to E, only regard $\varphi_k(x)$ and $\varphi'_k(x)$ as variable. We then find

$$f_E = \frac{1}{c^2}\int_\xi^x \left\{ \begin{vmatrix} \varphi_1(x) & \varphi_2(x) \\ \varphi_1(t) & \varphi_2(t) \end{vmatrix} \cdot \begin{vmatrix} \varphi_1(t) & \varphi_2(t) \\ \varphi'_1(\xi) & \varphi'_2(\xi) \end{vmatrix} - \begin{vmatrix} \varphi'_1(x) & \varphi'_2(x) \\ \varphi_1(t) & \varphi_2(t) \end{vmatrix} \cdot \begin{vmatrix} \varphi_1(t) & \varphi_2(t) \\ \varphi_1(\xi) & \varphi_2(\xi) \end{vmatrix} \right\} dt,$$

or

$$f_E = -\alpha_{21} I_{11} + (\alpha_{11} - \alpha_{22}) I_{12} + \alpha_{12} I_{22}, \qquad (23)$$

with

$$I_{ik} = \frac{1}{c}\int_\xi^x \varphi_k(t)\varphi_l(t)\,dt. \qquad (24)$$

From (23) we find for the second derivative:

$$f_{EE} = -\frac{\partial \alpha_{21}}{\partial E} I_{11} + \cdots - \alpha_{21}\frac{\partial I_{11}}{\partial E} + \cdots.$$

If we use (15) and (22) to evaluate $\partial \alpha_{kl}/\partial E$ we easily find that

$$f_{EE} = -f(I_{11}I_{22} - I_{12}^2) - \alpha_{21}\frac{\partial I_{11}}{\partial E} + (\alpha_{11} - \alpha_{22})\frac{\partial I_{12}}{\partial E} + \alpha_{12}\frac{\partial I_{22}}{\partial E}. \qquad (25)$$

We shall now use (1), (14), (23), and (25) to discuss the behaviour of f as an analytical function of E.

If E is very large and positive, we have the following approximate solution of (1):

$$\varphi_1(x) = \cos\left[(x-\xi)\sqrt{E}\right], \qquad \varphi_2(x) = \sin\left[(x-\xi)\sqrt{E}\right], \qquad (26)$$

and f is approximately given by

$$f = 2\cos\left[(x-\xi)\sqrt{E}\right]. \qquad (27)$$

If E is very large and negative, an approximate solution of (1) is

$$\varphi_1(x) = e^{-(x-\xi)\sqrt{-E}}, \qquad \varphi_2(x) = e^{+(x-\xi)\sqrt{-E}}, \qquad (28)$$

and f is approximately given by

$$f = e^{(x-\xi)\sqrt{-E}}. \qquad (29)$$

If E goes from $-\infty$ to $+\infty$, f will, to begin with, decrease from $+\infty$ until it reaches a minimum for the first time; it then increases for the first time

G. On the Eigenvalue Problem in a One-Dimensional Field of Force   201

to reach a maximum and after that it will infinitely often reach alternately minimum and maximum values which are asymptotically represented by (27).

We can now conclude from (23) and (6) that the absolute value of f is always larger than or equal to 2 for those values of E where f is stationary, that is, where f_E vanishes. The proof is easiest if we choose φ_1 and φ_2 to be real and such that the integral $I_{12} = (1/c) \int_\xi^x \varphi_1 \varphi_2 \, dx$ vanishes. Since I_{11} and I_{22} are always positive, it follows from (23) that the real quantities α_{12} and α_{21} have the same sign. It then follows from (6) that the product of α_{11} and α_{22} is not less than 1 and hence the absolute value of their sum f not less than 2.

We can also conclude from (25) that for a stationary value f_{st} of $f(E)$ we are dealing with a maximum when f_{st} is positive and with a minimum when f_{st} is negative.

We shall only prove this for $|f_{\text{st}}| > 2$. In this case we can choose immediately a real base φ_1, φ_2 such that both α_{12} and α_{21} vanish. Because of (23) it then follows from $f_E = 0$ that we also have $I_{12} = 0$. Equation (25) then simplifies to

$$f_{EE} = -f\, I_{11} I_{22} + (\alpha_{11} - \alpha_{22}) \frac{\partial I_{12}}{\partial E}. \tag{29}$$

Since we have $\alpha_{11} \alpha_{22} = 1$ it follows that $|\alpha_{11} - \alpha_{22}|$ is smaller than $|\alpha_{11} + \alpha_{22}| = |f|$. One also proves easily that

$$\left| \frac{\partial I_{12}}{\partial E} \right| < I_{11} I_{22}. \tag{30}$$

It then follows from (29) that in the stationary points f_{EE} is non-vanishing and has the opposite sign from f. Hence, $f_{\text{st}} > 2$, respectively, < 2 corresponds always to a maximum, respectively, a minimum.

The proof of (30) proceeds as follows:

$$\frac{\partial I_{12}}{\partial E} = \frac{1}{c} \frac{\partial}{\partial E} \int_\xi^x \varphi_1(t) \varphi_2(t) \, dt = \frac{1}{c} \int_\xi^x (\varphi_1 \varphi_{2E} + \varphi_{1E} \varphi_2) \, dt$$

$$= \frac{1}{c^2} \int_\xi^x dt \int_\xi^t dt' \begin{vmatrix} \varphi_1(t) & \varphi_2(t) \\ \varphi_1(t') & \varphi_2(t') \end{vmatrix} \{\varphi_1(t)\varphi_2(t') + \varphi_1(t')\varphi_2(t)\}$$

$$= \frac{1}{c^2} \int_\xi^x dt \int_\xi^t dt' \{\varphi_1^2(t)\varphi_2^2(t') - \varphi_1^2(t')\varphi_2^2(t)\},$$

$$\left| \frac{\partial I_{12}}{\partial E} \right| < \frac{1}{c^2} \int_\xi^x dt \int_\xi^t dt' \{\varphi_1^2(t)\varphi_2^2(t') + \varphi_1^2(t')\varphi_2^2(t)\}$$

$$= \frac{1}{c^2} \int_\xi^x \varphi_1^2(t) \, dt \int_\xi^x \varphi_2^2(t) \, dt = I_{11} I_{22}.$$

Finally we prove the earlier statement that the eigenvalues for which $f = +2$ (or -2) are, in general, non-degenerate. To do this we need only show that it is impossible to choose a base such that α_{11} and α_{22} are both equal to $+1$ (or -1) while simultaneously both α_{12} and α_{21} vanish. This follows, however, immediately from (23); only when f_E vanishes at the same time that $|f| = 2$ can and must α_{12} vanish at the same time as α_{21}.

H. The Use of Charge-Conjugated Wavefunctions in the Hole Theory of the Electron

From the 4-component Dirac wavefunction ψ_k ($k = 1, 2, 3, 4$) satisfying

$$\{(\boldsymbol{\alpha} \cdot [\mathbf{p} - e\mathfrak{A}]) + e\Phi + \beta m\} \psi = i\hbar \frac{\partial \psi}{\partial t} \tag{1}$$

another 4-component wavefunction ψ^L can be derived, which satisfies the same equation, but with the sign of e reversed:

$$\{(\boldsymbol{\alpha} \cdot [\mathbf{p} + e\mathfrak{A}]) - e\Phi + \beta m\} \psi^L = i\hbar \frac{\partial \psi^L}{\partial t}. \tag{2}$$

We call ψ^L the *charge-conjugated* function of ψ. The relation between ψ and ψ^L is a very simple one, when we choose for the matrices $\boldsymbol{\alpha}, \beta$ that particular representation, which causes the two components ψ_1, ψ_2 to transform like a relativistic spinor u, v, and ψ_3, ψ_4 like a spin-conjugated spinor $U^\dagger$ ($= -V^*$), $V^\dagger$ ($= U^*$), when a Lorentz transformation is applied.[1] In fact, writing

$$\psi = \begin{pmatrix} \psi_1 \\ \psi_2 \\ \psi_3 \\ \psi_4 \end{pmatrix} = \begin{pmatrix} u \\ v \\ -V^* \\ U^* \end{pmatrix}, \tag{3}$$

and defining the charge-conjugated wavefunction by

$$\psi^L = \begin{pmatrix} \psi_1^L \\ \psi_2^L \\ \psi_3^L \\ \psi_4^L \end{pmatrix} = \begin{pmatrix} U \\ V \\ -v^* \\ u^* \end{pmatrix} = \begin{pmatrix} \psi_4^* \\ -\psi_3^* \\ -\psi_2^* \\ \psi_1^* \end{pmatrix}, \tag{4}$$

it is easily verified that ψ^L satisfies (2) when ψ satisfies (1).

In order to prove this, consider for a moment ψ_k as a function of two variables s, r which each take only the values $+\frac{1}{2}$ and $-\frac{1}{2}$:

$$\psi = \psi_{s,r}, \psi_1 = \psi_{\frac{1}{2},\frac{1}{2}}, \psi_2 = \psi_{-\frac{1}{2},\frac{1}{2}}, \psi_3 = \psi_{\frac{1}{2},-\frac{1}{2}}, \psi_4 = \psi_{-\frac{1}{2},-\frac{1}{2}}.$$

---
[1] Compare for instance H. A. Kramers, *Hand u. Jahrb. d. Chem. Physik*, I, § 63, 64.

If $\boldsymbol{\sigma}(\sigma_x, \sigma_y, \sigma_z)$ and $\varrho_x, \varrho_y, \varrho_z$ are the Pauli matrices operating on s and r, the representation of $\boldsymbol{\alpha}, \beta$ under consideration takes the form:

$$\boldsymbol{\alpha} = \varrho_z \boldsymbol{\sigma}, \qquad \beta = \varrho_x. \tag{5}$$

Now the following formulæ are easily seen to hold for each of the three σ- and ϱ-operators:

$$(\sigma \psi)^L = -\sigma \psi^L, \qquad (\varrho \psi)^L = -\varrho \psi^L.$$

From this follows:

$$(\boldsymbol{\alpha} \psi)^L = \boldsymbol{\alpha} \psi^L, \qquad (\beta \psi)^L = -\beta \psi^L.$$

Since we have furthermore:

$$(\mathbf{p}\psi)^L = \hbar(-i\nabla \psi)^L = \hbar i \nabla \psi^L = -\mathbf{p}\psi^L, \qquad \left(i\hbar \frac{\partial \psi}{\partial t}\right)^L = -i\hbar \frac{\partial \psi^L}{\partial t},$$

$$(\mathfrak{A}\psi)^L = \mathfrak{A}\psi^L, \qquad (\Phi\psi)^L = \Phi\psi^L,$$

we see that the charge-conjugated of the left and right side of (1) are equal to the left and right side of (2), both multiplied by -1.

Any other representation of the $\boldsymbol{\alpha}, \beta$ matrices coresponds to a transformation of the wavefunction

$$\psi' = S\psi, \qquad \psi'^L = S\psi^L, \tag{6}$$

where S is an arbitrary non-singular matrix and we have

$$\boldsymbol{\alpha}' = S\boldsymbol{\alpha}S^{-1}, \qquad \beta' = S\beta S^{-1}.$$

In the particular case

$$S = \frac{e^{\pi i/4}}{\sqrt{2}}(1 + i\varrho_y \sigma_y), \qquad S^{-1} = \frac{e^{-\pi i/4}}{\sqrt{2}}(1 - i\varrho_y \sigma_y), \tag{7}$$

we find

$$\alpha'_x = \alpha_x = \varrho_z \sigma_x, \quad \alpha'_z = \alpha_z = \varrho_z \sigma_z, \quad \alpha'_y = -\beta = -\varrho_x, \quad \beta'_x = \alpha_y = \varrho_z \sigma_y.$$

The matrices $\boldsymbol{\alpha}'$ are now purely real whereas β' is purely imaginary. From (6) we find:

$$\begin{pmatrix} \psi'_1 \\ \psi'_2 \\ \psi'_3 \\ \psi'_4 \end{pmatrix} = \frac{e^{\pi i/4}}{\sqrt{2}} \begin{pmatrix} \psi_1 - i\psi_4 \\ \psi_2 + i\psi_3 \\ \psi_3 + i\psi_2 \\ \psi_4 - i\psi_1 \end{pmatrix}, \quad \begin{pmatrix} \psi'^L_1 \\ \psi'^L_2 \\ \psi'^L_3 \\ \psi'^L_4 \end{pmatrix} = \frac{e^{\pi i/4}}{\sqrt{2}} \begin{pmatrix} \psi^*_4 - i\psi^*_1 \\ -\psi^*_3 - i\psi^*_2 \\ -\psi^*_2 - i\psi^*_3 \\ \psi^*_1 - i\psi^*_4 \end{pmatrix} = \begin{pmatrix} \psi'^*_1 \\ \psi'^*_2 \\ \psi'^*_3 \\ \psi'^*_4 \end{pmatrix}$$

Thus, for this particular representation, charge-conjugation and complex-conjugation are identical:

H. The Use of Charge-Conjugated Wavefunctions

$$\psi'^L = \psi'^*. \tag{8}$$

It is immediately seen that Eq. (2) in this case is the conjugate complex of Eq. (1).

The property (8) must be invariant with respect to Lorentz transformations; this is directly verified, since — with the representation under consideration — the coefficients of the Lorentz transformation of $\psi_1', \psi_2', \psi_3', \psi_4'$ are real. In fact, an infinitesimal Lorentz transformation corresponds to the operator

$$I_{op} = \tfrac{1}{2} S \left(\mathrm{i} d_x \sigma_x + \mathrm{i} d_y \sigma_y + \mathrm{i} d_z \sigma_z + l_x \varrho_z \sigma_x + l_y \varrho_z \sigma_y + l_z \varrho_z \sigma_z \right) S^{-1},$$

where d_x, d_y, d_z (infinitesimal rotation) and l_x, l_y, l_z (infinitesimal pure Lorentz transformation) are all real. Inserting the value (7) for S we find for I_{op} a purely real matrix.

A representation where $\alpha, \mathrm{i}\beta$ are real has been used by Majorana[2] in his recent work on the quantum theory of the negaton and the positon. The concept of charge conjugation is, however, independent of the particular choice of representation. By its means we will in this article represent some results of the Majorana calculus (and, thereby, also of the Dirac-Heisenberg hole theory, with which it is practically equivalent) in a general form, which on account of its simplicity might be of some interest.

The Pauli principle is introduced in the usual way by promoting the wavefunction ψ to a "matrix operator" $\boldsymbol{\psi}$ satisfying

$$\boldsymbol{\psi}^*(q)\boldsymbol{\psi}(q') + \boldsymbol{\psi}(q')\boldsymbol{\psi}^*(q) = \delta(q,q'), \tag{9}$$

where q stands for the complete set of positional and spin coordinates. The same relation will hold for the charge-conjugated function:

$$\boldsymbol{\psi}^{L*}(q)\boldsymbol{\psi}^L(q') + \boldsymbol{\psi}^L(q')\boldsymbol{\psi}^{L*}(q) = \delta(q,q'). \tag{10}$$

In these expressions, the asterisk means hermitic conjugation.

In ordinary, non-relativistic quantum mechanics an operator F^{total}, which is symmetrical with respect to the electrons and which consists simply of a sum of identical hermitic operators $F(i)$,

$$F^{total} = \sum F(i)$$

($F(i)$ operates on the coordinates of the ith electron), is promoted to a matrix operator $\mathbf{F}$ by the well known formula:

$$\mathbf{F} = \int \boldsymbol{\psi}^* F \boldsymbol{\psi},$$

---

[2] *Nuovo Cimento* **14**, 171 (1937).

where $\int$ means integration over the three space coordinates and summation over the spin coordinates. With Dirac's equation (1) and its charge-conjugated equation (2), we would have the choice between the two completely analogous hermitic expressions:

$$\int \psi^* F \psi \quad \text{and} \quad \int \psi^{L*} F^L \psi^L, \tag{11}$$

where F^L is obtained from F by changing the quantity e representing the electric charge (and which in general may occur in F) into $-e$:

$$F^L(e) = F(-e). \tag{12}$$

Majorana's calculus can now be simply expressed by stating that the arithmetic mean of the two expressions (11) has to be taken:

$$\mathbf{F} = \tfrac{1}{2} \int \left(\psi^* F \psi + \psi^{L*} F^L \psi^L \right). \tag{13}$$

A particular representation of ψ is obtained by developing ψ in terms of the eigenfunctions of the free electron. Those corresponding to "positive" energies will be denoted by φ_λ; those corresponding to "negative" energies by φ_λ^L. For the latter we may indeed take the charge-conjugated of the former. The φ_λ may be taken to correspond to definite values of the impulse vector $\mathbf{p}$; for each $\mathbf{p}$ there will still be two φ_λ's, which may be taken to correspond to spin parallel (φ_λ^p) and antiparallel (φ_λ^a) to $\mathbf{p}$. Taking for simplicity the eigenfunctions to be normalised eigenfunctions in a big cubical space (volume Ω) with periodic boundary conditions, and using the representation (5), we may write for every possible impulse vector $\mathbf{p}_\lambda$ with components $\hbar \Omega^{-1/3} k_x$, $\hbar \Omega^{-1/3} k_y$, $\hbar \Omega^{-1/3} k_z$ (k_x, k_y, k_z integers)[3]

$$\left. \begin{array}{l} \varphi_\lambda^p = \dfrac{1}{\sqrt{\Omega}} \begin{pmatrix} C\alpha \\ C\beta \\ S\alpha \\ S\beta \end{pmatrix} e^{i(\mathbf{p}\cdot\mathbf{r})/\hbar}, \quad \varphi_\lambda^a = \dfrac{1}{\sqrt{\Omega}} \begin{pmatrix} -S\beta^* \\ S\alpha^* \\ -C\beta^* \\ C\alpha^* \end{pmatrix} e^{i(\mathbf{p}\cdot\mathbf{r})/\hbar}, \\[2em] \varphi_\lambda^{pL} = \dfrac{1}{\sqrt{\Omega}} \begin{pmatrix} S\beta^* \\ -S\alpha^* \\ -C\beta^* \\ C\alpha^* \end{pmatrix} e^{-i(\mathbf{p}\cdot\mathbf{r})/\hbar}, \quad \varphi_\lambda^{aL} = \dfrac{1}{\sqrt{\Omega}} \begin{pmatrix} C\alpha \\ C\beta \\ -S\alpha \\ -S\beta \end{pmatrix} e^{-i(\mathbf{p}\cdot\mathbf{r})/\hbar}, \end{array} \right\} \tag{14}$$

where C, S, α, β are defined by

$$\left. \begin{array}{l} C = \cos \tfrac{1}{2}\chi, \quad S = \sin \tfrac{1}{2}\chi, \quad \cot \chi = \dfrac{p}{m}, \\ p_z = p \cos \vartheta, \quad p_x + i p_y = p \sin \vartheta e^{i\psi}, \\ \alpha = \cos \tfrac{1}{2}\vartheta \, e^{-i\psi/2} \quad \beta = \sin \tfrac{1}{2}\vartheta \, e^{i\psi/2}. \end{array} \right\} \tag{15}$$

---

[3] Compare H. A. Kramers, loc. cit., p. 292.

H. The Use of Charge-Conjugated Wavefunctions

The representation in question can now be defined as

$$\psi = \sum_\lambda \mathbf{a}_\lambda \varphi_\lambda + \sum_\lambda \mathbf{b}_\lambda^* \varphi_\lambda^L. \tag{16}$$

To it corresponds the following representation for ψ^L

$$\psi^L = \sum_\lambda \mathbf{b}_\lambda \varphi_\lambda + \sum_\lambda \mathbf{a}_\lambda^* \varphi_\lambda^L. \tag{17}$$

In these formulæ the summation over λ includes summation over the two opposite spin directions. The $\mathbf{a}_\lambda, \mathbf{b}_\lambda$ and their hermitic conjugates $\mathbf{a}_\lambda^*, \mathbf{b}_\lambda^*$ are Wigner-Jordan matrices satisfying

$$\mathbf{a}_\lambda^* \mathbf{a}_\lambda + \mathbf{a}_\lambda \mathbf{a}_\lambda^* = 1, \quad \mathbf{b}_\lambda^* \mathbf{b}_\lambda + \mathbf{b}_\lambda \mathbf{b}_\lambda^* = 1, \quad \text{all other pairs anticommute.}$$

In a continuous description (16) and (17) have to be replaced by integrals. Such a description is particularly appropriate to the discussion of the Lorentz invariance of the calculus, but we will not enter upon it here.

We will now discuss some particular examples of the application of (13). Consider first the case of free electrons, that is, electrons in the absence of external fields and without interactions. We compute the energy operator $\mathbf{H}^0$

$$\mathbf{H}^0 = \tfrac{1}{2} \int \left(\psi^* H^0 \psi + \psi^{L*} H^0 \psi^L \right),$$

where

$$H^0 = H^{0L} = (\boldsymbol{\alpha} \cdot \mathbf{p}) + \beta m.$$

Now, since

$$H^0 \varphi_\lambda = E_\lambda \varphi_\lambda, \quad H^0 \varphi_\lambda^L = -E_\lambda \varphi_\lambda^L, \quad E_\lambda = +\sqrt{p_\lambda^2 + m^2},$$

we find

$$\left.\begin{aligned}
\mathbf{H}^0 &= \tfrac{1}{2} \sum_\lambda E_\lambda (\mathbf{a}_\lambda^* \mathbf{a}_\lambda - \mathbf{b}_\lambda \mathbf{b}_\lambda^*) + \tfrac{1}{2} \sum_\lambda E_\lambda (\mathbf{b}_\lambda^* \mathbf{b}_\lambda - \mathbf{a}_\lambda \mathbf{a}_\lambda^*) \\
&= \tfrac{1}{2} \sum_\lambda E_\lambda \{(\mathbf{a}_\lambda^* \mathbf{a}_\lambda - \mathbf{a}_\lambda \mathbf{a}_\lambda^*) + (\mathbf{b}_\lambda^* \mathbf{b}_\lambda - \mathbf{b}_\lambda \mathbf{b}_\lambda^*)\} \\
&= \sum_\lambda E_\lambda \{\mathbf{a}_\lambda^* \mathbf{a}_\lambda + \mathbf{b}_\lambda^* \mathbf{b}_\lambda - 1\},
\end{aligned}\right\} \tag{18}$$

where the -1 corresponds to the well known infinite negative zero-point energy of the hole theory. Generalising the $\mathbf{a}$ and $\mathbf{b}$ to time-dependent matrices $\mathbf{a}(t)$ and $\mathbf{b}(t)$, which for $t = 0$ become equal to $\mathbf{a}$ and $\mathbf{b}$, we find immediately from

$$\dot{\mathbf{a}}_\lambda = \frac{i}{\hbar} \left(\mathbf{H}^0 \mathbf{a}_\lambda - \mathbf{a}_\lambda \mathbf{H}^0 \right), \qquad \dot{\mathbf{b}}_\lambda = \frac{i}{\hbar} \left(\mathbf{H}^0 \mathbf{b}_\lambda - \mathbf{b}_\lambda \mathbf{H}^0 \right),$$

the well known time dependence

$$a_\lambda(t) = a_\lambda e^{-iE_\lambda t}, \qquad b_\lambda(t) = b_\lambda e^{-iE_\lambda t}.$$

Promoting also ψ to a time-dependent operator $\psi(t)$ the formula (16) takes the form:

$$\psi(x,y,z,t) = \sum_\lambda a_\lambda \varphi_\lambda^0 e^{i[(\mathbf{p}_\lambda \cdot \mathbf{r}) - E_\lambda t]/\hbar} + \sum_\lambda b_\lambda^* \varphi_\lambda^{0L} e^{-i[(\mathbf{p}_\lambda \cdot \mathbf{r}) - E_\lambda t]/\hbar},$$

where φ_λ^0 and φ_λ^{0L} are the values of φ_λ and φ_λ^L at the origin. It is of interest to compare this formula with the analogous formula for the quantized radiation field of vacuum electrodynamics,[4] where a similar formalism imposes itself automatically and ensures positive energy values for the light quanta.

The operator $\mathbf{e}$ of the total charge of the electrons will be given by choosing in (13):

$$F = -F^L = e,$$

and we find

$$\mathbf{e} = \frac{e}{2}\int(\psi^*\psi - \psi\psi^*) = \frac{e}{2}\sum_\lambda(a_\lambda^* a_\lambda + b_\lambda b_\lambda^*) - \frac{e}{2}\sum_\lambda(b_\lambda^* b_\lambda + a_\lambda a_\lambda^*)$$
$$= e\sum_\lambda(a_\lambda^* a_\lambda - b_\lambda^* b_\lambda) = -e\sum_\lambda(b_\lambda^* b_\lambda - a_\lambda^* a_\lambda). \qquad (19)$$

This leads us to consider $a_\lambda^* a_\lambda$ as the operator of the number of electrons with charge e in the state λ (we will call them *negatons*) and $b_\lambda^* b_\lambda$ as that of the electrons with charge $-e$ in this state (*positons*). This interpretation is in accordance with the usual interpretation of the formulæ to which one is led when external fields are taken into consideration and when the non-relativistic limiting case is discussed. It must, however, be remarked that the total number of electrons,

$$\mathbf{N} = \sum_\lambda(a_\lambda^* a_\lambda + b_\lambda^* b_\lambda), \qquad (20)$$

is not given by (13) when we put $F = F^L = 1$: we find indeed for $F = 1$ formally a constant:

$$1 = \sum_\lambda 1.$$

On the other hand, the simple but non-analytical operator

$$F = F^L = \frac{H^0}{|H^0|}$$

---

[4] H. A. Kramers, *loc. cit.*, p. 434, Eq. (103).

would give

$$\left(\frac{H^0}{|H^0|}\right) = \sum_\lambda (a_\lambda^* a_\lambda + b_\lambda^* b_\lambda - 1) = N - \sum_\lambda 1.$$

The last sum might be called the negative zero-point number of electrons; it corresponds to the negative zero-point energy in (18).

The operator of the charge density in space is given by

$$\varrho(x,y,z) = \sum_k \frac{e}{2}\left(\psi_k^* \psi_k - \psi_k^{L*}\psi_k^L\right). \tag{21}$$

If necessary it can be expressed explicitly in terms of the a, b, a^*, b^*. It is a particular case of the more general operator:

$$P(q',t';q,t) = \frac{e}{2}\left(\psi^*(q,t)\psi(q',t') - \psi^{L*}(q,t)\psi^L(q',t')\right), \tag{22}$$

where q and q' stand for two arbitrary, different choices of the space coordinates x, y, z and the spin coordinate k. The meaning of P is of course only well defined when the Hamiltonian H of the system is known, which governs the time dependence of all operators. In the particular case when the electrons are free, we have $H = H^0$. The corresponding P operator will be denoted by P^0. It is closely related to

$$R^0(q',t';q,t) = e\sum_\lambda \left(\varphi_\lambda^* \varphi_\lambda' e^{iE_\lambda(t-t')/\hbar} - \varphi_\lambda^{L*}\varphi_\lambda'^L e^{-iE_\lambda(t-t')/\hbar}\right). \tag{23}$$

This function is identical with the density matrix, which has been computed in Dirac's paper on the hole theory of 1934.[5] P in (22) may be called the general operator of the matrix density of the electric charge.

If the electrons are subject to an external field with potentials $\mathfrak{A}, \Phi$, we introduce in (13) for F, F^L the operators:

$$H = H^0 + e\left[\Phi - (\alpha \cdot \mathfrak{A})\right], \quad H^L = H^0 - e\left[\Phi - (\alpha \cdot \mathfrak{A})\right],$$

and we find for the energy operator of the system:

$$H = H^0 + H^1$$
$$H^1 = \frac{e}{2}\int \left\{\psi^*\left[\Phi - (\alpha \cdot \mathfrak{A})\right]\psi - \psi^{L*}\left[\Phi - (\alpha \cdot \mathfrak{A})\right]\psi^L\right\}. \tag{24}$$

If the action of the operator $\Phi - (\alpha \cdot \mathfrak{A})$ is described by means of the two-point operator function $\Omega(q,q')$:

$$\left[\Phi - (\alpha \cdot \mathfrak{A})\right] f(q) = \int_{q'} \Omega(q,q') f(q'),$$

---

[5] *Proc. Camb. Phil. Soc.* **30**, 150 (1934).

(24) can also be written in the form

$$\mathbf{H}^1 = \int_q \int_{q'} \mathbf{P}(q,q') \Omega(q,q'), \tag{25}$$

where $\mathbf{P}(q,q')$ is obtained from (22) by putting t and t' equal to zero. The $\mathbf{H}^1$ in (24) or (25) corresponds to the prescription of the Heisenberg-Dirac hole theory; expressed in the $\mathbf{a}_\lambda$ and $\mathbf{b}_\lambda$ it contains, on the one hand, the terms

$$\left. \begin{array}{l} \dfrac{e}{2} \sum_{\lambda\lambda'} (\Omega_{\lambda\lambda'} + \Omega^*_{\lambda\lambda'})(\mathbf{a}^*_\lambda \mathbf{a}_{\lambda'} - \mathbf{b}^*_\lambda \mathbf{b}_{\lambda'}), \\[1em] \Omega_{\lambda\lambda'} = \displaystyle\int \varphi^*_\lambda [\Phi - (\boldsymbol{\alpha}\cdot\mathfrak{A})] \varphi_{\lambda'}, \end{array} \right\} \tag{26}$$

which commute with $\mathbf{N}$ and, on the other hand, the terms

$$\left. \begin{array}{l} \dfrac{e}{2} \sum_{\lambda\lambda'} (\Omega_{\lambda\lambda'} + \Omega^*_{\lambda\lambda'})(\mathbf{a}^*_\lambda \mathbf{b}^*_{\lambda'} + \mathbf{b}_{\lambda'} \mathbf{a}_\lambda), \\[1em] \Omega_{\lambda\lambda'} = \displaystyle\int \varphi^*_\lambda [\Phi - (\boldsymbol{\alpha}\cdot\mathfrak{A})] \varphi^L_{\lambda'}. \end{array} \right\} \tag{27}$$

The terms (27) do not commute with $\mathbf{N}$; they correspond to pair formation ($\mathbf{a}^*\mathbf{b}^*$) and to pair annihilation ($\mathbf{b}\mathbf{a}$).

If the potentials of the external field are time-independent, the operator $H = H^0 + e[\Phi - (\boldsymbol{\alpha}\cdot\mathfrak{A})]$ will have eigenfunctions χ_e and corresponding eigenvalues E_e. The question arises if, and in what way, they correspond to stationary states of the one-electron problem in the field in question. It seems very difficult to give any definite answer at all to this question if no specification of the $\mathfrak{A}, \Phi$ field is given. This difficulty is related to the unsatisfactory and preliminary character of the hole theory in its present condition.

In the particular case, where the external field is due to a positive electric charge, smaller than $137|e|$, fixed at some point in space (hydrogen-like atom), the χ_e and E_e can be naturally divided into two groups. The first of these corresponds to positive values of E_e, which we will denote by E_m (eigenfunctions χ_m). For the second group the E_e's are negative; we denote them by $-\overline{E}_n$ (eigenfunctions $\overline{\chi}^L_n$). From comparison with the non-relativisttic treatment, we expect that if e is taken to be negative, the first group corresponds to one negative electron in the field of the nucleus (ordinary hydrogen atom), whereas the second group corresponds to one positive electron in this field. When the nuclear charge continuously decreases to zero, the set of χ_m functions merges continuously into the set of φ_λ functions defined by (14), whereas the set of $\overline{\chi}^L_n$ functions merges into the φ^L_λ functions. In this case we are therefore led to introduce the following representation of ψ and ψ^L:

H. The Use of Charge-Conjugated Wavefunctions 211

$$\left.\begin{array}{l}\psi = \sum_m c_m \chi_m + \sum_n \mathbf{d}_n^* \overline{\chi}_n^L, \quad \psi^L = \sum_n \mathbf{d}_n \overline{\chi}_n + \sum_m c_m^* \chi_m^L, \\ \mathbf{c}_m^* \mathbf{c}_m + \mathbf{c}_m \mathbf{c}_m^* = 1, \quad \mathbf{d}_m^* \mathbf{d}_m + \mathbf{d}_m \mathbf{d}_m^* = 1, \\ \text{all other pairs anticommute.}\end{array}\right\} \quad (28)$$

Comparing (28) with (16), we see that the c_m and $\mathbf{d}_n^*$ can be expressed in terms of the $\mathbf{a}_\lambda$ and $\mathbf{b}_\lambda^*$ and reversely.

The energy operator will now be given by

$$\mathbf{H} = \tfrac{1}{2} \int \left(\psi^* H \psi + \psi^{L*} H^L \psi^L \right),$$

$$= \sum_m E_m \mathbf{c}_m^* \mathbf{c}_m + \sum_n E_n \mathbf{d}_n^* \mathbf{d}_n - \tfrac{1}{2} \left[\sum_m E_m + \sum_n \overline{E}_n \right]. \quad (29)$$

The total charge is given by

$$\mathbf{e} = e \left[\sum_m \mathbf{c}_m^* \mathbf{c}_m - \sum_n \mathbf{d}_n^* \mathbf{d}_n \right].$$

Looking apart from the zero-point energy in (29), all stationary states have positive energy. Their Schrödinger functions, in the m, n representation, are given by functions

$$A(\ldots N_m \ldots ; \ldots \overline{N}_n \ldots), \quad N_m, \overline{N}_n = 1 \text{ or } 0, \quad (30)$$

which are zero for all $N_m, \overline{N}_n$ combinations with the exception of one particular combination $N_m^0, \overline{N}_n^0$, for which A equals 1. Every one of these states can also be interpreted in terms of "free electrons", but it appears not to be quite easy to determine how the $A(\ldots N_m \ldots ; \ldots \overline{N}_n \ldots)$ description is transformed into the $A(\ldots N_\lambda \ldots ; \ldots \overline{N}_\lambda \ldots)$ description, which would refer to free electrons.

One of the ordinary discrete states, say state m_0, of the hydrogen-like atom with one electron would correspond to $A(\ldots N_m \ldots ; \ldots \overline{N}_n \ldots)$ being different from zero only for that particular $N_m^0, \overline{N}_n^0$ combination for which all the $\overline{N}_n^0$ are zero, and for which also all N_m^0 are zero with the exception of $N_{m_0}^0$. In the description in terms of free electrons, the number of electrons in this state is of course not well defined, there being a probability of finding only one (negative) electron, a probability of finding three electrons (two negative, one positive), and so on.

Now, the electrons have up to this point been considered as independent, whereas in reality they act on each other through the medium of the electromagnetic field. It is of course possible to describe this interaction in a formal way, by introducing a quantised E, H field by the methods of quantum electrodynamics. In view of the unsatisfactory nature of these methods we might, as an approximation, try to introduce directly the Coulomb interaction between the electrons, in order to improve our scheme of calculating stationary states.

Now, in non-relativistic quantum mechanics, this Coulomb interaction would be represented by a matrix operator:

$$\mathbf{H}_2 = \frac{e^2}{2} \int_q \int_{q'} \psi^*(q)\psi^*(q')\frac{1}{r}\psi(q')\psi(q), \qquad r = |\mathbf{r}(q) - \mathbf{r}(q')|. \qquad (31)$$

The simplest but perhaps not correct way of generalizing this formalism in the hole theory would be:

$$\mathbf{H}_2 = \frac{e^2}{4} \int\int \left\{ \psi^*(q)\psi^*(q')\frac{1}{r}\psi(q')\psi(q) \right.$$
$$\left. + \psi^{L*}(q)\psi^{L*}(q')\frac{1}{r}\psi^L(q')\psi^L(q) \right\}. \qquad (32)$$

For large atomic number this energy might be considered as a perturbation. Its influence on the energy of the stationary states would be given by its expectancy value; this value does not vanish automatically in the case of the hydrogen-like atom (with one electron), in contrast to the result of applying (31) to such a state in non-relativistic quantum mechanics. One might say this is due to the fact that, in the hole theory, it can no longer be said that precisely one electron is present in the stationary states in question. As a result we expect that *a correction must be applied to the energy values of the stationary states of the hydrogen atom, as given by the Dirac theory of 1928.*

In a later paper we will discuss more closely the possibility of actually computing this correction.

I. Brownian Motion in a Field of Force and the Diffusion Model of Chemical Reactions

Abstract. A particle which is caught in a potential hole and which, through the shuttling action of Brownian motion, can escape over a potential barrier yields a suitable model for elucidating the applicability of the transition state method for calculating the rate of chemical reactions.

1. Introduction

In order to elucidate some points in the theory of the velocity of chemical reactions, the following problem is studied. A particle moves in an external field of force, but — in addition to this — is subject to the irregular forces of a surrounding medium in temperature equilibrium (Brownian motion). The conditions are such, that the particle is originally caught in a potential hole but may escape in the course of time by passing over a potential barrier. We want to calculate the probability of escape in its dependency on temperature and viscosity of the medium and to compare the values found with the results of the so-called "transition state method" for determining reaction velocities. The calculation rests on the construction and discussion of the equation of diffusion obeyed by a density distribution of particles in phase space.

For the sake of simplicity only a one-dimensional model is studied. Definite results could be obtained in the limiting cases of small and large viscosity; in both cases there exists a close analogy with Christiansen's treatment of chemical reactions as a diffusion problem. In the fairly general case where the potential barrier corresponds to a smooth maximum a reliable solution for any value of the viscosity is obtained. In that case the probability for escape is, for a large range of the value of the viscosity, practically equal to that computed by the transition method.

Our problem has also a direct bearing on the fission of an electrically charged hot drop of liquid, a question which was recently considered by Bohr and Wheeler in their discussion of the fission of uranium nuclei.

In § 2 the principles of Brownian motion are briefly discussed, and applied in order to set up an equation of diffusion in phase space.

In § 3 and § 4 the limiting cases of large and small viscosity are studied; they reduce both to a one-dimensional diffusion process.

In § 5 the formulæ found are applied to calculate the escape over a potential barrier. The results are compared with the transition method.

§ 6 discusses the relation of our model to actual problems of reaction velocity.

2. Principles of Brownian Motion in Phase Space

The equations of motion of a particle of mass 1 in a one-dimensional extension, where it is acted upon by the external field of force $K(q)$ and the irregular force $X(t)$ due to the medium, can be written as follows:

$$\dot{p} = K(q) + X(t), \qquad \dot{q} = p. \tag{1}$$

A theory of Brownian motion on the Einstein pattern can be set up if there exists a range of time intervals τ which has the following properties: On the one hand τ must be so short, that the change of velocity suffered in the course of τ may be considered as very small; on the other hand τ must be so large, that the chance for X to take a given value at the time $t + \tau$ is independent of the value which X possessed at the time t. We then consider the probability distribution of the quantity

$$B_\tau = \int_t^{t+\tau} X(t')\,dt', \tag{2}$$

which is assumed to be independent of t. Calling the distribution function $\varphi_\tau(B; p, q)$ — besides on τ and on temperature it may depend on the velocity p and the position q of the particle — it is further assumed that the moments,

$$\overline{B_\tau^n} = \int_{-\infty}^{+\infty} B^n \varphi\, dB, \qquad (\overline{B_\tau^0} = 1) \tag{3}$$

depend on τ in such a way that, practically, they can be represented by the first non-vanishing term of a development $a\tau + b\tau^2 + \cdots$. The possibility of a term proportional to τ in the expression for $\overline{B_\tau^n}(n > 1)$ is clearly due to the fact that the values which X takes at moments $t_1, t_2, \ldots t_n$ which lie sufficiently close together are no longer independent; in fact $\overline{B_\tau^n}$ is represented by a volume integral $\int \cdots \int X(t_1)X(t_2) \cdots X(t_n)\,dt_1 \cdots dt_n$ over an n-dimensional cube; the contribution to this integral due to a narrow cylinder extending along the diagonal $t_1 = t_2 = \cdots = t_n$ may give a term proportional to τ.

Einstein's original theory can be expressed by the assertions

$$\begin{aligned}
\overline{B_\tau^n} &= -\eta p \tau, \\
\overline{B_\tau^2} &= \nu\tau + \ldots, \\
\overline{B_\tau^n} &= 0 \cdot \tau + \ldots, \qquad n > 2,
\end{aligned} \tag{4}$$

where the "viscosity" η and the constant ν may still depend on temperature and position. Between η and ν the relation

$$\nu = 2\eta T \tag{5}$$

must hold, where T is the absolute temperature (defined in such a way that Boltzmann's constant equals 1); this is most easily seen by remarking that the Brownian motion does not disturb the equipartition of energy (Langevin). Expressing this fact by

$$\overline{p(t+\tau)^2} = \overline{p(t)^2},$$

we get immediately from

$$p(t+\tau) = p(t) + B$$

that

$$\overline{2p(t)B} + \overline{B^2} = 0 \quad \to \quad -2\overline{p^2}\eta + \nu \quad \to \quad -2T\eta + \nu = 0.$$

From this proof we see too that $\overline{B_\tau}$ never can vanish. We will see presently, however, that (4) represents by no means the only possible dependence of $\overline{B_\tau}/\tau$ and $\overline{B_\tau^2}/\tau$ on p, and also that there is no a priori reason why higher moments of B_τ should contain no term linear in τ.

Writing generally

$$\overline{B_\tau^n} = \mu_n \tau, \tag{6}$$

we will now derive the equation of diffusion for an ensemble of particles with density $\rho(p,q)$ in p,q-space. The density at a point $A(p_1, q_1)$ at time $t + \tau$ may be thought of as being derived from the densities at a previous moment t along the straight line for which $q = q_2 = q_1 - p_1 \tau$. Denoting by $p_2 = p_1 - K\tau$ the value which p would have taken at the time t if no Brownian forces had acted, we may write

$$\rho(p_1, q_1, t+\tau) = \rho(p_2 + K\tau, q_2 + p_2\tau, t+\tau)$$
$$= \int_{-\infty}^{+\infty} \rho(p_2 - B, q_2)\varphi(B; p_2 - B, q_2)\, dB.$$

Developing with respect to the first power of τ and to the first and higher powers of B (as far as it appears in $p_2 - B$) we get

$$\rho(p_2, q_2) + \frac{\partial \rho}{\partial t} + \frac{\partial \rho}{\partial p}K\tau + \frac{\partial \rho}{\partial q}p\tau$$
$$= \int_{-\infty}^{+\infty} \left(\rho\varphi - B\frac{\partial}{\partial p}(\rho\varphi) + \frac{B^2}{2}\frac{\partial^2}{\partial p^2}(\rho\varphi) - \cdots \right) dB.$$

Using (6) we obtain the following equation of the Fokker-Planck type:

$$\frac{\partial \rho}{\partial t} = -K(q)\frac{\partial \rho}{\partial p} - p\frac{\partial \rho}{\partial q} - \frac{\partial}{\partial p}(\mu_1 \rho) + \frac{1}{2}\frac{\partial^2}{\partial p^2}(\mu_2 \rho) - \cdots. \tag{7}$$

This is the well known Gibbs equation completed with terms, due to the Brownian motion. The current density has a q-component equal to $p\rho$ and a p-component equal to

$$K\rho + \mu_1\rho - \frac{1}{2}\frac{\partial}{\partial p}(\mu_2\rho) \cdots$$

The fundamental condition to be imposed on the μ's states that the Boltzmann distribution

$$\rho_B = e^{-(\frac{1}{2}p^2 + U(q))/T}, \qquad K = -\frac{\partial U}{\partial q}, \tag{8}$$

should be stationary. This gives

$$\frac{\partial}{\partial p}\left\{-\mu_1 e^{-p^2/2T} + \frac{1}{2}\frac{\partial}{\partial p}\left(\mu_2 e^{-p^2/2T}\right) - \frac{1}{6}\frac{\partial^2}{\partial p^2}\left(\mu_3 e^{-p^2/2T}\right) \cdots\right\} = 0,$$

$$-\mu_1 - \frac{p}{2T}\mu_2 + \frac{1}{2}\frac{\partial \mu_2}{\partial p} - \cdots = F(q,T)e^{p^2/2T}.$$

From physical considerations we expect that $\mu_1, \mu_3, \ldots$ are odd functions of p whereas $\mu_2, \mu_4, \ldots$ are even, so that F should be zero. The simplest possibility is Einstein's case:

$$\mu_1 = -\eta p, \qquad \mu_2 = 2\eta T, \qquad \mu_3 = \mu_4 = \cdots = 0.$$

Other laws of friction should also be possible, however; for instance

$$\mu_1 = -\eta p - \zeta p^3, \qquad \mu_2 = 2\eta T + 4\zeta T^2 + 2\zeta T p^2, \qquad \mu_3 = \mu_4 = \cdots = 0.$$

An example of a case where μ_3 does not vanish would be

$$\mu_1 = -\eta p - \zeta p^3, \qquad \mu_2 = 2\eta T + 6\zeta T^2, \qquad \mu_3 = 6\zeta T^2 p, \qquad \mu_4 = \cdots = 0.$$

I do not know if the more complicated cases might have some physical application. In the following we will restrict ourselves to the simple Einstein case, where the diffusion equation takes the form (η is taken independent of q)

$$\frac{\partial \rho}{\partial t} = -K(q)\frac{\partial \rho}{\partial p} - p\frac{\partial \rho}{\partial q} + \eta\frac{\partial}{\partial p}\left(p\rho + T\frac{\partial \rho}{\partial p}\right). \tag{9}$$

For previous work on the Brownian motion, the reader is referred to Uhlenbeck and Ornstein's article of 1930,[1] in which also diffusion equations of the Fokker-Planck type are discussed (although not in phase space) and in which references to the earlier work are to be found. I have found no discussion of the possibility that $\mu_3, \mu_4, \ldots$ should not be negligible.

---

[1] G. Uhlenbeck and L. S. Ornstein, *Phys. Rev.* **38**, 823 (1930).

3. The Case of Large Viscosity

Large viscosity means, that the effect of the Brownian forces on the velocity of the particle is much larger than that of the external force $K(q)$. Assuming that K does not change very much over a distance $\sqrt{T}/\eta$, we expect that, starting from an arbitrary initial ρ distribution, a Maxwell velocity distribution will be established very soon (i.e., after a time lapse of the order of $1/\eta$) for every value of q:

$$\rho(q,p,t) \cong \sigma(q,t)\,e^{-p^2/2T}. \tag{10}$$

From that time on a slow diffusion of the density distribution σ in the q-coordinate will take place, which may be expected to satisfy the Smoluchowski diffusion equation

$$\frac{\partial \sigma}{\partial t} = -\frac{\partial}{\partial q}\left(\frac{K}{\eta}\sigma - \frac{T}{\eta}\frac{\partial \sigma}{\partial q}\right), \tag{11}$$

where T/η represents the diffusion constant. As long as no perfect temperature equilibrium is attained, (10) holds only approximately. This is even the case when the external force is zero. In that case it is true enough that the velocities, which a particle takes in the course of time, are exactly distributed in the Maxwell way; the velocity distribution of the particles at a given point q deviates, however, from Maxwell, since otherwise there would be no diffusion current ($= \int \rho p\, dp$).

In order to derive (11) from (9), we rewrite (9) in the following form:

$$\frac{\partial \rho}{\partial t} = \eta\left[\frac{\partial}{\partial p} - \frac{1}{\eta}\frac{\partial}{\partial q}\right]\left[p\rho + T\frac{\partial \rho}{\partial p} - \frac{K}{\eta}\rho + \frac{T}{\eta}\frac{\partial \rho}{\partial q}\right] - \frac{\partial}{\partial q}\left[\frac{K}{\eta}\rho - \frac{T}{\eta}\frac{\partial \rho}{\partial q}\right].$$

We now integrate the right and left side of this equation along a straight line $q + p/\eta = \text{const} = q_0$ from $p = -\infty$ to $+\infty$. Denoting the integral of ρ along this line by $\sigma(q_0)$, we obtain:

$$\frac{\partial \sigma}{\partial t} = -\int_{q+p/\eta=q_0} \frac{\partial}{\partial q}\left[\frac{K}{\eta}\rho - \frac{T}{\eta}\frac{\partial \rho}{\partial q}\right] dp \cong -\frac{\partial}{\partial q_0}\left[\frac{K(q_0)}{\eta}\sigma(q_0) - \frac{T}{\eta}\frac{\partial \sigma(q_0)}{\partial q_0}\right].$$

The approximate validity of this equation is a consequence of the approximate validity of (10), if it is also allowed to assume that in the region of p-values which contributes essentially to the value of the integral (i.e., for $|p| \lesssim \sqrt{T}$), the variation of q (which is of the order $\sqrt{T}/\eta$) is small compared to the q-distances over which K and σ (and η) undergo marked variations. These are, however, precisely the conditions which a priori have to be imposed in order to ensure the applicability of (17).

A stationary diffusion current obeys the law

$$w = \frac{K}{\eta}\sigma - \frac{T}{\eta}\frac{\partial \sigma}{\partial q} = \text{const}. \tag{12}$$

Since it can also be written in the form

$$w = -\frac{T}{\eta} e^{-U/T} \frac{\partial}{\partial q} \left(\sigma e^{U/T} \right),$$

we obtain, on integration between two points A and B on the q-coordinate:

$$w = \left[T \middle| \sigma e^{U/T} \middle|_B^A \right] \bigg/ \left[\int_A^B \eta e^{U/T} \, dq \right]. \tag{13}$$

This result will enable us to derive in §5 an expression for the escape of a particle from a potential hole over a potential barrier.

4. The Case of Small Viscosity

We will restrict ourselves to the case where the particle would perform a motion of oscillatory type if no Brownian forces were present. By small viscosity is meant that the latter forces cause only a small variation of the energy during the time of an oscillation. The effect of the Brownian motion will therefore, in its main aspect, consist in the gradual change of the distribution of the ensemble over the different energy values. Denoting by I the area inside a curve of constant energy:

$$I(E) = \oint p \, dq,$$

and denoting by $\bar{\rho} \, dI$ the fraction of the ensemble lying inside the ring-shaped area dI, we find an equation for $\bar{\rho}(E,t)$ by averaging (9) over such a ring dI. The first two members on the right-hand side of (9) give zero, since in the absence of Brownian forces the distribution in energy is maintained. As regards the third member, we may write, on account of $\partial E/\partial p = p$

$$\overline{\frac{\partial}{\partial p} \left(p\rho + T \frac{\partial \rho}{\partial p} \right)} = \overline{\frac{\partial}{\partial p} \left(p\bar{\rho} + T \frac{\partial E}{\partial p} \frac{\partial \bar{\rho}}{\partial E} \right)}$$
$$= \bar{\rho} + T \frac{\partial \bar{\rho}}{\partial E} + \overline{p^2} \frac{\partial}{\partial E} \left(\bar{\rho} + T \frac{\partial \bar{\rho}}{\partial E} \right)$$

The mean value $\overline{p^2}$ is equal to the action integral over one period (i.e., I) divided by the period of oscillation. Denoting the frequency by ω:

$$\omega = \frac{dE}{dI},$$

we have therefore $\overline{p^2} = I\omega$, and the diffusion equation takes the form:

$$\frac{\partial \bar{\rho}}{\partial t} = \eta \left[1 + \omega I \frac{\partial}{\partial E} \right] \left[\bar{\rho} + T \frac{\partial \bar{\rho}}{\partial E} \right] = \eta \left[1 + I \frac{\partial}{\partial I} \right] \left[\bar{\rho} + T \frac{\partial \bar{\rho}}{\partial E} \right],$$

or, writing ρ for $\bar{\rho}$:

$$\frac{\partial \rho}{\partial t} = \eta \frac{\partial}{\partial I}\left(I\rho + TI\frac{\partial \rho}{\partial E}\right). \qquad (14)$$

This corresponds to a diffusion along the I (or E) coordinate; the diffusion term proper is given by

$$\eta \frac{\partial}{\partial I}\left(TI\frac{\partial \rho}{\partial E}\right) = \eta \frac{\partial}{\partial I}\left(T\frac{I}{\omega}\frac{\partial \rho}{\partial I}\right),$$

and corresponds to a diffusion coefficient $\eta TI/\omega$.

A stationary state of diffusion with current density w corresponds to

$$w = -\eta\left(I\rho + TI\frac{\partial \rho}{\partial E}\right) = -\eta T I e^{-E/T} \frac{\partial}{\partial E}\left(\rho e^{E/T}\right).$$

Integration between two points A and B along the E (I) coordinate gives

$$w = \left[\eta T \left|\rho e^{E/T}\right|_B^A\right] \Big/ \left[\int_A^B \frac{1}{I} e^{E/T}\, dE\right]. \qquad (15)$$

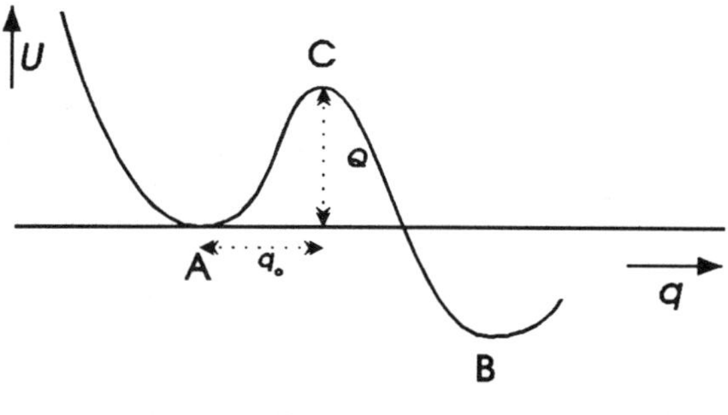

Fig. I.1. Potential field with smooth barrier.

5. Escape over a Potential Barrier

Let the potential function U be of the type illustrated in Fig. I.1. The particle is originally caught at A. The height Q of the potential barrier is supposed to be large compared with T. The discussion of the escape from A over C to B follows closely the line of thought in Christiansen's treatment[2] of chemical reaction as a diffusion problem.[3] Our ensemble in phase space is thought of as illustrating the phases of a great number of similar particles each in its

---
[2] J. A. Christiansen, *Zs. Phys. Chem.* **B33**, 145 (1936).
[3] The author is very much indebted to Prof. Christiansen for the privilege of discussing this treatment with him some years ago.

own field U. Our treatment corresponds obviously to the assumption that quantum theory effects are negligible.

At A the particle can be in a bound state. In the figure, B corresponds to another state of binding but of lower energy. If our system of particles were in thermodynamical equilibrium, the ensemble density would be proportional to $e^{-E/T}$ and the net number of particles passing from A to B would vanish. If, however, to start with, the number of particles bound at A is larger than would correspond to thermal equiibrium with the number at B, a diffusion process will be started, tending to establish equilibrium. This process will be slow if, as we have already assumed, the potential barrier Q is large compared with T. We may therefore expect that — at any moment — it can be compared with a stationary diffusion process. Even if, to start with, the particles in the hole A should not be Boltzmann-like distributed, a Boltzmann distribution near A (and also near B) will have been established a long time before an appreciable number of particles have escaped; our quasi-stationary diffusion will thus correspond to a flow from a quasi-infinite supply of Boltzmann-distributed particles at A to the region B.

I have not been able to solve exactly the problem of this quasi-stationary diffusion in p,q-space for arbitrary values of the viscosity η. In the limiting cases of large and small η, however, it reduces to a one-dimensional problem.

Large Viscosity

Assuming a quasi-stationary state, in which practically no particle has yet arrived at B, whereas near A thermal equilibrium has practically been established, application of (13) gives

$$w = \frac{T}{\eta}\sigma_A \left\{\int_A^B e^{U/T}\,dq\right\}^{-1}, \qquad \sigma_A = \left[\sigma e^{U/T}\right]_{\text{near }A}.$$

The number n_A of particles near A can be calculated, if we assume that U near A can be represented by $\frac{1}{2}(2\pi\omega)^2 q^2$ (harmonic oscillator of frequency ω):

$$n_A = \int_{-\infty}^{+\infty} \sigma_A\, e^{-(2\pi\omega)^2 q^2/2T}\,dq = \frac{\sigma_A}{\omega}\sqrt{\frac{T}{2\pi}}.$$

The reaction velocity $r = w/n_A$ denotes the chance in unit time that a particle which originally was caught at A escapes to B. It is given by

$$r = \frac{w}{n_A} = \frac{\omega}{\eta}\sqrt{2\pi T}\left\{\int_A^B e^{U/T}\,dq\right\}^{-1}.$$

The main contribution to the integral is due to a small region near C. Assuming, as would correspond to the drawing of U in Fig. I.1, that there is no sudden jump in the curvature, we may write

$$U_{\text{near } C} = Q - \tfrac{1}{2}(2\pi\omega')^2(q-q_C)^2, \tag{16}$$

whence

$$\int_A^B e^{U/T}\,dq \cong e^{Q/T}\int_{-\infty}^{+\infty} e^{-(2\pi\omega')^2(q-q_C)^2/2T}\,dq = \frac{1}{\omega'}\sqrt{\frac{T}{2\pi}}\,e^{Q/T},$$

and for the reaction velocity we find

$$r \cong \frac{2\pi\omega\omega'}{\eta}\,e^{-Q/T}. \tag{17}$$

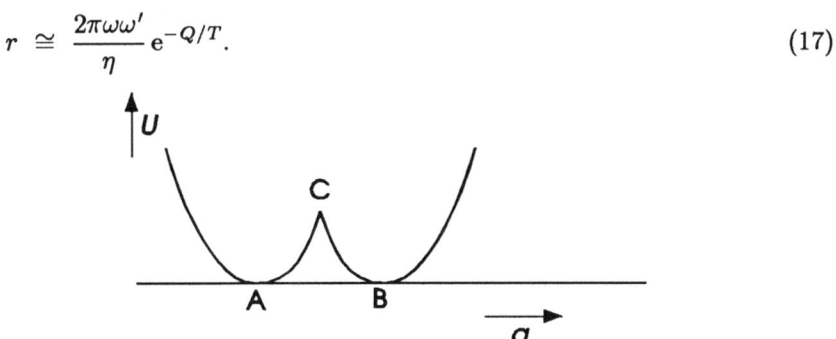

Fig. I.2. Potential field with edge-shaped barrier.

As an example of another behaviour of $U(q)$ near C, let us consider the case where U between A and C can be exactly represented by $U = \tfrac{1}{2}(2\pi\omega)^2 q^2$, and that the curve B is obtained by reflection of AC with respect to C (compare Fig. I.2). We now find

$$\int_A^B e^{U/T}\,dq = 2e^{Q/T}\int_0^\infty e^{-(2\pi\omega)^2 q_C(q-q_C)/T}\,d(q-q_C)$$

$$= \frac{2T}{(2\pi\omega)^2 q_C}\,e^{Q/T} = \frac{2T}{2\pi\omega\sqrt{2Q}}\,e^{Q/T},$$

$$r \cong \frac{2\pi\omega 2}{\eta}\sqrt{\pi}\sqrt{\frac{Q}{T}}\,e^{-Q/T}.$$

The fact that the drop of the value of $\sigma e^{U/T}$ from σ_A to zero occurs in a small q region near C (at C itself $\sigma e^{U/T}$ practically equals $\tfrac{1}{2}$) means that the resistance against escape is practically localized in this region. Assuming again the validity of (16) in this region, the corresponding diffusion process in phase space can even be described exactly for any value of η. In fact, considering the situation as stationary, Eq. (9) now assumes the form

$$0 = -(2\pi\omega')^2 q'\frac{\partial\rho}{\partial p} - p\frac{\partial\rho}{\partial q'} + \eta\frac{\partial}{\partial p}\left(p\rho + T\frac{\partial\rho}{\partial p}\right),$$

where, for simplicity, we have written q' instead of $q-q_C$. With the substitution

$$\rho = \zeta e^{-\left(p^2 - (2\pi\omega')^2 q'^2\right)/2T}$$

we get

$$0 = -(2\pi\omega')^2 q' \frac{\partial \zeta}{\partial p} - p \frac{\partial \zeta}{\partial q'} - \eta p \frac{\partial \zeta}{\partial p} + \eta T \frac{\partial^2 \zeta}{\partial p^2}. \tag{18}$$

The solution $\zeta = $ const. corresponds to thermal equilibrium. The equation (18) admits also a solution where ζ is a function of a linear combination u of p and q':

$$\zeta = \zeta(u), \qquad u = p - aq'. \tag{19}$$

Inserting (19) in (18) we obtain

$$0 = \left(ap - (2\pi\omega')^2 q' - \eta p\right) \zeta' + \eta T \zeta''. \tag{20}$$

This equation can be fulfilled if

$$(a - \eta)p - (2\pi\omega')^2 q' = (a - \eta)(p - aq'),$$

which gives for a the condition

$$(2\pi\omega')^2 = a(a - \eta), \qquad a = \frac{\eta}{2} \pm \sqrt{\frac{\eta^2}{4} + (2\pi\omega')^2}. \tag{21}$$

Equation (20) now takes the form

$$0 = (a - \eta)u \frac{d\zeta}{du} + \eta T \frac{d^2\zeta}{du^2},$$

which is solved, apart from by $\zeta = $ const., by

$$\zeta = K \int^u e^{-(a-\eta)u^2/2\eta T} \, du. \tag{22}$$

If in (21) the upper sign is chosen, $a - \eta$ will be positive and (22) just represents a diffusion of the desired type. It will correspond to a situation where practically no particles are yet to be found in the region to the right of C, if we take $-\infty$ for the lower limit of the integral. In the region well to the left of C we find then

$$\zeta = K \sqrt{\frac{2\pi \eta T}{a - \eta}},$$

and the density in phase space near A will become equal to

$$\zeta = K \sqrt{\frac{2\pi \eta T}{a - \eta}} \, e^{Q/T} \, e^{-\left(p^2 + (2\pi\omega)^2 q^2\right)/2T}. \tag{23}$$

The number w of particles passing in unit time the point C will be obtained by integrating ρp over p from $-\infty$ to $+\infty$ for $q' = 0$:

$$w = \int_{-\infty}^{+\infty} \rho p \, dp = K \int_{-\infty}^{+\infty} dp \cdot p e^{-p^2/2T} \int_{\infty}^{p} e^{-(a-\eta)p^2/2\eta T} \, dp = KT\sqrt{\frac{2\pi\eta T}{a}}.$$

On the other hand, the number of particles caught near A equals

$$n_A = K\sqrt{\frac{2\pi\eta T}{a-\eta}} e^{Q/T} \iint_{-\infty}^{+\infty} e^{-\left(p^2 + (2\pi\omega)^2 q^2\right)/2T} \, dp \, dq$$

$$= K\sqrt{\frac{2\pi\eta T}{a-\eta}} e^{Q/T} \frac{T}{\omega}. \tag{24}$$

The probability of escape is therefore now given by

$$r = \frac{w}{n_A} = \omega \sqrt{\frac{a-\eta}{a}} e^{-Q/T},$$

which by (21) (upper sign) gives

$$r = \frac{\omega}{2\pi\omega'} \left(\sqrt{\tfrac{1}{4}\eta^2 + (2\pi\omega')^2} - \tfrac{1}{2}\eta \right) e^{-Q/T}. \tag{25}$$

For $\eta/2 \gg 2\pi\omega'$ this formula reduces to the formula (17) previously found, whereas for small viscosity, $\eta/2 \ll 2\pi\omega'$, it reduces to

$$r \cong \omega e^{-Q/T} = r_{tr} \quad \text{(transition state method value)}. \tag{26}$$

The probability of escape represented by this formula corresponds exactly to the value which would be given by the *transition state method* of calculating reaction velocities.[4] In fact, according to this method, one considers the particles near A to be in perfect temperature equilibrium with those near B so that we have thermal equilibrium also at C, and one calculated the number of particles w which in unit time pass through the transition point C from the left to the right. This number is of course equal to that passing C from the right to the left and is given by

$$w = \left[\int_0^\infty \rho_0 p \right]_{(C)} dp = e^{-Q/T} \int_0^\infty p e^{-p^2/2T} \, dp = T e^{-Q/T},$$

where

$$\rho_0 = e^{-E/T}$$

---

[4] H. Pelzer and E. Wigner, *Zs. Phys. Chem.* **B15**, 445 (1932); H. Eyring, *J. Chem. Phys.* **3**, 107 (1935); M. C. Evans and M. Polanyi, *Trans. Faraday Soc.* **31**, 875 (1935).

represents the Boltzmann-Gibbs distribution in phase space. The number n_A of particles caught near A is now given by the value T/ω of the double integral occurring in (24), and for the reaction velocity we get exactly the value (26). This result can be interpreted by stating, that the region near C offers no additional resistance in the case of small η. It would, however, be wrong to conclude from this that the transition method is unconditionally applicable in this case, since now the shuttling effect of the Brownian motion is so small that the subsequent delivery of particles with energy Q (or more) near C may be insufficient. The rate of escape is now determined by a diffusion process along the energy coordinate, and can be calculated as follows.

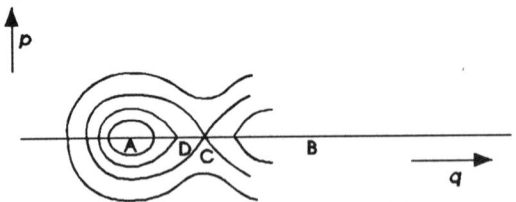

Fig. I.3. Phase space for smooth potential barrier.

SMALL VISCOSITY

We consider a quasi-stationary diffusion in phase space of the type investigated in §4. The density ρ is practically constant along curves of constant energy over a range which starts at A (energy zero) and which extends to energy curves cutting the q-axis not in the transition point C itself but in points D very near to C (compare Fig. I.3). This last restriction has to be made, if the U function near C has a smooth maximum as in Fig. I.1, because in this case the frequency tends to zero, when the energy tends to Q; for energies near Q the viscosity is therefore no longer small in the sense used in §4. If, on the other hand, the U function behaves near C in the way illustrated by Fig. I.2, such a restriction is not necessary and the stationary diffusion in energy extends up to the energy curve passing through C itself. Points in phase space which are shuttled up to an energy larger than Q will immediately travel to the region to the right of C. As our calculation will anyhow be somewhat less exact than in the case of large viscosity, we will moreover make the simplifying assumption that particles leaving the A region near C will practically never return (as for instance would be the case if the escape corresponded to dissociation, and if at the time considered practically no dissociated states of the system are yet at hand).

We can now apply formula (15) and write it in the form:

$$w = \eta T \left\{ \left(\rho e^{E/T}\right)_{\text{near } A} - \left(\rho e^{E/T}\right)_C \right\} \Big/ \left\{ \int_{\text{near } A}^{C} \frac{1}{I} e^{E/T} dE \right\}. \qquad (27)$$

The subscript "near A" is necessary, because if we integrated strictly from the point A, i.e., from energy zero ($E = I = 0$), the integral, would diverge.

We may take it, that "near A" means an energy of the order T and thus corresponds to points in phase space, where ρ is still of the same order as in A itself, i.e., to points within the region in which practically all particles of the ensemble are assembled. Furthermore, the assumption that particles leaving at C will practically never reenter the A region means that $(\rho e^{E/T})_C$ may be equalled to zero and that the upper limit in the integral could be taken to correspond to the energy at C. It is easily seen that, in the case of a symmetrical U function of the type illustrated in Fig. I.2, the value for w thus calculated will need a correction in the form of a factor $\frac{1}{2}$, since $(\rho e^{E/T})_C$ will then be $\frac{1}{2}$ of $(\rho e^{E/T})_{\text{near } A}$.

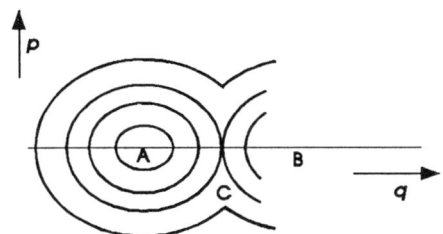

Fig. I.4. Phase space for edge-shaped potential barrier.

Denoting $(\rho e^{E/T})_{\text{near } A}$ by ρ_A, we have therefore

$$w \cong \eta T \rho_A \left\{ \int_T^Q \frac{1}{I} e^{E/T} dE \right\}^{-1}.$$

Since the main contribution to the integral is due to E-values which differ from Q by a quantity of the order of magnitude T, we may for I take the value I_C corresponding to the energy curve through C:

$$\int_T^Q \frac{1}{I} e^{E/T} dE \cong \frac{1}{I_C} e^{Q/T} \int_0^\infty e^{-(Q-E)/T} d(Q-E) = \frac{T}{I_C} e^{Q/T}.$$

Since I_C will be of the order of magnitude Q/ω, where ω is the proper frequency of particles near A, we get

$$w \cong \frac{\eta \rho_A Q}{\omega} e^{-Q/T}.$$

The probability of escape is again obtained by dividing w by the number of particles assembled near A, i.e., by $n_A = \rho_A T/\omega$. Thus we get

$$r = \frac{w}{n_A} \cong \eta \frac{Q}{T} e^{-Q/T}. \tag{28}$$

If the U function to the left of C could be represented exactly by $\frac{1}{2}(2\pi\omega)^2 q^2$, the error in this approximation for r would only be of the relative order of magnitude T/Q. In a case like that corresponding to Fig. I.3, it might be wrong by a factor of a few units; not only the estimate for the value of I_C is too low, but also the upper limit of the integral in (27) is too high so that the estimate for r is too small.

The validity of (28) implies that η is small compared with ω, i.e., that the motion of the oscillator is a periodically damped one. Aperiodic damping would mean $\eta = 4\pi\omega$, and one may expect that in this case the rate of escape is now fairly well described by (25). I have not been able to find a trustworthy method for extending (28) to η-values which are not small compared to $4\pi\omega$. One might hope to find an exact solution in the case where U to the left of C is exactly equal to $\frac{1}{2}(2\pi\omega)^2 q^2$: it would depend on the solution of the fundamental diffusion equation with particularly chosen boundary conditions.

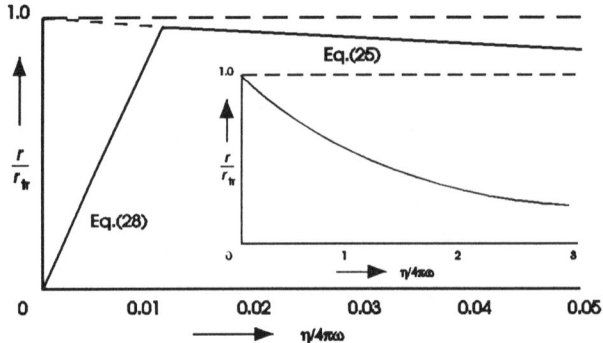

Fig. I.5. Probability of escape as a function of viscosity ($\omega' = \omega$, $Q/T = 10$).

Due to our assumption that Q/T is large (at least of the order of magnitude, say, 5; without this assumption the problem of the probability of escape is not well defined anyhow, since the conception of a Boltzmann distribution near A would no longer be valid), it is hardly of any importance, however, to try to improve on (28) in the region of small η. In fact, even for a very small value of η ($\eta/\omega \cong T/Q$), formula (28) yields a value for r equal to that given by (25), i.e., practically equal to that given by the transition state method. Figure I.5 illustrates this in the particular case $Q/T = 10$, $\omega' = \omega$. We conclude that formula (28) or (25) applies according as η/ω is smaller or larger than T/Q. *In particular we expect that the transition state method gives results which are correct, say, within 10% in a rather wide range of η-values ($\omega T/Q \lesssim \eta \lesssim 1.2\omega'$).*

In the region of small η-values this is supported by a consideration of what happens to particles which enter the A-region through C with a velocity of the order $\sqrt{T}$. If such particles are practically caught all of them after a time of the order of $1/\omega$, it means that the transition method is very reliable. Now it is clear that such a capture will take place not only when η is equal to $4\pi\omega$ (aperiodic damping), but also when η is smaller even by a factor of the order

T/Q. In fact, this means that after a time $1/\omega$ the systematic friction has been sufficient to lower the energy by an amount of the order T.

6. Relation of Our Model to Actual Reaction Processes

As well known, the appearance of the Arrhenius factor $e^{-Q/T}$ in the expression for a reaction velocity can often be interpreted by assuming that in the chain of elementary reactions which lead from an initial state A to a final state B there occurs an intermediate product in such a state C, that a (free) energy Q would be required in order to reach C from A. In simple cases C may just correspond to an activated state of a molecule the normal state of which corresponds to state A. The *activation energy* Q should be the highest energy barrier through which the system should have to pass, even if it chooses the energetically most economic way. If the molecules in the *transition state* C (concentration N_C were in temperature equilibrium with those in state A (concentration N_A), the temperature dependence of their relative concentration N_C/N_A would be given mainly by $e^{-Q/T}$. They would therefore correspond to the rarest of all intermediate products occurring in the reaction, and it is plausible that their concentration mainly determines the rate of the reaction. The so-called *transition state method* calculates the "flow" of reaction products on their way from A through C to B, which in thermodynamical equilibrium between A and C would be present at C, and declares this flow to be a reliable value for the actual reaction velocity r. In some cases this method is not unambiguous, because it is difficult to decide what assumptions have to be made about the concentration of products between state C and the final state B; in many cases it yields, however, definite results.

When discussing the reliability of this transition method, reactions of an outspoken autocatalytic character (chain reactions) must as a rule be discarded, since here the necessary activation energy can be supplied by reaction products appearing between C and B; the mechanism of reactions which lead from A to C and which are responsible for the temperature equilibrium between A and C is now no longer the only one which counts.[5] But even if autocatalytic reactions are excluded, there are several reasons why the transition method might give wrong results. The very fact that a reaction is going on means that state C will never be in exact temperature equilibrium with state A; there may be a one-sided supply of molecules in state C from state A: those in state C which pass in the direction of B may never be compensated by a backward flow $B \to C \to A$. Thus the problem is raised if the restocking $A \to C$ is sufficient in order to maintain the equilibrium concentration at C.

---

[5] Still, chain reactions may, in particular cases, just lead to the transition-method value for r. Compare J. A. Christiansen and H. A. Kramers, *Zs. Phys. Chem.* **104**, 451 (1923).

The simplest example where the restocking gets too small is that of a homogeneous gas reaction, where the pressure is so low that a sufficient number of activated molecules is no longer supplied by the collisions. But even if there is no difficulty as regards the concentration at C, the reaction may be much slower than that given by the transition method, due to the fact that the molecules in C are still practically in temperature equilibrium with stages in the reaction D, E, ... on the way from C to B, so that the flow at C in the direction B is compensated to a smaller or larger extent by a flow from E, D through C backward in the direction of A.[6]

Our problem of the escape of a particle in a potential hole A over a barrier C clearly supplies a simplified classical model, in which it is possible to judge the reliability of the transition method. The Brownian forces of the medium illustrate the mechanism which strives to bring about temperature equilibrium. The value of the viscosity coefficient η (which may depend on T even in the manner of an exponential function) is a measure for the intensity with which the molecules in the different states react with the surrounding medium. If η is too small, the restocking is too small, and the reaction velocity falls beneath the transition-method value; if it is too large, the net flow at the transition point will be much smaller than the flow in the direction $A \to C \to B$ alone: in a fairly large region of η-values, however, the transition method holds good ($\omega T/Q \lesssim \eta \lesssim 1.2\omega'$).

The model illustrates also the ambiguity involved in the conception "transition state". For small η, it should be a state of definite energy Q, and one imagines easily in what manner the model might be generalized, in order to obtain a closer resemblance to actual processes, especially as regards the introduction of discrete quantized energy states. For larger η the transition state in our model is, however, mainly characterized by the value of the spatial coordinate q_C; in actual problems we may indeed meet cases where the main characteristic of the potential barrier to be overcome corresponds to a definite spatial arrangement of atoms in a molecule, while their velocities can be treated more or less on classical lines. If the latter condition is far from fulfilled, quantummechanical "tunnel effects" for which there is no room in our model, could also play a part.[7]

In order to get some insight in the virtues and the defects of our model, let us consider three examples of reactions.

Consider first a simple reaction of molecules in a solvent, for instance the slow racemisation of an optically active substance in water. The transition from the left-handed to the right-handed configuration comes mainly about through the existence of states in which some atomic vibrations are so strongly excited that the activated molecule no easily flaps over from the

---

[6] In *Trans. Faraday Soc.* **34**, 3–81 (1938) a general discussion on reaction kinetics is given, in which the question of the transition state method is frequently touched upon. For quantummechanical effects compare J. O. Hirschfelder and E. Wigner, *J. Chem. Phys.* **7**, 616 (1939).

[7] J. O. Hirschfelder and E. Wigner, *loc.cit.*

left-handed into the right-handed state, or that even the difference between these two states is obliterated. It is true that the flapping over is also possible in non-activated states (tunnel effect) but its probability is negligible then. The activation comes about through the interaction with the water molecules and this interaction is symbolized in our model by the Brownian forces. This symbolisation is more or less justified when we compare our model with a sort of pulsating structure in a homogeneous medium, but it may of course be that actualy the activation is better described by comparing it with short-timed interactions ("collisions" with a water molecule) which cause sudden considerable changes in the vibrational state. Anyhow, an estimate, based on the value of the coefficient of viscosity in water, of the effective η to be inserted in our formulæ in this case suggests that the restocking is well provided for and that we are probably justified in applying the transition method.

In homogeneous gas reactions (one might think of the racemisation of an evaporated optically active substance) our Brownian forces would represent the action of the separate collisions between the molecules. Such a representation would only be of quantitative value in the extreme case where the change in the state of motion of a reactive molecule due to a single collision with another gas molecule is — on the average — only very small. This case would be realised if the vibrating atoms in the molecule were heavy compared with the mass of the majority in the gas (say, dissociation of diatomic iodine molecules in helium). It reminds us of Lorentz' calculation[8] of the Brownian motion of a spherical particle in a highly rarefied gas, although here, of course, only the effect on the translational (and rotational) motion of a free particle was considered. In general, the collisions may induce sudden jumps of the reactive molecule to states in which the energy is quite different and the Brownian motion model has, at its best, no more than a qualitative significance.

In this connection it might be of interest to calculate the probability of escape of a particle in a potential hole A over a barrier C under the assumption that the particle is subject to the collisions with the molecules of a Maxwell-distributed gas. If the mass of such a molecule is small compared to that of the particle, the Brownian motion model holds good; if the ratio of the masses is not small, the diffusion equation in phase space is no longer a differential equation but an integro-differential equation. Still in the limit of large viscosity the Smoluchowski equation (11) would apply.

Our second example is that of a polyatomic molecule with a great number of vibrational degrees of freedom, one of which may lead to reaction (say, dissociation of N_2O_5).[9] Here the coupling between the different modes of vibration (due to terms in the potential energy higher than quadratic) influence the "reactive" vibration in a similar way as the Brownian forces influence the motion of the particles in our model. It may of course be that in

---
[8] H. A. Lorentz, *Les théories statistiques en thermodynamique*, Teubner (1912) p. 47.
[9] Compare L. S. Kassel, *The Kinetics of Homogeneous Gas Reactions*, Ch. V.

the activated state the distinction of one particular reactive vibration, which is subject to perturbations from the other vibrations, is no longer sound and that consequently a one-dimensional model of the transition state is largely at fault.

As a third example consider the fission of heavy nuclei in the way this phenomenon was treated by Bohr and Wheeler.[10] They consider the nucleus as a hot drop of homogeneously charged liquid; such a drop has five modes of vibration which may lead to fission. Bohr and Wheeler justify the assumption that the problem may be treated as a classical one and they calculate the probability of fission by the transition method. The temperature of the nucleus is not introduced explicitly, but the calculation is in essence not different from a classical application of the transition method. The occurrence of h in Bohr and Wheeler's formula is only due to the circumstance that the excitation of the nucleus is described in terms of a density of levels.

As a matter of fact, our model corresponds rather closely to the Bohr-Wheeler model. Our q corresponds to their fission coordinate and our η corresponds to the resistance to which the vibration of the drop is subject as a consequence of the "viscosity" of the nuclear matter. The transition method in this case is therefore justified if the friction is not very small, nor so large that the drop vibrations are over-aperiodically damped to a high degree. Of course, at the present state of our knowldege, a marker error in Bohr and Wheeler's estimate would only be due to the friction being abnormally small or abnormally large. Still it is not uninteresting to consider the question of the coefficient of viscosity of nuclear matter somewhat more closely. Even if a nucleus in its normal state behaved as a perfectly hard, non plastic crystal, there is no reason to exclude the possibility that the excited nucleus possesses a finite coefficient of internal friction. In view of the surprising properties of He II it is even dangerous to assert that this coefficient cannot be extremely small; this assumption would not necessarily contradict Bohr's assumption that a single neutron impinging on a nucleus is in first instance captured. This assumption is, however, not well reconcilable with the idea that the nuclear matter should behave as a perfectly hard crystal, i.e., a certain amount of plasticity is anyhow to be expected.

---

[10]N. Bohr and J. A. Wheeler, *Phys. Rev.* **56**, 426 (1939).

J. Statistics of the Two-Dimensional Ferromagnet

Abstract. In an effort to make statistical methods available for the treatment of cooperational phenomena, the Ising model of ferromagnetism is treated by rigorous Boltzmann statistics. A method is developed which yields the partition function as the largest eigenvalue of some finite matrix, as long as the manifold is only one dimensionally infinite. The method is carried out fully for the linear chain of spins which has no ferromagnetic properties. Then a sequence of finite matrices is found whose largest eigenvalue approaches the partition function of the two-dimensional square net as the matrix order gets large. It is shown that these matrices possess a symmetry property which permits location of the Curie temperature if it exists and is unique. It lies at

$$\frac{J}{kT_c} = 0.8814$$

if we denote by J the coupling energy between neighboring spins. The symmetry relation also excludes certain forms of singularities at T_c, as, e.g., a jump in the specific heat. However, the information thus gathered by rigorous analytic methods remains incomplete.

Transition temperatures of various types are a well-known phenomenon in the study of matter, and the statistical distribution laws form a generally accepted piece of theory. It is also generally believed that the former are a consequence of the latter. This is, however, by no means immediately obvious and an examination of the literature shows that there is not more than one case in which a proof of this fact has been attempted. The case which has received successful treatment is the condensation of vapors.[1] This paper wishes to carry out a similar treatment for the Curie transition of ferromagnets.

The problem has a mechanical and a statistical aspect. On the mechanical side we wish to improve our understanding of the responsible coupling forces.

---

[1] J. E. Mayer, *J. Chem. Phys.* **5**, 67 (1937); J. E. Mayer and Ph. G. Ackermann, *J. Chem. Phys.* **5**, 74 (1937); M. Born, *Physica* **4**, 1034 (1937); B. Kahn, Dissertation Utrecht (1938). The last paper has the most rigorous treatment of the matter.

On the statistical side we wish to derive with certainty the thermal properties from a reasonable accurate mechanical model. Both aspects have received extensive attention. Quantum theory has explained satisfactorily the origin and nature of the coupling forces. There are also several theories available which explain in terms of them the thermal behavior of ferromagnets. Not one, however, applies just straight statistics to the mechanical data.[2] Generally some simplifying assumption is introduced to facilitate the evaluation of the partition function. It follows that the results obtained are not necessarily a consequence of the mechanical model, but may well be due to the statistical approximation.

The present paper is an attempt to gain sound statistical information about some model of a ferromagnet. The Ising model has been chosen because its extreme simplicity makes it particularly suitable for such a purpose. In Part I, we shall show that the task of finding the state sum can be reduced to finding the largest eigenvalue of some matrix. The matrix will be very simple in the case of the linear chain and we shall re-derive by this method the results of Ising. No such simple solution will be possible for the two-dimensional net where the matrix is infinite. However, some precise information can still be gained which will be collected in the latter sections of the paper.

In Part II, we shall complement this knowledge by approximate treatments. Some of them are already well known, as for instance, the power series approximation, the Heisenberg method, the order-disorder method of Bethe. In addition, we wish to add two treatments of our own. Both are based on the matrix method. One will be a semi-numerical treatment to answer a specific question left open in Part I, the other a new approximation method giving results in closed form. It will be shown that this latter is very much superior to the older procedures.

1. The Mechanical Model

The Ising model can be explained as follows. Assume a set of spins μ_1, μ_1, $\mu_3 \cdots \mu_N$ arranged in some regular order. Let each of the spins be capable of two orientations which we characterize by $\mu_i = +1$ and $\mu_i = -1$. Then the Ising model assumes that the forces on each spin depend only on the orientation of its immediate neighbors in addition to an eventually applied magnetic field. In particular, if all direct neighbors of a given spin are equivalent the model contains only two parameters, namely the magnetic moment m of each spin and a quantity J which is the energy gained if two neighbors change from an antiparallel to a parallel position. With these two definitions the total energy E takes the form

$$E = -\tfrac{1}{2} \sum_{\langle i,k \rangle} \mu_i \mu_k - mH \sum_i \mu_i, \qquad (1)$$

---

[2] For a detailed discussion of various approximations see Part II.

where here as in the future $\sum_{(i,k)}$ shall mean that the sum is carried out over all pairs (i, k) which are direct neighbors.

Most statistical questions concerning (1) can be considered solved if we can evaluate the so-called partition function f

$$f = \underset{\mu_i=\pm 1}{\boldsymbol{\Sigma}} \exp\left[K \sum_{(i,k)} \mu_i \mu_k + C \sum_i \mu_i\right] \tag{2}$$

with

$$K = \frac{J}{2kT} \tag{3}$$

and

$$C = \frac{mH}{kT}. \tag{4}$$

The bold face summation sign $\boldsymbol{\Sigma}_{\mu_i=\pm 1}$ is to be understood to extend over all possible states of the system, i.e., it would have to be written explicitly as

$$\underset{\mu_i=\pm 1}{\boldsymbol{\Sigma}} \sim \sum_{\mu_1=\pm 1} \sum_{\mu_2=\pm 1} \sum_{\mu_3=\pm 1} \cdots \sum_{\mu_N=\pm 1}.$$

Once f is obtained most important physical consequences can be derived from it. We obtain for instance the total magnetization M and the total energy E:

$$M = m\frac{\partial \log f}{\partial C} \tag{5}$$

and

$$E = -MH - \tfrac{1}{2}H\frac{\partial \log f}{\partial K}. \tag{6}$$

2. The Linear Chain

Ising himself carried out the calculation (2) for the linear chain.[3] However, we shall reconsider this problem since it forms an easy introduction for the eigenvalue method of evaluating f.

Let us suppose first the chain to be finite with n members $\mu_0, \mu_1, \mu_2, \cdots \mu_{n-1}$, as indicated in Fig. J.1. Then, by Boltzmann's theorem, the probability for a particular arrangement of spins $\mu_0 = +1$, $\mu_1 = +1$, $\mu_2 = -1$, $\mu_3 = +1$, $\cdots \mu_{n-1} = -1$ is proportional to $e^{-E/kT}$, because every arrangement has weight 1. Since the energy E is given by (1) this probability is proportional to

---

[3] E. Ising, *Zs. Phys.* **31**, 253 (1925).

$$\exp\left[K(\mu_0\mu_1+\mu_1\mu_2+\mu_2\mu_3+\cdots+\mu_{n-2}\mu_{n-1})\right.$$
$$\left.+C(\mu_0+\mu_1+\mu_2+\cdots+\mu_{n-2}+\mu_{n-1})\right],$$

where K and C are given by (3) and (4). Exactly the same consideration is possible if we add one extra spin in the position $[P]$ of Fig. J.1. The resulting expression is the same as above except that both sums extend to μ_n. It follows that the two probabilities differ from each other only through the factor

$$\exp\left[K\mu_{n-1}\mu_n + C\mu_n\right].$$

From these all-over probabilities others answering more simple questions can be obtained by summation. Let us determine for example from our first expression the probability $P(\mu_{n-1})$ that μ_{n-1} has either value regardless of the values of $\mu_0, \mu_1, \mu_2, \cdots \mu_{n-2}$. This is easily found to be

$$P(\mu_{n-1}) \sim \sum_{\mu_0,\mu_1,\cdots\mu_{n-2}=\pm 1} \exp\left[K(\mu_0\mu_1+\mu_1\mu_2+\cdots\mu_{n-2}\mu_{n-1})\right.$$
$$\left.+C(\mu_0+\mu_1+\cdots\mu_{n-1})\right].$$

By summing the second probability containing μ_n over the same μ's a probability $P(\mu_{n-1},\mu_n)$ can be obtained giving the chance for any of the four combinations $++, +-, -+, --$. The quantities $P(\mu_{n-1})$ and $P(\mu_{n-1},\mu_n)$ still differ by the same factor, *viz.*,

$$\lambda P(\mu_{n-1},\mu_n) = P(\mu_{n-1})\exp\left[K\mu_{n-1}\mu_n + C\mu_n\right],$$

the unknown factor λ entering because the Boltzmann exponentials are only proportional to probabilities.

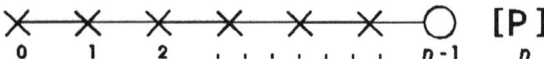

Fig. J.1. Building up the linear chain. The elementary step consists in placing a spin in position $[P]$, whose sign depends only on the spin $\bigcirc$.

If we sum both sides in the above expression over $\mu_{n-1} = \pm 1$ we get the probability $P(\mu_n)$ for μ_n having either value in terms of the same probabilities for μ_{n-1} before μ_n was added. However, *if the chain is very long*, the physical situation described by the two P's is identical. Hence $P(\mu_n)$ and $P(\mu_{n-1})$ must be the same mathematical functions of their argument

$$\lambda P(\mu_n) = \sum_{\mu_{n-1}=\pm 1} P(\mu_{n-1})\exp\left[K\mu_{n-1}\mu_n + C\mu_n\right].$$

These two linear equations have the form of a matrix eigenvalue problem. If we symmetrize the matrix by the substitution

$$a(\mu) = P(\mu)\exp\left[\tfrac{1}{2}C\mu\right]$$

the problem takes the form

$$\sum_{\mu'=\pm 1}\mathcal{H}(\mu,\mu')a(\mu') = \lambda a(\mu) \tag{7a}$$

with

$$\mathcal{H}(\mu,\mu') = \exp\left[K\mu\mu' + \tfrac{1}{2}C\mu + \tfrac{1}{2}C\mu'\right]. \tag{7b}$$

We shall consider the solution of (7) later on.

Before doing so let us clear up the significance of the two eigenvalues λ_1 and λ_2.[4] This can be done most easily by using the fundamental theorem which develops any matrix in terms of its eigenvectors:

$$\mathcal{H}(\mu_1,\mu_2) = \lambda_1 a_1(\mu_1)a_1(\mu_2) + \lambda_2 a_2(\mu_1)a_2(\mu_2),$$

where $a_1(\mu)$ and $a_2(\mu)$ denote the eigenvectors belonging to λ_1 and λ_2, respectively. They are orthogonal and may be assumed normalized

$$\sum_{\mu=\pm 1} a_i(\mu)a_k(\mu) = \delta_{ik}.$$

This permits us to unite two $\mathcal{H}$'s as follows

$$\sum_{\mu_2}\mathcal{H}(\mu_1,\mu_2)\mathcal{H}(\mu_2,\mu_3) = \lambda_1^2 a_1(\mu_1)a_1(\mu_3) + \lambda_2^2 a_2(\mu_1)a_2(\mu_3)$$

and next

$$\sum_{\mu_2}\sum_{\mu_3}\mathcal{H}(\mu_1,\mu_2)\mathcal{H}(\mu_2,\mu_3)\mathcal{H}(\mu_3,\mu_4) = \lambda_1^3 a_1(\mu_1)a_1(\mu_4) + \lambda_2^3 a_2(\mu_1)a_2(\mu_4)$$

and so on until finally

$$\sum_{\mu_2,\mu_3\cdots\mu_N}\mathcal{H}(\mu_1,\mu_2)\mathcal{H}(\mu_2,\mu_3)\mathcal{H}(\mu_3,\mu_4)\cdots\mathcal{H}(\mu_N,\mu_{N+1})$$
$$= \lambda_1^N a_1(\mu_1)a_1(\mu_{N+1}) + \lambda_2^N a_2(\mu_1)a_2(\mu_{N+1}).$$

If we close the ring of spins by the assumption that $\mu_{N+1} = \mu_1$ and sum over this last spin we get

$$\sum_{\mu_i}\mathcal{H}(\mu_1,\mu_2)\mathcal{H}(\mu_2,\mu_3)\cdots\mathcal{H}(\mu_N,\mu_1) = \lambda_1^N + \lambda_2^N.$$

Substituting for the matrices $\mathcal{H}$ their values as given by (7) we verify that the expression on the left-hand side is exactly the partition function f, as defined by (2), for a closed ring of N spins

---

[4] The elegant form of procedure used here is due to Mr. E. Montroll who applied it first to the theory of molecular chains.

$$f_N = \lambda_1^N + \lambda_2^N. \tag{8}$$

Finally, if the length N of the chain tends to infinity the smaller root λ_2 may be neglected.

What is the value of the larger root? The eigenvalue problem (7) reads

$$\begin{pmatrix} e^{K+C} & e^{-K} \\ e^{-K} & e^{K-C} \end{pmatrix} \begin{pmatrix} a(+) \\ a(-) \end{pmatrix} = \lambda \begin{pmatrix} a(+) \\ a(-) \end{pmatrix} \tag{7c}$$

and hence

$$\lambda = e^K \cosh C + \sqrt{e^{2K} \sinh^2 C + e^{-2K}}. \tag{9}$$

It has already been pointed out by Ising himself[5] that a linear chain of spins is not ferromagnetic. This can easily be verified by calculating the total magnetization with the help of (5) and (8):

$$M = \frac{mN \sinh C}{\sqrt{\sinh^2 C + e^{-4K}}}, \tag{10a}$$

an expression which, because of (4), vanishes with H. The initial molecular paramagnetic susceptibility comes out to be

$$\chi = \frac{m^2}{kT} e^{J/kT}. \tag{10b}$$

At low temperatures this value is very much larger than without the cooperative coupling J, but it is still finite.

In the absence of a magnetic field we have

$$\lambda = 2 \cosh K,$$

which gives for the energy as a function of temperature

$$E = -\tfrac{1}{2} NJ \tanh K = -\tfrac{1}{2} NJ \tanh \frac{J}{kT}. \tag{11}$$

This is a smooth increase from $-\tfrac{1}{2} NJ$ to 0 as the temperature rises.

3. Matrix Form of the Square Net Problem

The successful calculation of Section 2 has unfortunately no bearing upon ferromagnetism since the model proves to be paramagnetic only. The situation is entirely different if we consider the still simplified case of a square net of spins having two infinite dimensions. Peierls[6] has proved that this model

---

[5] *Loc.cit.*
[6] R. Peierls, *Proc. Camb. Phil. Soc.* **32**, 477 (1936). Ising's erroneous conclusions for the two-dimensional case are still quoted in some papers, e.g., Lamek Hulthén, *Arkiv för Matematik, Astronomi och Fysik* **26A**, No. 11 (1940).

J. Statistics of the Two-Dimensional Ferromagnet

is ferromagnetic in the sense that it has a non-zero magnetization at absolute zero. We can conclude from that result that the spontaneous magnetization cannot possibly be represented by a single analytic function of temperature since it vanishes identically in the high temperature range. This in turn would make us suspect the existence of a singular point at which the magnetization ceases to vanishes identically. It follows that the two-dimensional Ising model is a fair test for the general statistical theory of ferromagnets.

The reduction of the linear chain problem can be described in a qualitative way as follows. It is possible to build up a chain by repeating constantly one and the same operation, namely, adding another spin beyond the one just placed previously. In fact, if the chain is really very long no physical change takes place through the addition of one more spin. The successful mathematical treatment is based on this identity on the one hand and on the other on the fact that the *state of the last spin μ_n is only dependent upon the state of its predecessor μ_{n-1}*. It follows that the function $P(\mu_n)$ depends operationally on $P(\mu_{n-1})$, yet is the same mathematical function of its argument. Exactly this is expressed in Eq. (7).

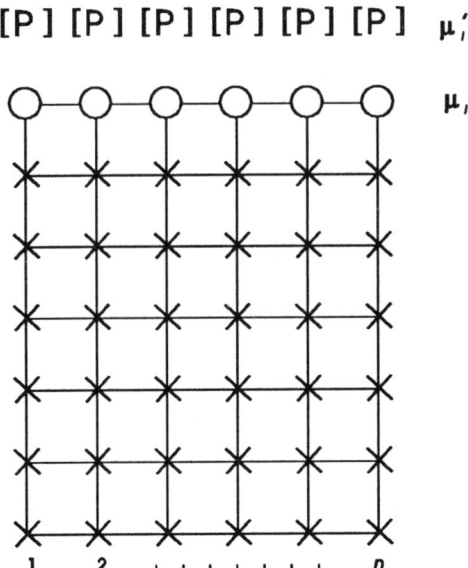

Fig. J.2. Building up the infinite strip. Elementary step consists in filling all positions [P] whose situation depends on each other and the spins ◯.

A strip of spins having infinite length but finite width can obviously be built up in an analogous manner. Let us recall briefly the steps with the help of Fig. J.2. First we write down two probability expressions, one referring to all spins marked by crosses and rings (we assume the positions [P] to be vacant), and the other to all crosses, rings, and positions [P]. Both expressions follow directly from Boltzmann's theorem, that is they are proportional to $e^{-E/kT}$, E being given by (1). In interpreting (1) we must remember that the neighborhood in $\sum_{(i,k)}$ is now both horizontal and vertical, as expressed by the connecting rods in Fig. J.2. As a consequence, the factor by which the probabilities differ is equal to

$$\exp\Big[K\left(\mu'_1\mu'_2 + \mu'_2\mu'_3 + \cdots + \mu'_{n-1}\mu'_n\right)$$
$$+ K\left(\mu_1\mu'_1 + \mu_2\mu'_2 + \cdots + \mu_n\mu'_n\right) + C\left(\mu'_1 + \mu'_2 + \cdots + \mu'_n\right)\Big],$$

where the indices of the μ's refer to the column number. Only the μ's of the two top lines are involved in the expression, the very top being indicated by μ'_i, the next one by μ_i (see Fig. J.2). Probabilities referring only to the two lines just mentioned can be gained by summing over all crossed spins. We are still left with a function $P(\mu_i)$ of n variables and $P(\mu_i, \mu'_i)$ of $2n$ variables. The two probabilities still differ from each other by the same factor

$$\rho P(\mu_i, \mu'_i) = P(\mu_i)\exp\left[K\sum_{i=1}^{n-1}\mu'_i\mu'_{i+1} + K\sum_{i=1}^{n}\mu_i\mu'_i + C\sum_{i=1}^{n}\mu'_i\right],$$

the constant ρ appearing for the same reason as above (§2). Finally, if the strip is very long $P(\mu'_i) = \sum_{\mu_i} P(\mu_i, \mu'_i)$ must be the same function of its variables as $P(\mu_i)$. The substitution

$$a(\mu_i) = P(\mu_i)\exp\left[\tfrac{1}{2}K\sum_{i=1}^{n-1}\mu_i\mu_{i+1} + \tfrac{1}{2}C\sum_{i=1}^{n}\mu_i\right]$$

will make the resulting eigenvalue problem symmetrical. It reads then

$$\sum_{\mu'_i}\mathcal{H}(\mu_i, \mu'_i)a(\mu'_i) = \rho a(\mu_i) \tag{12a}$$

with

$$\mathcal{H}(\mu_i, \mu'_i) = \exp\Big[K\sum_{i=1}^{n}\mu_i\mu'_i + \tfrac{1}{2}K\sum_{i=1}^{n-1}\mu_i\mu_{i+1}$$
$$+ \tfrac{1}{2}K\sum_{i=1}^{n-1}\mu'_i\mu'_{i+1} + \tfrac{1}{2}C\sum_{i=1}^{n}\mu_i + \tfrac{1}{2}C\sum_{i=1}^{n}\mu'_i\Big]. \tag{12b}$$

To find the significance of ρ we turn again to the matrix development theorem. We have this time 2^n roots ρ_i since the matrix is of that order. As a starting point, we use two basic statements

$$\sum_{\mu_i} a_p(\mu_i)a_q(\mu_i) = \delta_{pq},$$

$$\mathcal{H}(\mu_i, \mu_i') = \sum_{p=1}^{2^n} \rho_p a_p(\mu_i) a_p(\mu_i').$$

If we consider now a finite ring which is n spins wide and m spins in circumference we find that its partition function can be written in terms of the "nuclei" $\mathcal{H}$ thus

$$f = \sum_{\mu_i^k} \mathcal{H}(\mu_i^1, \mu_i^2)\mathcal{H}(\mu_i^2, \mu_i^3)\mathcal{H}(\mu_i^3, \mu_i^4) \cdots \mathcal{H}(\mu_i^{m-1}, \mu_i^m)\mathcal{H}(\mu_i^m, \mu_i^1).$$

Putting in the development for each matrix $\mathcal{H}$, interchanging summations and remembering the orthogonality conditions we find

$$f = \sum_{p=1}^{2^n} \rho_p^m. \tag{13}$$

If we make the strip very long while keeping the width the same, m will become very large and all but the largest root ρ can be neglected. It is well known from the theory of matrices that the largest root has an eigenvector $a(\mu)$ with only positive components. This must be so, of course, because of their probability significance.

The doubly infinite square net results from (12) and (13) only in the limit when the width n of the strip tends to infinity as well as m. It has therefore been our endeavor to find a way of connecting n and $n+1$ just as our Eq. (12) carries out the reduction from $m+1$ to m. It has been impossible so far to do this in a rigorous manner. However, a very powerful approximation method can be based upon Eq. (12) which will be discussed in Part II, Section 7.

In spite of this failure some exact results can be obtained for the two-dimensional case; they are based on certain matrix identities. In order to obtain them we have to take up once more our basic steps which led to the eigenvalue problem (12). For it is possible to use instead of the matrix $\mathcal{H}$ another equivalent one which does not have some of its inconveniences. The matrix $\mathcal{H}$ is of order 2^n; it is filled solidly with elements which take up various values in a more or less haphazard way. In addition, its largest eigenvalue changes its meaning as n increases because it does not refer to one spin but to a whole line of them, i.e., we do not have $\rho^{nm} = f$, but only $\rho^m = f$. The reason for these features is that we built up the strip in large steps, filling a large number of vacant positions [P] at one time (Fig. J.2).

This situation can be amended partly by arranging the spins along the thread of a screw instead of a simple strip. For this purpose we dispose of the free right- and left-hand edges in Fig. J.2 by bringing togethe each left end spin with the right end spin of the next line; the resulting order is visualized in Fig. J.3. Its advantage is that the whole net can now be built up through the simple operation of placing a new spin immediately beside its predecessor as we move along the thread of the screw. Figure J.3 indicates this operation by a [P] as in the two previous figures.

The mathematical build-up of the eigenvalue problem follows the two previous procedires. Boltzmann's theorem may be used to give the probabilities for all the μ's with or without a spin being placed at $[P]$. As a second step we may eliminate, just as before, all explicit references to the crossed spins by summation. If we denote by n the number of spins making up one pitch of the screw we can re-label the ring spins as indicated on Fig. J.4. Let us call $A(\mu_{n-1}, \mu_{n-2} \cdot \cdot \mu_1, \mu_0)$ the probability referring to them alone and $P(\mu_n, \mu_{n-1} \cdot \cdot \mu_0)$ the one including μ_n in position $[P]$ as well. The factor by which the two quantities differ is now much simpler than it was in our previous treatment in §3. It contains only the couplings which link up μ_n with μ_{n-1} and μ_0 in addition to the action of the field H on μ_n

$$\lambda P(\mu_n, \cdot \cdot \mu_0) = A(\mu_{n-1}, \cdot \cdot \mu_0) \exp\left[K(\mu_n(\mu_{n-1} + \mu_0) + C\mu_n)\right].$$

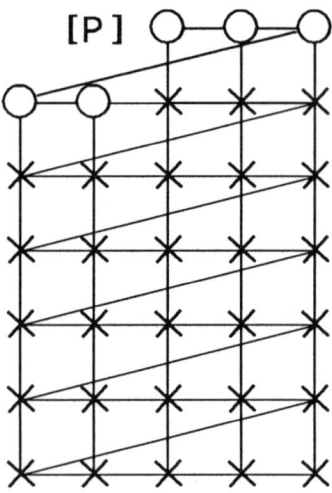

Fig. J.3. Building up the infinite screw. Elementary step $[P]$ demands knowledge of spins $\bigcirc$, partly for the step itself, but mostly for later similar steps to follow.

The eigenvalue problem follows from this equation if we sum $P(\mu_n, \cdot \cdot \mu_1, \mu_0)$ over μ_0 and notice that the resulting situation is identical with the one described by $A(\mu_{n-1}, \cdot \cdot \mu_0)$ provided the screw is very long. The only difference is that μ_1 now occupies the place of μ_0, μ_2 of μ_1, and so forth. It follows that we get the equation

$$\sum_{\mu_0} \exp\left[K(\mu_n(\mu_{n-1} + \mu_0) + C\mu_n\right] A(\mu_{n-1}, \cdot \cdot \mu_0) = \lambda A(\mu_n, \cdot \cdot \mu_1). \quad (14)$$

If we compare (14) with (12) we notice that the matrix has become essentially asymmetric and that we have not achieved any reduction of its order.

But we have realized two improvements: Each line contains now only two non-zero elements and the eigenvalue λ is now the partition function per individual spin regardless of the order of the matrix.

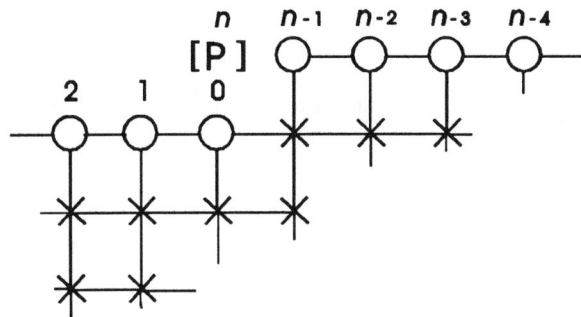

Fig. J.4. Numbering of spins in the infinite screw problem.

The proof of the latter proposition repeats the pattern outlined after Eq. (12), except that we have to take into account the asymmetry of the matrix. Let λ_i be its 2^n eigenvalues and $A_i(\mu_n \cdots \mu_1)$ the corresponding eigenvectors to the right. Then we have to consider in addition the eigenvectors to the left which we call $B_i(\mu_n \cdots \mu_1)$ and which satisfy the equations

$$\sum_{\mu_0} \exp\left[K(\mu_n(\mu_{n-1}+\mu_0)+C\mu_n\right] B(\mu_{n-1}, \cdots \mu_0) = \lambda B(\mu_n, \cdots \mu_1). \quad (15)$$

Before applying the basic equations it is advantageous to eliminate the appearance of identical μ's on either side of our Eqs. (14) and (15). This can be done by repeating n times the matrix operation indicated. If we discard the terms in C, i.e., let the magnetic field be zero for simplicity, we get

$$\sum_{\mu_1,\mu_2\cdots\mu_n} \exp\left[K\sum_{i=n}^{2n-1}\mu_i\mu_{i+1} + K\sum_{i=1}^{n}\mu_i\mu_{i+n}\right] A(\mu_n \cdots \mu_1)$$
$$= \lambda^n A(\mu_{2n} \cdots \mu_{n+1})$$

and

$$\sum_{\mu_{n+1},\mu_{n+2}\cdots\mu_{2n}} \exp\left[K\sum_{i=n}^{2n-1}\mu_i\mu_{i+1} + K\sum_{i=1}^{n}\mu_i\mu_{i+n}\right] B(\mu_{2n} \cdots \mu_{n+1})$$
$$= \lambda^n B(\mu_n \cdots \mu_1).$$

It follows that with proper normalization

$$\sum_{\mu_i} B_p(\mu_n \cdots \mu_1) A_q(\mu_n \cdots \mu_1) = \delta_{pq}$$

and

$$\exp\left[K\sum_{i=n}^{2n-1}\mu_i\mu_{i+1}+K\sum_{i=1}^{n}\mu_i\mu_{i+n}\right]$$
$$=\sum_{p=1}^{2^n}\lambda_p^n A_p(\mu_{2n}\cdot\cdot\mu_{n+1})B_p(\mu_n\cdot\mu_1).$$

We can write down this same formula for the next pitch of the screw

$$\exp\left[K\sum_{i=2n}^{3n-1}\mu_i\mu_{i+1}+K\sum_{n+i=1}^{2n}\mu_i\mu_{i+n}\right]$$
$$=\sum_{p=1}^{2^n}\lambda_p^n A_p(\mu_{3n}\cdot\cdot\mu_{2n+1})B_p(\mu_{2n}\cdot\mu_{n+1}).$$

Now we mulitply these two formulas with each other and sum over the spins of the middle row $\mu_{n+1},\mu_{n+2}\cdot\cdot\mu_{2n}$. We get

$$\sum_{\mu_{n+1},\mu_{n+2}\cdots\mu_{2n}}\exp\left[K\sum_{i=n}^{3n-1}\mu_i\mu_{i+1}+K\sum_{i=1}^{2n}\mu_i\mu_{i+n}\right]$$
$$=\sum_{p=1}^{2^n}\lambda_p^{2n}A_p(\mu_{3n}\cdot\cdot\mu_{2n+1})B_p(\mu_n\cdot\mu_1).$$

We can continue building up that way on the left the complete partition function of the problem. If we dispose of the ends of our screw by making it endless on a torus-shaped body, i.e., if we put

$$\mu_{mn+i}=\mu_i,$$

then we get finally

$$f=\sum_{p=1}^{2^n}\lambda_p^{mn}=\sum_{p=1}^{2^n}\lambda_p^N, \tag{16}$$

where $mn=N$ is the total number of spins present. If m is made sufficiently large while keeping n fixed, all but the largest eigenvalue λ may be neglected.

It has proved necessary in the study of the matrix (14) to avoid all necessary complications of the problem. One such complication is the dependence of λ on two parameters K and C, corresponding physically to the temperature and the magnetic field (Eqs. (3), (4)). It is an obvious simplification to study the square net at zero field only. Unfortunately, the disadvantage of such a step is greater than is immediately obvious, for Eq. (5) shows us that it will be impossible under those circumstances to study even zero-field magnetization as a function of temperature. This reduces us henceforth entirely to the study of the energy and the specific heat. It is fortunate that these quantities show singularities concurrently with the magnetization at the Curie point, and their behavior as functions of T should be sufficient to discern the complete behavior of our model. From (6) and (16) we can thus deduce the following simplified expressions for the energy E and the molar specific heat C

$$E = -\tfrac{1}{2}NJ\frac{d\log\lambda}{dK} \tag{17}$$

and

$$\frac{C}{R} = K^2\frac{d^2\log\lambda}{dK^2}, \tag{18}$$

where R is the gas constant — $R = Nk$.

If we make this assumption of zero magnetic field the parameter C in (14) is zero and the matrix operator is invariant with respect to inversion of all spins. It follows that the eigenvectors are either symmetric or antisymmetric with respect to that operation. Since the largest eigenvalue has only positive components it must belong to the symmetric class

$$A(\mu_n, \mu_{n-1} \cdots \mu_1) = A(-\mu_n, -\mu_{n-1}, \cdots -\mu_1). \tag{19}$$

This follows also from the significance of the A's as probabilities. Equation (19) reduces the number of components of $\mathbf{A}$ from 2^n to 2^{n-1}.

It has proved to be convenient to write the matrix in (14) as a thing with lines and columns. This implies some ordering of the components of $\mathbf{A}$. The following scheme has been adopted.

Take a given arrangement of μ's, for instance

$$+ + - - + + - +,$$

replace all the plus signs by zeros and all the minus signs by ones, thus:

0 0 1 1 0 0 1 0;

read the number thus obtained as if 2 were our decimal unit and let it be the order number of the arrangement. The above arrangement for instance has the order number 50. As an example let us write down the components for the case $n = 5$:

+ + + + + 0	+ + − + + 4	+ − + + + 8	+ − − + + 12
+ + + + − 1	+ + − + − 5	+ − + + − 9	+ − − + − 13
+ + + − + 2	+ + − − + 6	+ − + − + 10	+ − − − + 14
+ + + − − 3	+ + − − − 7	+ − + − − 11	+ − − − − 15

Because of the symmetry condition (19) the components 16 to 31 are identical with 15 to 0. We find generally

component (i) = component $(2^n - i - 1)$.

From here on it is a straightforward matter to bring the matrix 14 into its customary square form. In the case of $n = 5$ we find, for example,

$$\mathfrak{M}(K) = \begin{vmatrix} \alpha & 1 & 0 & 0 & 0 & 0 & 0 & 0 & 0 & 0 & 0 & 0 & 0 & 0 & 0 & 0 \\ 0 & 0 & \alpha & 1 & 0 & 0 & 0 & 0 & 0 & 0 & 0 & 0 & 0 & 0 & 0 & 0 \\ 0 & 0 & 0 & 0 & \alpha & 1 & 0 & 0 & 0 & 0 & 0 & 0 & 0 & 0 & 0 & 0 \\ 0 & 0 & 0 & 0 & 0 & 0 & \alpha & 1 & 0 & 0 & 0 & 0 & 0 & 0 & 0 & 0 \\ 0 & 0 & 0 & 0 & 0 & 0 & 0 & 0 & \alpha & 1 & 0 & 0 & 0 & 0 & 0 & 0 \\ 0 & 0 & 0 & 0 & 0 & 0 & 0 & 0 & 0 & 0 & \alpha & 1 & 0 & 0 & 0 & 0 \\ 0 & 0 & 0 & 0 & 0 & 0 & 0 & 0 & 0 & 0 & 0 & 0 & \alpha & 1 & 0 & 0 \\ 0 & 0 & 0 & 0 & 0 & 0 & 0 & 0 & 0 & 0 & 0 & 0 & 0 & 0 & \alpha & 1 \\ 0 & 0 & 0 & 0 & 0 & 0 & 0 & 0 & 0 & 0 & 0 & 0 & 0 & 0 & \beta & 1 \\ 0 & 0 & 0 & 0 & 0 & 0 & 0 & 0 & 0 & 0 & 0 & 0 & \beta & 1 & 0 & 0 \\ 0 & 0 & 0 & 0 & 0 & 0 & 0 & 0 & 0 & 0 & \beta & 1 & 0 & 0 & 0 & 0 \\ 0 & 0 & 0 & 0 & 0 & 0 & 0 & 0 & \beta & 1 & 0 & 0 & 0 & 0 & 0 & 0 \\ 0 & 0 & 0 & 0 & 0 & 0 & \beta & 1 & 0 & 0 & 0 & 0 & 0 & 0 & 0 & 0 \\ 0 & 0 & 0 & 0 & \beta & 1 & 0 & 0 & 0 & 0 & 0 & 0 & 0 & 0 & 0 & 0 \\ 0 & 0 & \beta & 1 & 0 & 0 & 0 & 0 & 0 & 0 & 0 & 0 & 0 & 0 & 0 & 0 \\ \beta & 1 & 0 & 0 & 0 & 0 & 0 & 0 & 0 & 0 & 0 & 0 & 0 & 0 & 0 & 0 \end{vmatrix}, \quad (20)$$

where $\alpha = e^{2K}$ and $\beta = e^{-2K}$. For other values of n the order of the matrix is a different power of two. Apart from that the matrix stays the same. The non-zero elements occupy a characteristic, turned-over V-shape. On the top arm e^{2K} and 1 are alternating, on the bottom e^{-2K} and 1. Altogether each line and each column has two non-zero elements.

Our statistical problem is now reduced to finding the largest eigenvalue $\lambda(K)$ of the matrix $\mathfrak{M}$:[7]

$$\mathfrak{M}(K)\mathbf{A}(K) = \lambda(K)\mathbf{A}(K) \tag{21}$$

and to studying the behavior of $\lambda(K)$ as the order of $\mathfrak{M}(K)$ increases indefinitely. For only in that limit does the solution correspond to the doubly infinite net. The parameter K represents the temperature variable as defined in Eq. (3).

4. Properties of the Square Net Solution

It has been mentioned already that we have failed to find an exact solution in closed form for our problem. This does not mean, however, that no information concerning λ can be obtained. For some properties of the matrix $\mathfrak{M}$ have a bearing upon the properties of λ.

It is possible, for instance, to replace K by $-K$ in (20) through a simple re-ordering of terms. Using the case $n = 5$ again as a test case this re-ordering matrix $\mathfrak{R}$ takes the following form

---

[7] When using matrix notation we shall use uniformly the following convention: matrices will be printed in German type, vectors in bold face Latin type.

$$\mathfrak{R} = \begin{vmatrix} 0&0&0&0&0&0&0&0&0&0&1&0&0&0&0&0\\ 0&0&0&0&0&0&0&0&0&0&0&1&0&0&0&0\\ 0&0&0&0&0&0&0&0&1&0&0&0&0&0&0&0\\ 0&0&0&0&0&0&0&0&0&1&0&0&0&0&0&0\\ 0&0&0&0&0&0&0&0&0&0&0&0&0&1&0&0\\ 0&0&0&0&0&0&0&0&0&0&0&0&0&0&1&0\\ 0&0&0&0&0&0&0&0&0&0&0&0&0&0&0&1\\ 0&0&0&0&0&0&0&0&0&0&0&1&0&0&0&0\\ 0&0&0&0&0&0&0&0&0&0&0&0&1&0&0&0\\ 0&0&1&0&0&0&0&0&0&0&0&0&0&0&0&0\\ 0&0&0&1&0&0&0&0&0&0&0&0&0&0&0&0\\ 1&0&0&0&0&0&0&0&0&0&0&0&0&0&0&0\\ 0&1&0&0&0&0&0&0&0&0&0&0&0&0&0&0\\ 0&0&0&0&0&0&1&0&0&0&0&0&0&0&0&0\\ 0&0&0&0&0&0&0&1&0&0&0&0&0&0&0&0\\ 0&0&0&0&1&0&0&0&0&0&0&0&0&0&0&0\\ 0&0&0&0&0&1&0&0&0&0&0&0&0&0&0&0 \end{vmatrix}, \qquad (22)$$

and its effect on $\mathfrak{M}(K)$ is given by

$$\mathfrak{R}\mathfrak{M}(K)\mathfrak{R} = \mathfrak{M}(-K). \qquad (23)$$

It is clear that the transformation does not alter the eigenvalue spectrum and hence

$$\lambda(K) = \lambda(-K). \qquad (24)$$

The substitution $K \to -K$ means replacing the ferromagnetic coupling J by an antiferromagnetic coupling of equal strength. It follows that such a coupling would also produce the thermal effects of ferromagnetism. This holds in particular for an eventual Curie point. Many approximate treatments destroy this basic symmetry.

It is easy to extend the definition (22) of $\mathfrak{R}$ to other orders. For if we compare that definition with the list of vector components at the end of §3, we see that it simply changes the sign of every other spin. It should be mentioned in this connection that this invariance does not exist for $n = 2, 4, 6 \cdots$ because in those cases a completely antiferromagnetic pattern cannot be fitted into our screw arrangement of Fig. J.3. For $n = 1, 3, 5 \cdots$ the difficulty does not arise and this partial sequence is sufficient to establish the invariance of λ for $n \to \infty$.

A more interesting result is obtained by operating on $\mathfrak{M}(K)$ with the unitary symmetric matrix $\mathfrak{T}$ which for $n = 5$ has the following form:

$$\mathfrak{T} = \frac{1}{4}\begin{vmatrix}
1 & 1 & 1 & 1 & 1 & 1 & 1 & 1 & 1 & 1 & 1 & 1 & 1 & 1 & 1 & 1 \\
1 & 1 & 1 & 1 & 1 & 1 & 1 & 1 & -1 & -1 & -1 & -1 & -1 & -1 & -1 & -1 \\
1 & 1 & 1 & 1 & -1 & -1 & -1 & -1 & -1 & -1 & -1 & -1 & 1 & 1 & 1 & 1 \\
1 & 1 & 1 & 1 & -1 & -1 & -1 & -1 & 1 & 1 & 1 & 1 & -1 & -1 & -1 & -1 \\
1 & 1 & -1 & -1 & -1 & -1 & 1 & 1 & 1 & 1 & -1 & -1 & -1 & -1 & 1 & 1 \\
1 & 1 & -1 & -1 & -1 & -1 & 1 & 1 & -1 & -1 & 1 & 1 & 1 & 1 & -1 & -1 \\
1 & 1 & -1 & -1 & 1 & 1 & -1 & -1 & -1 & -1 & 1 & 1 & -1 & -1 & 1 & 1 \\
1 & 1 & -1 & -1 & 1 & 1 & -1 & -1 & 1 & 1 & -1 & -1 & 1 & 1 & -1 & -1 \\
1 & -1 & -1 & 1 & 1 & -1 & -1 & 1 & 1 & -1 & -1 & 1 & 1 & -1 & -1 & 1 \\
1 & -1 & -1 & 1 & 1 & -1 & -1 & 1 & -1 & 1 & 1 & -1 & -1 & 1 & 1 & -1 \\
1 & -1 & -1 & 1 & -1 & 1 & 1 & -1 & -1 & 1 & 1 & -1 & 1 & -1 & -1 & 1 \\
1 & -1 & -1 & 1 & -1 & 1 & 1 & -1 & 1 & -1 & -1 & 1 & -1 & 1 & 1 & -1 \\
1 & -1 & 1 & -1 & -1 & 1 & -1 & 1 & 1 & -1 & 1 & -1 & -1 & 1 & -1 & 1 \\
1 & -1 & 1 & -1 & -1 & 1 & -1 & 1 & -1 & 1 & -1 & 1 & 1 & -1 & 1 & -1 \\
1 & -1 & 1 & -1 & 1 & -1 & 1 & -1 & -1 & 1 & -1 & 1 & -1 & 1 & -1 & 1 \\
1 & -1 & 1 & -1 & 1 & -1 & 1 & -1 & 1 & -1 & 1 & -1 & 1 & -1 & 1 & -1
\end{vmatrix}$$

(25)

For arbitrary values of n the matrix can be formed with the help of the following prescription: Suppose you have the matrix for the case $n-1$. Then you get the matrix for the case n by writing down twice each line of the previous matrix and then continuing on the right-hand side alternately in a symetric and antisymmetric fashion. In addition, divide by $\sqrt{2}$ to preserve unitary character. The matrix of order 1 you start out with is the number 1.

If we transform $\mathfrak{M}(K)$ with this matrix $\mathfrak{T}$ we find that it goes over from its form (20) into a matrix having upright V-shape. Upon closer inspection we can write it as follows

$$\mathfrak{T}\mathfrak{M}(K)\mathfrak{T} = \sinh 2K \; \mathfrak{M}^+(K^*),$$

where the cross on $\mathfrak{M}$ stands for transposition, i.e., exchange of lines and columns and K^* is an auxiliary quantity defined through

$$e^{2K^*} = \coth K$$

or more symmetrically

$$\sinh 2K \sinh 2K^* = 1 \qquad (26a)$$

or also

$$\frac{\sinh 2K}{\cosh^2 2K} = \frac{\sinh 2K^*}{\cosh^2 2K^*}. \qquad (26b)$$

The matrix relation can also be brought in a more symmetric form through the substitution

$$\mathfrak{V}(K) = \frac{1}{\cosh 2K} \mathfrak{M}(K), \qquad (27)$$

which yields

$$\mathfrak{T}\mathfrak{V}(K)\mathfrak{T} = \mathfrak{V}^+(K^*), \qquad (28)$$

a relation which, because of

$$\mathfrak{T} = \mathfrak{T}^+, \quad \text{and} \quad \mathfrak{T}^2 = 1$$

is obviously reversible.

Equation (28) has an important bearing upon λ. Neither transformation with $\mathfrak{T}$ nor transposition of $\mathfrak{M}$ does change its eigenvalues and hence we find for λ

$$\frac{\lambda(K)}{\cosh 2K} = \frac{\lambda(K^*)}{\cosh 2K^*}.$$

We shall have occasion later to study this invariant quotient which we define as $x(K)$:

$$x(K) = \frac{\lambda(K)}{\cosh 2K} \tag{29}$$

and for which

$$x(K) = x(K^*). \tag{30}$$

If we interpret (26) with the help of (3) we see that it associates two temperatures with each other. As one of them rises from 0 to ∞ the other one drops from ∞ to 0. The significance of Eq. (30) is then that singular temperatures can only arise in pairs, since every singularity at K will be matched by one at K^*. The only exception to this rule is the temperature for which

$$\sinh 2K_c = 1, \tag{31a}$$

since K_c is its own mate. The numerical value of K_c is found to be

$$K_c = 0.44069. \tag{31b}$$

We conclude therefrom that if λ possesses one singularity only it must occur at the temperature given by (31). It is therefore the only possible location of the Curie point.

It is, of course, impossible to determine by such a symmetry argument the nature of the singularity, since we are not even certain of its existence. For the symmetry property in question is common to the whole sequence of matrices of order $1, 2, 4, 8, 16 \cdots$. All these finite matrices have solutions λ which are continuous throughout. But we can use our information in a negative way to exclude with certainty certain types of possible singularities. Using the definition (29) we find from (17) and (18)

$$E = -NJ \left\{ \tanh 2K - \tfrac{1}{2} \frac{d \log x}{dK} \right\} \tag{32}$$

and

$$\frac{C}{R} = K^2 \left\{ \frac{4}{\cosh^2 2K} + \frac{d^2 \log x}{dK^2} \right\}. \tag{33}$$

Now from (26) we can derive the following relations at the Curie point

$$\left(\frac{dK^*}{dK} \right)_{K=K_c} = -1, \qquad \left(\frac{d^2 K^*}{dK^2} \right)_{K=K_c} = 2\sqrt{2}$$

and therefore from (30)

$$\left(\frac{dx}{dK} \right)_{K_c+0} + \left(\frac{dx}{dK} \right)_{K_c-0} = 0 \tag{34}$$

and

$$\left(\frac{d^2 x}{dK^2} \right)_{K_c+0} - \left(\frac{d^2 x}{dK^2} \right)_{K_c-0} = -\sqrt{2} \left\{ \left(\frac{dx}{dK} \right)_{K_c+0} - \left(\frac{dx}{dK} \right)_{K_c-0} \right\}. \tag{35}$$

Because of (32) Eq. (34) tells us that if the energy is continuous at the Curie point it must have the value

$$E(K_c) = -\tfrac{1}{2}\sqrt{2} NJ \tag{36}$$

which is a rather slow growth from the low temperature minimum $-NJ$ to the high temperature value $E = 0$. In case of a phase transition (36) would at least represent the arithmetic mean of the values it has in the two phases.

Equation (35) tells us that *if the energy at K_c is continuous then the specific heat is also continuous unless it is infinite.* This infinity, if existing, would have to be of a rather symmetric nature since $C(K) - C(K^*)$ must tend to zero as we approach the Curie point. This result will be of great importance in Part II since it is in contradiction to most approximate solutions, which show a jump of the specific heat at the Curie point.

Both these results can be united into a single statement. If we study the sequence of solutions x_n for $n = 1, 2, 3 \cdots$ the specific heat as given by (33) must either tend to infinity at $K = K_c$ or else both energy and specific heat are continuous. The question thus formulated is specific enough to permit numerical treatment. It will occupy Section 6 in Part II and will give strong evidence that the specific heat is actually infinite at the Curie temperature.

We hope that the matrix method of solving statistical problems will be of use to other workers in this field. It may be mentioned in this connection that the treatment of the three-dimensional Ising model can also be reduced to the solution of a sequence of V-shaped matrices. It seems altogether as if a better understanding of such V-matrices might be helpful for statistics. One would expect this not to be too difficult in view of the small number of non-zero elements and their periodic structure. Their main drawback as compared to other simple matrices seems to be that the linear system to which they belong does not have the structure of a system of recursion relations.

In conclusion we want to express our thanks to Mr. E. Montroll for a helpful discussion.

K. Fundamental Difficulties of a Theory of Particles

1. Difficulties in the Classical Electron Theory

We may make a distinction at least for practical purposes between "real" (A) and "formal" (B) difficulties of a theory:
A. A number of facts are not explained, although they are related to explained facts, or they may even be in contradiction with the theory.
B. Lack of coherence, which may become apparent either in theoretical incompleteness, or in logical inconsistency.

Both real and formal incompleteness (which often go together, but not always) are a difficulty, but not an objection. A relatively "open" theory more easily avoids the danger of dogmatising than a relatively "closed" theory. No small part of the resistance which the theory of relativity had to overcome was due to the beautiful closed form of classical mechanics.

As an introduction we first think of the difficulties of the classical electron theory, some 40 or 50 years ago. Dispersion (γ), Fresnel's æther drag, Zeeman effect (α), cathode rays (β) and secondary electrons were interpreted by Lorentz on the basis of the field equations and the equations of motion:[1]

$$m_{\exp}\ddot{x} = K_x + e\left[E_x + \frac{1}{c}(\dot{y}H_z - \dot{z}H_y)\right] + \frac{2e^2}{3c^3}\dddot{x}. \tag{1}$$

The effects α (1896), β (1897) and γ (in 1898) led in the course of this development to the establishment of the existence of the *negative electron*. In Eq.(1) m is the experimental mass, K the binding force (Lorentz usually puts $K_x = -\alpha x$); E and H are the "incident" external electromagnetic field. The last term is the radiation reaction (computed with retardation).

Concerning A: The spectral laws of Balmer and Rydberg were not explained. Are they perhaps even in contradiction with the theory? Nor was the anomalous Zeeman effect explained; the attempts of Lorentz and Voigt were somewhat at variance with the "spirit" of the theory. Concerning B: the theory was certainly incomplete, in view of the unanswerable questions with respect to the structure of the electron. Is the theory perhaps even inconsistent? One should bear in mind that (1) arose from Lorentz's model in the following way (we omit the H term):

---

[1] H. A. Lorentz, *The Theory of Electrons*, Teubner, Leipzig (1909) p. 251ff.

$$m_0\ddot{x} = K_x + eE_x - f\frac{e^2}{ac^2}\ddot{x} + \frac{2e^2}{3c^3}\dddot{x} + g\frac{ae^2}{c^4}\ddddot{x} + \cdots. \qquad (2)$$

Here m_0 is the inertial mass, a the electron radius, while f and g are numerical factors depending on the structure. To obtain Eq.(1), which is the basis of practically all interpretable phenomena, one had to put

1)
$$m_{\text{exp}} = m_0 + f\frac{e^2}{ac^2}. \qquad (3)$$

If the second term on the right, the electromagnetic mass, is of the same order as m_{exp}, then a becomes of the order 10^{-13} cm (conventional electron radius). That this is small compared to atomic dimensions gave rise to a certain amount of satisfaction.[2]

2) E (and H) are practically constant inside the electron. Even to-day the wavelength of the hardest of the known γ rays ($h\nu \sim$ 15 MeV, $\lambda \sim 10^{-11}$ cm) is large compared to 10^{-13} cm.

3) The $\dddot{x}$ term and the following terms are negligible when applied to interpretable phenomena.

The hope of obtaining information about m_0 itself (from the change of mass with velocity) vanished, when relativity theory, and with it Lorentz's contractible electron, were adopted. This suggested that the inscrutability of the structure of the electron should be established as a principle. For instance, it inspired the various modifications of the original theory, in which the electron was "really" a point. One should recall, on the one hand, the attempts and considerations of Mie, Hilbert, Born-Infeld, et al.[3] and, on the other hand, the idea, resumed in 1938 by Dirac,[4] to let a tend to zero and m_0 to minus infinity, in such a way that m_{exp} remains finite and positive. Apart from the fact that so far all these theories have yielded no concrete physical result, it seems to me, that they do not do justice to the "spirit" of the original theory (i.e., the part of "truth" contained in the theory).

Theories like that of Poincaré, with its stress field inside the electron, which keeps the charge together, still make use of a structure but try to give a relativistic justification for it. I do not wish to point out the difficulties encountered in this direction. They are just as "academic" as the questions relating to the point electron. For there was already an approximate, asymptotic theory (starting from (1)), which is structure independent: in this theory the electron was characterized by the charge e and the mass m_{exp} and it was possible to interpret the facts to a certain extent.

---

[2] V. F. Weisskopf, *Phys. Rev.* **56**, 72 (1939).
[3] M. Born and L. Infeld, *Proc. Roy. Soc.* **A143**, 410 (1934); **A144**, 425 (1934); **A147**, 522 (1934); E. Feenberg, *Phys. Rev.* **47**, 148 (1935).
[4] P. A. M. Dirac, *Proc. Roy. Soc.* **A167**, 148 (1938); M. H. L. Pryce, *Proc. Roy. Soc.* **A168**, 389 (1938); L. Infeld and P. R. Wallace, *Phys. Rev.* **57**, 797 (1940).

2. Difficulties in the Modern Theory of Particles

Many new particles have been added since the discovery of the negaton. First of all appeared the positive nuclei in 1911, of which, however, only the proton could claim the title of elementary particle. The whole edifice of quantum theory arose. The fact that the conventional electron radius is small compared to atomic dimensions turned out to be a consequence of the smallness of $e^2/\hbar c \cong 1/137$. It is true that wavefunctions ψ are used, but the fact that these are functions of the "coordinates of the particles" indicates that in quantum theory, just as previously in Eq. (1), one operates with point particles: one has to do — at least that is what one thinks — with a consistent quantization of the above mentioned asymptotic theory. A huge quantity of experimental material is interpreted by the modern theory. The particles are characterized not only by their charge and mass, but also by their spin; the latter shows up in the appearance of a set of wavefunctions $\psi_1, \psi_2, \ldots$ instead of a single one, more or less in analogy with the classical description of the electromagnetic field by six space-time functions. We ask again about the real and formal difficulties A and B.

A. Are there facts with which the theory cannot cope, which may even be in contradiction with it, and a closer investigation of which will perhaps throw light on the nature of the particles? The last question implies for example the question whether the value of $e^2/\hbar c$ can be "understood", and whether it will be possible to say more concerning the spin multiplicities and the masses than is known at present. The somewhat surprising answer to this question is: *probably*. The situation appears to be as follows.

α). Facts that do not involve explicitly the structure of the nucleus. The results from double electron scattering[5] seem to be in contradiction with the theory. Repetition of these experiments is desirable. Furthermore: it is well known that so far great mathematical difficulties made a precise calculation of the wavelengths and intensities of spectral lines for atoms or molecules well-nigh impossible, even if the corresponding Schrödinger equation were exactly known. For light atoms the latter is known well enough. Computations like those of Hylleraas make a contradiction between facts and theory improbable in this case. For heavy atoms, however, one is faced with a many-body problem, in which relativistic effects yield an important contribution. Now the theory is not yet able to state the Hamiltonian more exactly than to $(v/c)^2$ terms. Only when the reduction to a one-electron problem is justifiable can one probably rely on Dirac's theory. This means that $(v/c)^4$ effects in the wavelengths of the spectral lines (and occasionally even $(v/c)^2$ effects in their intensities) must exist, against which the theory is still powerless. Unfortunately, because of the above mentioned mathematical difficulties, it has so far been impossible to use this experimental source to obtain more data about the relativistic interactions of electrons.

---

[5] See L. Rosenfeld, *Ned. Tijds. Natuurk.* **10**, 53 (1943).

β). There are many facts which have a bearing on the nuclear structure and which the theory cannot deal with. From Rosenfeld's lecture we know, however, that the theory is here—even in non-relativistic approximation—still far from unambiguous. Was it not possible recently for Pais to improve the theory of the photo-effect of the deuteron? Hence it may be presumed, but it is not certain, that the known facts are already a source from which one may draw essentially new knowledge about the theory of particles; it may be possible to include them all within the prevailing scheme of wavefunctions and Hamiltonian. It must be granted that for problems like the one of the magnetic moment of the proton such an optimistic, or rather pessimistic, expectation is likely to be wrong.

B. In its theoretical aspect the present-day quantum theory of particles is not only open, and unfinished, nay, there is even a lack of logical consistency. This lies in the notorious *divergencies* of sums or integrals, which occur in the calculation of certain effects and clearly demonstrate that the theory is wrong. But still not wholly useless—which is the remarkable thing in the present situation. Guided by physical or other considerations, one has been able to draft so-called *subtraction prescriptions*, which make the mentioned sums or integrals convergent. Example: in a situation in which there is only one light quantum present in space the expression for E^2 in a space point diverges. One subtracts the expression for E^2 at that same point for the case where no light quantum is present. The procedure to accomplish this subtraction comes readily to mind; the result is a convergent expression, which we use confidently, not least of all because it exhibits automatically the desired correspondence with the analogous classical situation. In many other cases, though not in this case, the situation in quantum theory resembles very much the divergence of the electromagnetic mass of an electron if one lets its radius tend to zero. Concerning a famous divergence noticed by Dirac (viz., in the expression for $E_1 - E_2 - h\nu$ in a Bohr radiation jump) it was shown by Serpe,[6] that it is exactly paralel to the divergence of the electromagnetic mass; but for other divergences, in particular those which occur in the further elaboration of hole theory, such a correspondence is no longer present. Here the electromagnetic mass, for instance, does not tend to infinity as $1/a$, but as $\log(1/a)$. One assertion can be made: the hope that the difficulties of a classical relativistic electron theory (dualism: particle-field) would disappear of themselves through the mathematical mechanism of the quantization prescriptions has turned out to be in vain. One may say that it is the "fault" of the relativity theory, which only permits "contact forces" (and no direct forces at distance) between particles, for this is the quantum theoretical translation of the idea of the relativistic field theory of classical physics (where the interaction between two fields is described by their values and the values of their derivatives in one and the same space-time point).

---

[6] J. Serpe, *Physica* **7**, 133 (1940); **8**, 226 (1941); W. Opechowski, *Physica* **8**, 161 (1941).

The unfortunate thing is that this relativistic character of the theory is often jeopardized, or parhaps destroyed, by the subtraction prescriptions now in vogue.

3. Attempts to Solve the Difficulties

At present three currents can be distinguished among the attempts to attack the difficulties methodically. Among the methods I do not include the well-known program: the x, y, z, t description of the relativity theory is to be dropped and replaced with something more profound involving some "smallest length" ($\sim 10^{-13}$ cm) and a "smallest time". That is not to say that it should not be valuable to keep this program in mind, and that the eventual solution might not be regarded as a realisation of it, but I have (in spite of certain authors) the feeling that it does not exhibit the necessary promise as a starting point for a methodical investigation; it is for me, so to say, too mathematical, has too few contacts with experiment.

These three currents are thus as follows:

I. The introduction of more new fields (with their associated particles), with the purpose that the nasty diverging expressions will cancel each other without additional subtraction prescriptions, so that the result is finite. Bopp and Stueckelberg[7] worked in this direction, by considering the electron as source not merely of an electromagnetic field, but besides of a new field, of the Yukawa type, with short range. On the one hand these theories are somewhat fictitious and, not to mention other difficulties, not all divergencies are cut out. On the other hand, this kind of investigation yields various suggestions, which might turn out of great value. For instance, they may throw some light on the question of the difference between the proton and the neutron masses.

II. The contemplation of the classical particle theory, in order to derive from the correspondence postulate starting points for an improved quantum theory. To this belong in the first place the above mentioned theories of Born-Infeld and Dirac, which modify or re-interpret the classical theory in such a way that they permit to work with a real point particle. These classical theories, however, turn out to be extremely difficult, if not impossible, to quantize and I suspect that their lack of success is because they violate the spirit of the original classical theory. In the second place I here wish to mention the investigations of Opechowski and myself. These investigations brought to light the fact that the quantized interaction between an electrically charged particle and the electromagnetic field, which is in fact the prototype after which all later descriptions of interactions have been modelled, does not exhibit the full correspondence with Lorentz' classical electron theory, which after all one ought to have required. When comparing Eq. (1) with Eq. (2), one may say that the factor m_{exp} in (1) arose from a classical subtraction prescription

---

[7] F. Bopp, *Ann. Phys.* **38**, 345 (1940); **42**, 573 (1942); E. C. G. Stueckelberg, *Helv. Phys. Acta* **14**, 51 (1941).

relative to the electromagnetic momentum, which is embodied in Eq. (3). The feature of the classical theory, which is closely related that E and H in (1) represent the external and not the total electromagnetic field (which diverges at the point of the electron), has been neglected, so to say, in the literature, with the consequence that the usually advocated quantum theory of electron and radiation is not directly related to the asymptotic theory of Lorentz, whose results one would have liked to carry over into a quantized form. The fact that in spite of this numerous results could be derived in a satisfactory way (Dirac's derivation of Einstein's A's and B's and of the scattering formulæ) is due to details into which I cannot now enter. Typical for the conventional quantum theory is that it is incapable of formulating a stationary state in which purely monochromatic light is being scattered by a bound or free electron, whereas in the classical electron theory this is a very simple problem. It also fails in other respects (example: Dirac's calculation about the shift of spectral lines, i.e., the already mentioned calculation of $E_1 - E_2 - h\nu$). We have been able to give, in a modest (among other things non-relativistic) approximation, a formulation of the interaction of electron and radiation which in our opinion is more correct, and which satisfies the correspondence postulate. Further investigations must decide whether, by continuing in this direction, one can develop a theory which yields a satisfactory, relativistically invariant formulation of the radiation-electron problem.

III. Heisenberg's recent investigations[8] concerning the possibility of a relativistic description of the interaction that is not based on the use of a Hamiltonian with interaction terms in a Schrödinger equation. Heisenberg considers only free particles and introduces a formalism ("scattering matrix") by means of which the result of a short interaction (scattering) between these particles can be described. Formerly the scattering matrix could be derived from the Hamiltonian, but now we are to consider the scattering matrix as fundamental. We do not care whether a Schrödinger equation for particles in interaction exists; we do not care which correspondence requirements exist and how the scattering matrix can obey them. It is interesting that the scattering matrix is also able in principle to answer the question, in which stationary states the particles considered can be bound together. These are related to the existence and the position of zeros and poles of the eigenvalues of the scattering matrix, considered as a complex function of its arguments. Heisenberg could already give a (very simple) model of a two-particle system, in which a perfectly sharply relativistically determined stationary state occurs, while there are no divergence difficulties whatsoever.

However promising, this is still only a beginning, and in particular with regard to a correct description of the electromagnetic fields of photons I expect difficulties, which the investigations in this direction will have to overcome. Fortunately, Heisenberg's program is still open in several respects, and one

---

[8] W. Heisenberg, *Zs. Phys* **120**, 513, 673 (1942).

may perhaps expect a great deal from a fortunate combination with further ideas.

L. The Behavior of Macromolecules in Inhomogeneous Flow

Abstract. Following up the ideas of W. Kuhn's skein theory, J. J. Hermans has recently elaborated a theory of the properties of dilute solutions of highly polymerized molecules. Hermans describes the behavior of linear molecules by the diffusion of their end points. He finds that the contribution of a single molecule to the viscosity coefficient as well as to the double refraction of the solution is proportional to the square of the degree of polymerization, in agreement with the experiments. In the present article we investigate the statistical behavior of the individual links of the solved molecules. In this way we arrive at results equivalent with those of Hermans. Our method, however, can also be applied to molecules which possess branching points and rings.

1. Introduction

In connection with Kuhn's skein theory, Hermans[1] developed a simplified method to describe the influence of macromolecules on viscosity and double refraction. In this method the behavior of the molecule is characterized by the diffusion of the end points. This can be described as if two mass points which exert a fictive elastic force upon each other are diffusing in the liquid. The equilibrium distribution of the vector joining the end points, showing spherical symmetry in a liquid at rest, is deformed under the influence of the flow. It is assumed that this flow does not alter the elastic force. By means of a relatively simple calculation Hermans shows that the contributions of a molecule both to the viscosity and to the birefringence of flow are proportional to the *square* of the number of elementary links, in conformity with experiments (Staudinger, Signer).

In the following we attempt to develop a more detailed theory. Our method also enables us to give a treatment of macromolecules which possess branching points and rings.

---

[1] J. J. Hermans, *Physica* **10**, 777 (1943).

2. The Behavior of a System of Mutually Connected Particles in a Flow Derived from a Velocity Potential

Let the systematic force which the liquid exerts on the ith particle be $-\zeta_i \mathbf{v}_i$, where $\mathbf{v}_i$ is the velocity of the particle with respect to the liquid. This velocity can be written $-\mathbf{v}_i' + \mathbf{v}_i''$, if $\mathbf{v}_i'$ represents the velocity of the liquid and $\mathbf{v}_i''$ that of the particle. If the streaming is irrotational and stationary, and if its velocity potential is $\Psi(\mathbf{r})$, we have $\mathbf{v}_i' = -\nabla \Psi(\mathbf{r}_i)$. Consequently, the force which the liquid exerts on the ith particle can be written

$$X_i = -\frac{\partial U}{\partial x_i} - \zeta_i \dot{x}_i + X_i', \qquad (y), \qquad (z), \tag{1}$$

$$U = \sum_i \zeta_i \Psi(\mathbf{r}_i). \tag{2}$$

Here X_i' is a random force of such a nature that for a free particle it would maintain the equipartition. The influence of the pressure forces which has here been neglected will be considered in Section 6.

The equations of motion are the same as those which would apply to our system in a liquid at rest if it were subject to a field of force with potential energy U. Introducing generalized coordinates $q^1, q^2 \cdots q^s$ and conjugated momenta $p_1; p_s$ and writing

$$T = \tfrac{1}{2} a_{kl} \dot{q}^k \dot{q}^l = \tfrac{1}{2} a^{kl} p_k p_l, \qquad (p_k = a_{kl} \dot{q}^l) \tag{3}$$

for the kinetic energy T, the molecules will be distributed in p, q-space according to Boltzmann:

$$W \prod_k dp_k\, dq_k = \text{const.}\, e^{-(T+U)/\theta} \prod_k dp_k\, dq_k, \tag{4}$$

where U must be regarded as a function of the q's. Boltzmann's constant times the absolute temperature has been called θ.

Average values of $\dot{q}$ and q can now be calculated in principle. A function which is linear in the $\dot{q}$'s (or p's) will on the average be zero. The average value of a function which is quadratic in $\dot{q}$ can simply be found by first reducing it to the average function of a function $F(q_1 \cdots q_s)$ of the q's alone by means of the relation

$$\langle \dot{q}^k \dot{q}^l \rangle_{\text{Av}} = a^{kl} \theta. \tag{5}$$

Averaging a function $F(q_1 \cdots q_s)$ of the q's alone amounts, as is well known,[2] to computing the ratio

$$\langle F(q_1 \cdots q_s) \rangle_{\text{Av}} = \int F e^{-U/\theta} (A)^{1/2} \prod_k dq^k \bigg/ \int e^{-U/\theta} (A)^{1/2} \prod_k dq^k, \tag{6}$$

---

[2] J. W. Gibbs, *Elementary Principles of Statistical Mechanics*, Chap. 6.

where A is the determinant of the coefficients a_{kl}:

$$A = \text{Det}|a_{kl}|. \tag{7}$$

3. Pearl-Necklace Model of a Molecule

As a model for a macromolecule we consider a number of particles with coordinates $\mathbf{r}_1(x_1 y_1 z_1) \cdots \mathbf{r}_i \cdots \mathbf{r}_N$, in which each two successive particles are connected by a weightless rod of length L. Let the mass m and the frictional constant ζ be the same for all these particles. The direction cosines of the ith "link", i.e., of the rod which connects the ith particle with the $(i+1)$st particle, will be called $\alpha_i, \beta_i, \gamma_i$ (vector $\boldsymbol{\omega}_i$). If x, y, z ($\mathbf{r}$) are the coordinates of the center of gravity of our model, one has

$$x_i = x + \frac{L}{N}\Big(\alpha_1 + 2\alpha_2 + \cdots i\alpha_i - [N-i-1]\alpha_{i+1}$$
$$- [N-i-2]\alpha_{i+2} \cdots - \alpha_{N-1}\Big), \tag{8}$$

$$\sum_1^N \dot{x}_i^2 = \frac{L^2}{N} \sum_{\mu,\nu} g_{\mu\nu} \dot{\alpha}_\mu \dot{\alpha}_\nu + N\dot{x}^2, \tag{9}$$

$$g_{\mu\nu} = \begin{cases} \mu(N-\nu), & \nu \geq \mu, \\ \nu(N-\mu), & \mu \geq \nu, \end{cases} \tag{10}$$

$$T = \frac{mL^2}{2N} \sum_{\mu\nu} g_{\mu\nu}(\dot{\boldsymbol{\omega}}_\mu \cdot \dot{\boldsymbol{\omega}}_\nu) + \tfrac{1}{2}Nm(\dot{x}^2 + \dot{y}^2 + \dot{z}^2). \tag{11}$$

The simplest model is obtained if it is assumed that the two rods which meet in a particle are completely free to rotate with respect to each other; a link would then be more or less analogous to W.Kuhn's "statistical chain-element". In that case we introduce, for instance, in addition to x, y, z the polar angles $\vartheta_\mu, \varphi_\mu$:

$$\alpha_\mu = \cos\vartheta_\mu = c_\mu, \quad \beta_\mu = \sin\vartheta_\mu \cos\varphi_\mu = s_\mu C_\mu, \quad \gamma_\mu = \sin\vartheta_\mu \sin\varphi_\mu = s_\mu S_\mu.$$

The total number of coordinates is $2N+1$. The coefficients a_{kl} in the kinetic energy are easily derived from (11) using the following formulas:

$$(\dot{\boldsymbol{\omega}}_\mu \cdot \dot{\boldsymbol{\omega}}_\nu) = (s_\mu s_\nu + c_\mu c_\nu C_{\mu\nu})\dot{\vartheta}_\mu \dot{\vartheta}_\nu + s_\mu s_\nu C_{\mu\nu}\dot{\varphi}_\mu \dot{\varphi}_\nu$$
$$+ c_\mu s_\nu S_{\nu\mu}\dot{\vartheta}_\mu \dot{\varphi}_\nu + S_\mu c_\nu S_{\mu\nu}\dot{\varphi}_\mu \dot{\vartheta}_\nu, \tag{12}$$
$$C_{\mu\nu} = \cos(\varphi_\mu - \varphi_\nu), \qquad S_{\mu\nu} = \sin(\varphi_\mu - \varphi_\nu).$$

With 2, 3, 4, 5 links, these formulas lead to the following expressions for the determinant A of the coefficients a_{kl} (the angle between the μth and the $(\mu+1)$st link is denoted by ψ_μ):

$N-1$	A (apart from a numerical factor)
2	$s_1^2 s_2^2 (1 - \frac{1}{4}\cos^2\psi_1)$
3	$s_1^2 s_2^2 s_3^2 (1 - \frac{1}{4}[\cos^2\psi_1 + \cos^2\psi_2])$
4	$s_1^2 s_2^2 s_3^2 s_4^2 (1 - \frac{1}{4}[\cos^2\psi_1 + \cos^2\psi_2 + \cos^2\psi_3]$ $+ \frac{1}{16}\cos^2\psi_1 \cos^2\psi_3)$
5	$s_1^2 s_2^2 s_3^2 s_4^2 s_5^2 (1 - \frac{1}{4}[\cos^2\psi_1 + \cos^2\psi_2 + \cos^2\psi_3 + \cos^2\psi_4]$ $+ \frac{1}{16}[\cos^2\psi_1 \cos^2\psi_3 + \cos^2\psi_1 \cos^2\psi_4 + \cos^2\psi_2 \cos^2\psi_4])$

(13)

For $N - 1 = 5$ we have calculated A only for the special case that all links are lying in the same plane; for $N - 1 = 2, 3, 4$, however, the formulas (13) are of general applicability. We have not yet succeeded in deriving a general expression for A holding for arbitrary values of N, but its form can be surmised.

In a liquid at rest the probability that the probability that ϑ_μ and φ_μ have values between θ_μ and $\vartheta_\mu + d\vartheta_\mu$ and φ_μ and $\varphi_\mu + d\varphi_\mu$, respectively, is

$$\text{const.} A^{1/2} \prod_\mu d\vartheta_\mu \, d\varphi_\mu. \tag{14}$$

From the formulas (13) it is seen how the orientations of the various links are correlated: The correlation is strongest between successive links. The probability that they are perpendicular to each other is somewhat larger than that they are parallel or antiparallel; in the case of two links these probabilities are in the ratio $(4/3)^{1/2} \cong 1.15$.

A somewhat less artificial model would be one in which a given constant value ψ_0 (for instance the supplement of the tetrahedron angle 109.5°) is assigned to the angle between two successive links. The number of degrees of freedom in that case amounts to $3 + 2 + N - 2 = N + 3$, but it is difficult to introduce generalized coordinates in a simple manner. If desired, however, one can start from the former case, i.e., retain the complete number of $2N+1$ degrees of freedom, but complete this by introducing a potential energy

$$U' = \tfrac{1}{2} K \left\{ (\psi_1 - \psi_0)^2 + (\psi_2 - \psi_0)^2 + \cdots (\psi_{N-2} - \psi_0)^2 \right\}. \tag{15}$$

We would thus introduce explicitly the bending vibrations of the valence bonds instead of replacing these by rigid bonds. It is clear that the incomplete freedom of rotation of a link with respect to another link from which it is separated by one neighbor could, if desired, also be accounted for by a suitable potential energy.

4. Internal Friction in a Non-Uniform Flow

In a volume element of sufficiently small size the Poiseuille flow which occurs in the experiments is of the well-known laminar type

$$u = \kappa y, \quad v = 0, \quad w = 0. \tag{16}$$

This flow, however, is not irrotational and we will rather replace it by

$$u = -\frac{\partial \Psi}{\partial x} = \tfrac{1}{2}\kappa y, \quad v = -\frac{\partial \Psi}{\partial y} = \tfrac{1}{2}\kappa x, \quad w = -\frac{\partial \Psi}{\partial z} = 0, \quad \Psi = \tfrac{1}{2}\kappa xy. \tag{17}$$

Superposing a uniform rotation $u = \tfrac{1}{2}\kappa y, v = -\tfrac{1}{2}\kappa x$ on this flow, one obtains the laminar flow (16). If κ is not too large, the effect of this rotation can be neglected, as will be shown in Section 6.

According to the considerations of Section 2, the effect of the flow (17) can be described by a potential energy U (compare Eq. (2)):

$$U = -\tfrac{1}{2}\zeta\kappa \sum_{1}^{N} x_i y_i. \tag{18}$$

Omitting the coordinates of the center of gravity, this becomes

$$U = -\frac{\kappa \zeta L^2}{2N} \sum_{\mu\nu} g_{\mu\nu} \alpha_\mu \beta_\nu,$$

since it is clear that the coefficients $g_{\mu\nu}$ which occur in the dependency of $\sum xy$ on the α's and β's are the same as those which occur in the dependency of $\sum \dot{x}^2$ on the $\dot{\alpha}$'s (see Eq. (9)).

The influence of the flow is described by the Boltzmann factor $e^{-U/\theta}$ which for sufficiently small κ can be replaced by

$$e^{-U/\theta} \cong 1 - \frac{U}{\theta}. \tag{19}$$

If $P_{xx}, P_{xy}, \cdots$ represents the stress tensor in the liquid, the coefficient of internal friction η is defined by

$$P_{xy} = \eta \left(\frac{\partial u}{\partial y} + \frac{\partial v}{\partial x} \right) = \kappa \eta.$$

We will now calculate the average contribution δP_{xy} of a single macromolecule per unit volume to the value of P_{xy}. The contribution $\delta \eta$ per molecule to the viscosity will then be given by

$$\delta \eta = \frac{\delta P_{xy}}{\kappa}. \tag{20}$$

Two effects contribute to δP_{xy}, the motion of the particles and the tension in the links. The former amounts to

$$\delta_I = -m \sum \langle \dot{x}_i \dot{y}_i \rangle_{\text{Av}},$$

because the probability that in unit time a plane of unit area perpendicular to the y axis is traversed by the ith particle in the direction of the negative y axis is $-\dot{y}_i$, and the momentum transferred in this process is $m\dot{x}_i$.

The second contribution results from the tensions in the links. If S_μ is the tension in the μth link, this tension contributes an amount $S_\mu \alpha_\mu$ to δP_{xy} if the link intersects the plane mentioned. The probability for the occurrence of such an intersection is $L\beta_\mu$ per molecule; hence

$$\delta_{II} = L \sum_\mu \langle S_\mu \alpha_\mu \beta_\mu \rangle_{\text{Av}}.$$

If the total force on the ith mass point resulting from the tension is called X_i'', Y_i'', Z_i'', we have

$$X_i'' = \sum_j S_i^{(j)} \alpha_i^{(j)}, \qquad Y_i'' = \sum_j S_i^{(j)} \beta_i^{(j)}, \tag{21}$$

where the sum extends over all links j which meet in the ith particle. Usually there will be only two of such links, but in a ramified molecule — for which this calculation also holds good — there can be three or more. Considering now the virial-like expression

$$V = \sum_i X_i'' y_i \left(= \sum_i Y_i'' x_i \right),$$

in which we sum over all the particles, and substituting (21), each $S_i^{(j)} \alpha_i^{(j)}$ will occur twice, viz., for those two particles which form the ends of the links concerned, but in these two cases the $\alpha_i^{(j)}$'s are of opposite sign. Consequently we may write

$$V = \sum_\mu S_\mu \alpha_\mu (y_i - y_i') = -\sum_\mu S_\mu \alpha_\mu \cdot L\beta_\mu,$$

where y_i and y_i' characterize the two ends of the μth link, and where α_μ is the direction cosine of the link $i \to i'$. Now, if X_i represents the force which the ith particle experiences from the liquid, we have according to Eq. (1):

$$X_i'' = m_i \ddot{x}_i - X_i = m_i \ddot{x}_i + \frac{\partial U}{\partial x_i} + \zeta_i \dot{x}_i - X_i'.$$

In the average value of $\sum_i X_i'' y_i$ in Boltzmann's distribution the contributions of the last two terms in the third member vanish, and we finally obtain

L. The Behavior of Macromolecules in Inhomogeneous Flow

$$\delta_{II} = -\sum_i \langle X_i'' y_i \rangle_{\text{Av}} = -m \sum_i \langle \ddot{x}_i y_i \rangle_{\text{Av}} - \sum_i \left(\frac{\partial U}{\partial x_i} y_i \right)_{\text{Av}},$$

$$\delta P_{xy} = \delta_I + \delta_{II} = -m \sum_i \left(\langle \dot{x}_i \dot{y}_i \rangle_{\text{Av}} + \langle \ddot{x}_i y_i \rangle_{\text{Av}} \right) - \sum_i \left(\frac{\partial U}{\partial x_i} y_i \right)_{\text{Av}}. \quad (22)$$

The sum of the first two terms in the third member represents the derivative with respect to time of the average value of $-m \sum_i \dot{x}_i y_i$ and therefore contributes nothing. For the remaining expression one can write $\sum_i \langle X_i''' u_i \rangle_{\text{Av}}$, if X_i''' represents the systematic force resulting from the streaming liquid. That this expression is equal to $\kappa \delta \eta$ also follows immediately from the considerations given by J. M. Burgers[3] in connection with Einstein's work. Burgers studied the changes which forces acting on the liquid in a certain region (here $-X_i'''$) will bring about in the flow of the liquid outside that region, and arrived at a formula which is equivalent to ours.

With the aid of (18) and (20) it follows from (22) that

$$\delta\eta = \tfrac{1}{2}\zeta \sum_i \langle y_i^2 \rangle_{\text{Av}} = \tfrac{1}{6}\zeta \langle (x_i^2 + y_i^2 + z_i^2) \rangle_{\text{Av}} = \frac{\zeta L^2}{6N} \sum_{\mu\nu} g_{\mu\nu} \langle (\mu\nu) \rangle_{\text{Av}}. \quad (23)$$

To a sufficient approximation we may here take the average over the Boltzmann distribution *in the liquid at rest* ($U = 0$ in Eq. (4)). In the third member we have introduced the abbreviated notation

$$(\omega_\mu \omega_\nu) = (\mu\nu)$$

for the cosine of the angle between the μth and the νth link.

An accurate calculation of the expression (23) would require a precise knowledge of the correlation between the orientations of two different links. Now, in the free pearl necklace these correlations are small for neigboring links and negligible for more remote links. Moreover, in those cases with a small number of links where the calculation was carried out, the correlation only depends on the second power of the cosine of the angle between the two links, which means that $\langle (\mu\nu) \rangle_{\text{Av}}$ becomes zero. It would seem that this also applies to arbitrary N-values. Since $\langle (\mu\mu) \rangle_{\text{Av}} = 1$, we may therefore write

$$\delta\eta = \frac{\zeta L^2}{6N} \sum_{\mu=1}^{N-1} g_{\mu\nu} = \frac{\zeta L^2}{6N} \sum \mu(N-\mu) = \frac{\zeta L^2}{6N} \frac{(N-1)N(N+1)}{6},$$

or, if N is assumed to be large:

$$\delta\eta \cong \frac{\zeta L^2 N^2}{36}. \quad (24)$$

---

[3] J. M. Burgers, *"Viscosity Report II"*, Proc. Roy. Acad. Amsterdam [1] **16**, 128 (1938).

As in the analogous formula derived by Hermans, this equation shows $\delta\eta$ to be proportional to the square of the number of links, in conformity with Staudinger's viscosity rule.

If the free rotation of the links is of a nature such as that known from organic chemistry, i.e., if two successive links always enclose a given angle α, we get

$$\langle(\mu\mu)\rangle_{\text{Av}} = 1, \qquad \langle(\mu, \mu \pm 1)\rangle_{\text{Av}} = \cos\alpha,$$
$$\langle(\mu, \mu \pm 2)\rangle_{\text{Av}} = \cos^2\alpha, \qquad \langle(\mu\nu)\rangle_{\text{Av}} = \cos^{|\mu-\nu|}\alpha.$$

To a sufficient approximation we may then write

$$\delta\eta \cong \frac{\zeta L^2 N^2}{36} \cot^2 \tfrac{1}{2}\alpha. \tag{25}$$

If the links are connected by ordinary C–C bonds, α is the supplement of the tetrahedron angle, $\cos\alpha$ amounts to $\tfrac{1}{3}$, $\cot^2 \tfrac{1}{2}\alpha$ equals 2, and the influence on the viscosity becomes, therefore, twice as large as in the case of the free pearl necklace.

5. Double Refraction in Non-Uniform Flow

The birefringence is described by a dielectric tensor ϵ_{xx}, $\epsilon_{xy} = \epsilon_{yx}$, etc., which is related to the polarization tensor $\bar{p}_{xx}$, $\bar{p}_{xy}$ resulting from the influence of a single molecule per unit volume according to the formulas

$$\epsilon_{xx} = \epsilon_0 + 4\pi G \bar{p}_{xx}, \qquad \epsilon_{xy} = 4\pi G \bar{p}_{xy}, \qquad \text{etc.} \tag{26}$$

Here $\epsilon_0^{1/2}$ is the refractive index of the liquid and G represents the number of macromolecules in unit volume. To simplify matters, we assume that the contribution of each link to the polarization tensor shows cylindrical symmetry round the direction of the link and can therefore be written

$$p_\lambda = a \begin{pmatrix} 1 & 0 & 0 \\ 0 & 1 & 0 \\ 0 & 0 & 1 \end{pmatrix} + b \begin{pmatrix} \alpha_\lambda^2 - \tfrac{1}{3} & \alpha_\lambda \beta_\lambda & \alpha_\lambda \gamma_\lambda \\ \beta_\lambda \alpha_\lambda & \beta_\lambda^2 - \tfrac{1}{3} & \beta_\lambda \gamma_\lambda \\ \gamma_\lambda \alpha_\lambda & \gamma_\lambda \beta_\lambda & \gamma_\lambda^2 - \tfrac{1}{3} \end{pmatrix}. \tag{27}$$

The total polarization tensor of the molecule becomes

$$p = \sum_{1}^{N-1} p_\lambda.$$

To a first approximation the average contribution resulting from a is independent of κ and simply amounts to

$$(N-1)a \begin{pmatrix} 1 & 0 & 0 \\ 0 & 1 & 0 \\ 0 & 0 & 1 \end{pmatrix}.$$

The average contribution which results from the anisotropy constant b will to a first approximation be proportional to κ and originate from the term $-U/\Theta$ in the expression (19).

From considerations of rotational invariance it is at once obvious that only the xy component of p in (27) gives a contribution which differs from zero. Its average value per molecule per unit volume will be

$$\bar{p}_{xy} = b \sum_{1}^{N-1} \langle (\alpha_\lambda \beta_\lambda)^2_{\text{Av}} \rangle_{\text{Av}} = b \frac{\kappa \zeta L^2}{2\Theta N} \left\langle \left(\sum_\lambda \alpha_\lambda \beta_\lambda \right) \left(\sum_{\mu\nu} g_{\mu\nu} \alpha_\mu \beta_\nu \right) \right\rangle_{\text{Av}}. \quad (28)$$

The angular brackets enclosing the product in the third member of (28) mean that we are to integrate over a Boltzmann distribution $e^{-T/\Theta} \prod dp\, dq$. Since no $\dot{q}$ occurs, we can at once average over the q's with $A^{1/2}$ as weight factor:

$$\langle F \rangle_{\text{Av}} = \int F A^{1/2} \prod dq \bigg/ \int A^{1/2} \prod dq. \quad (29)$$

It is suitable first to use the following formula which follows immediately from rotational invariance and in which the choice of the coordinate system no longer plays a part:

$$\tfrac{1}{2} \langle \alpha_\lambda \beta_\lambda (\alpha_\mu \beta_\nu + \alpha_\nu \beta_\mu) \rangle_{\text{Av}} = \frac{1}{15} \langle \{ \tfrac{3}{2}(\lambda\mu)(\lambda\nu) - \tfrac{1}{2}(\mu\nu) \} \rangle_{\text{Av}}. \quad (30)$$

We thus obtain for the polarization tensor per molecule:

$$\bar{p}_{xx} = \bar{p}_{yy} = \bar{p}_{zz} \cong Na, \quad \bar{p}_{xz} = \bar{p}_{yz} = 0, \quad \bar{p}_{xy} \cong C\kappa b,$$

$$C = \frac{\zeta L^2}{60 \Theta N} \sum_{\lambda \mu \nu} g_{\mu\nu} \{ 3 \langle (\lambda\mu)(\lambda\nu) \rangle_{\text{Av}} - \langle (\mu\nu) \rangle_{\text{Av}} \}. \quad (31)$$

This shows that the polarization tensor is deformed as a result of the flow; it has become "tri-axial." For electrical oscillations parallel to z the polarization has remained NA, as in the liquid at rest. In the directions perpendicular to z which make angles of $45°$ and $-45°$ with the x and y axis, however, the polarization has respectively increased and decreased by an amount $C\kappa b$.

The constant C is proportional to ζ, and therefore to the viscosity of the liquid, and inversely proportional to the absolute temperature. Above all, however, we are interested in its dependence on the number of links in the molecule. To discuss this dependence, we note that according to (31) everything depends on the correlation between the orientations of different links, in as far as λ, μ, and ν are different. For links which are far apart, this correlation is practically negligible. In the model of completely free rotation of neighboring links with respect to each other only the correlation between two adjacent links differs perceptibly from zero. The main terms are then doubtless those for which μ and ν in (31) are exactly or almost equal. In this case the average value of $3(\lambda\mu)(\lambda\nu)$ practically cancels that of $(\mu\nu)$ if λ differs perceptibly from μ, and the sum in formula (31) becomes approximately

$$2 \sum_{1}^{N-1} g_{\mu\nu} \cong \tfrac{1}{3}N^3,$$

while for the constant C we obtain the approximate expression

$$C \cong \frac{\zeta L^2}{180\Theta} N^2. \tag{32}$$

If in (28) we had only retained the term with $\lambda = \mu = \nu$, we would have arrived at the same result. Thus the double refraction by a single molecule becomes proportional to the square of the number of links, as was also found by Hermans. It is to be expected that in a more accurate calculation applied to the pearl necklace with completely free rotation, the result will have to be corrected by a factor which does not differ much from unity. With the partially free rotation of actual molecules the analogous correction factor will be larger than 1; for C–C bonds a preliminary estimate yielded a value between 2 and 3.

6. Remarks on the Approximations Involved

I. Influence of the Pressure Forces

In formula (1) the force which results from the non-uniform pressure in the flow of the liquid has been neglected. If a_{xi} represents the x component of the acceleration $\mathbf{a}$ at the center of the ith particle and if m'_i is the mass of the liquid replaced by the particle, the x component of the force concerned is $m'_i a_{xi}$. Since in the stationary flow we have

$$a_x = \frac{1}{2}\frac{\partial}{\partial x}(\nabla\Psi)^2, \qquad (y), \qquad (z),$$

the influence of the pressure would result in an additional term

$$-\tfrac{1}{2}\sum_i m'_i (\nabla\Psi_i)^2$$

in U. If Ψ is considered small of first order, our approximation amounts to omitting a term of the second order. With the potential $\Psi = -\tfrac{1}{2}\kappa xy$ used by us, the ratio between the pressure force ma and the frictional force $\zeta v''$ is of the order $m'\kappa/\zeta$, which for a sphere of radius 10^{-7} cm in water ($\eta = 0.01$) is about $10^{-13}\kappa$, i.e., negligible.

II. Influence of Rotation

So far we have not accounted for the rotation of the liquid. If the flow of the liquid consists exclusively of a uniform rotation, the average behavior of our system can still be described on the basis of an assembly in temperature equilibrium if we insert the distribution

$$W \prod dp\, dq = e^{-(T-(\boldsymbol{\Omega}\cdot \mathbf{I}))/\Theta} \prod dp\, dq. \tag{33}$$

Here $\boldsymbol{\Omega}$ is the angular velocity of the liquid and $\mathbf{I}$ the angular momentum of the system. On account of the fact that in the absence of forces $\mathbf{I}$ would be an additive integral of motion, such a distribution would be stationary in an ideal gas whose particles are systems of the type considered here. Maxwell already pointed out, that with free particles this distribution corresponds to a uniform rotation of a gas with angular velocity $\boldsymbol{\Omega}$. Similarly, in the present case, where each particle possesses in addition internal degrees of freedom, the distribution mentioned includes a superposed uniform rotation of each of the mass-points constituting the systems. This is at once evident from the identity

$$\begin{aligned} T - (\boldsymbol{\Omega}\cdot \mathbf{I}) &= \sum \tfrac{1}{2} m_i (\dot{x}_i^2 + \dot{y}_i^2 + \dot{z}_i^2) - \sum m_i (\boldsymbol{\Omega}\cdot [\mathbf{r}_i \wedge \dot{\mathbf{r}}_i]) \\ &= \sum_i \left\{ \tfrac{1}{2} m_i (\dot{\mathbf{r}}_i - [\boldsymbol{\Omega}\wedge \mathbf{r}_i])^2 - \tfrac{1}{2} m_i ([\boldsymbol{\Omega}\wedge \mathbf{r}_i])^2 \right\}, \end{aligned}$$

since it shows that the distribution (33) is equivalent to the distribution

$$W = e^{-(T'+U')/\Theta},$$

if T' is considered as the kinetic energy referred to the coordinate system rotating with angular velocity $\boldsymbol{\Omega}$, while

$$U' = -\tfrac{1}{2} \sum_i m_i ([\boldsymbol{\Omega}\wedge \mathbf{r}_i])^2$$

represents the potential of a centrifugal force acting on all the particles. The influence of the pressure forces can be accounted for by the potential

$$U'' = \tfrac{1}{2} \sum_i m'_i ([\boldsymbol{\Omega}\wedge \mathbf{r}_i])^2.$$

If the system consists of spheres with the same specific weight as that of the liquid, $m_i = m'_i$, and U' and U'' exactly cancel.

If now the system is placed in a liquid which itself rotates with angular velocity $\boldsymbol{\Omega}$, the system and the random forces exerted by the liquid will be the same in the rotating coordinate system as in a liquid at rest: the distribution (33) will, therefore, be left undisturbed.

If the flow is stationary but of an arbitrary nature and not irrotational, the equilibrium distribution can no longer be described by one of the simple assemblies known in the kinetic theory of gases. Using a method known from the theory of the Brownian movement,[4] it is then still possible to construct

---

[4] H. A. Kramers, *Physica* **7**, 287 (1940).

an equation of diffusion which describes the change with time of an arbitrary distribution in phase space. We would then have to determine the stationary solution of this equation. In the special case of the flow (16), representing the superposition of a pure rotation and an irrotational flow, both of which are proportional to κ, the mathematical problem becomes more simple. Yet it would not be permissible to assume that the stationary distribution is now described by the simple combination of (33) and (4), i.e., by

$$W = e^{-(T+\frac{1}{2}\kappa I_z - \frac{1}{2}\kappa\zeta \sum xy)/\Theta}.$$

This is seen, for instance, from the fact that with increasing κ the system will gradually be oriented parallel to the direction of the flow, whereas according to this formula there can only be question of a preference orientation at 45°. Each of the two terms following T in the exponent applies only in the absence of the other one. However, if their influence on the distribution is small, errors will only occur if one would be interested in effects of the order κ^2. Effects of the order κ will be described by the distribution

$$W = e^{-T/\Theta} \left\{ 1 - \tfrac{1}{2}\kappa \frac{I_z}{\Theta} + \tfrac{1}{2}\kappa\zeta \frac{\sum xy}{\Theta} \right\}.$$

As the term with I_z cannot have any influence on the double refraction (since it corresponds to a uniform rotation of the liquid) we are led to conclude that it was actually permissible to neglect the rotation in the preceding section. The rotation can also be neglected in the viscosity. The terms which must be added to the sum $\sum X''y$ in Section 4 in order to account for the rotation in the determination of δP_{xy}, exactly cancel in as far as quantities $\sim \kappa$ are concerned. This follows at once from the fact, that $\delta P_{xy} = 0$ in pure rotation, but it can, of course, also be verified by direct calculation (the term with $-\tfrac{1}{2}\kappa I_z/\Theta$ in the last mentioned formula will then also play a part).

A simple estimate shows that even with molecules of length 10^{-4} cm, it will not be simple to realize κ values of such a magnitude that κ^2 effects would be perceptible in dilute solutions.

7. Ramified Molecules

In virtue of the derivations of the formulas (23) and (31) for viscosity and birefringence, these formulas also hold good for branched molecules; we will only have to insert other expressions for $g_{\mu\nu}$. These $g_{\mu\nu}$'s occur, for instance, in the expression

$$T = \frac{mL^2}{2N} \sum_{\mu\nu} g_{\mu\nu}(\dot\mu\dot\nu), \quad (\dot\mu\dot\nu) = (\dot\alpha_\mu\dot\alpha_\nu + \dot\beta_\mu\dot\beta_\nu + \dot\gamma_\mu\dot\gamma_\nu),$$

representing the kinetic energy of the molecule if the movement of the center of gravity is omitted. The value of $g_{\mu\nu}$ can now be determined in the following manner.

L. The Behavior of Macromolecules in Inhomogeneous Flow

Cut the molecule at the μth link and at the νth link; it will then fall apart in three pieces, to wit two end pieces which contain, respectively, half the μth and half the νth link, and a middle piece. The number of particles in these three parts is called N_1, N_2, and N_3, respectively; their sum is N. With $\mu = \nu$ we get $N_3 = 0$.

If $\mu \neq \nu$, the expression

$$T_{\mu\nu} = \frac{mL^2}{2N} \{g_{\mu\mu}(\dot\mu\dot\mu) + 2g_{\mu\nu}(\dot\mu\dot\nu) + g_{\nu\nu}(\dot\nu\dot\nu)\}$$

will be equal to the kinetic energy of a macromolecule, in which only the links μ and ν possess non-zero angular velocities $\dot\omega_\mu$ and $\dot\omega_\nu$ while all other links perform translations only, in such a way that the center of gravity remains fixed in space. $T_{\mu\nu}$ therefore also represents the kinetic energy of a system of three particles with masses mN_1, mN_2, and mN_3, of which both 1 and 3 and 3 and 2 are connected by a rod of length L. Since the g's are numerical coefficients, it will suffice to consider the case where the particles momentarily are lying on a straight line. Introducing the momentary velocities v_1, v_2, and v_3:

$$N_1 v_1 + N_2 v_2 + N_3 v_3 = 0, \quad v_3 - v_1 = L|\dot\omega_\mu|, \quad v_2 - v_2 = L|\dot\omega_\nu|,$$

we get

$$T_{\mu\nu} = \tfrac{1}{2}m(N_1 v_1^2 + N_2 v_2^2 + N_3 v_3^2)$$
$$= \frac{mL^2}{2N}\{N_1(N_3 + N_2)(\dot\mu\dot\mu) + 2N_1 N_2(\dot\mu\dot\nu) + (N_1 + N_3)N_2(\dot\nu\dot\nu)\}. \quad (34)$$

This shows that the value of $g_{\mu\nu}$ is equal to the product of the number of particles in the end pieces which are formed when cutting the μth and the νth link:

$$g_{\mu\nu} = N_1 N_2. \tag{35}$$

Specializing $\mu = \nu$, or also directly from (34), we obtain

$$g_{\mu\mu} = N_1 N_2, \tag{36}$$

where N_1 and N_2 are the number of mass-points in the two pieces in which the molecule falls apart when cutting the μth link.

We will now restrict ourselves to the viscosity. In order to compute the friction from (23), we may assume that with completely free rotation of two adjacent links the average value $\langle(\mu\nu)\rangle_{\text{Av}}$ for two different links ($\mu \neq \nu$) is almost always zero, as it was in the unbranched pearl necklace. For the effect of a single molecule on the friction this gives

$$\delta\eta = \frac{\zeta L^2}{6N} \sum_\mu g_{\mu\mu} = \frac{\zeta L^2}{6} \langle N_1 N_2\rangle_{\text{Av}}, \tag{37}$$

where the angular brackets with subscript $_{Av}$ now indicate the average over all possibilities of cutting.

For an unbranched molecule this average value becomes

$$\langle N_1 N_2 \rangle_{Av} = \frac{1}{N} \sum_{1}^{N-1} N_1(N - N_1) \cong N^2 \int_0^1 x(1-x)\,dx = \tfrac{1}{6} N^2,$$

in conformity with the formula (24). For a ramified molecule $\langle N_1 N_2 \rangle_{Av}$ will always be smaller. For instance, if the molecule consists of s branches of N/s links each, all starting from a single point:

$$\langle N_1 N_2 \rangle_{Av} = \left\{ \frac{1}{2s} - \frac{1}{3s^2} \right\} N^2.$$

For $s = 1$ and $s = 2$ this is equal to $\tfrac{1}{6} N^2$, as it should be, but for $s > 2$ it is always smaller than $\tfrac{1}{6} N^2$.

8. Viscosity Contribution of a Macromolecule Closed to a Ring

Let a linear molecule consisting of N particles be closed to a ring by joining the Nth mass-point to the 1st particle by means of an Nth link of length L. The kinetic energy will still be given by

$$T = \frac{mL^2}{2N} \sum_{\mu,\nu=1}^{N-1} g_{\mu\nu}(\dot{\mu}\dot{\nu}), \tag{38}$$

but as the result of the new kinematic bond the possible values of T will be independent of the link which is assumed to have closed the ring. The coefficients $g_{\mu\nu}$ may, therefore, be replaced by the average values of all the $g_{\mu\nu}$'s for which the μth and the νth link are separated by the same number of links in the ring:

$$\bar{g}_{\mu\nu} = \frac{1}{N} \sum_{\rho=0}^{N-1} g_{\mu+\rho,\nu+\rho} \cong \frac{1}{6N} \left\{ |\mu - \nu|^3 + (N - |\mu - \nu|)^3 \right\}.$$

The link between N and 1 will be equivalent to the others; the numbers μ and ν will vary from 1 to N. In the sum the numbers $\mu + \rho$ (and $\nu + \rho$) must be considered as identical with $\mu + \rho - N$ (and $\nu + \rho - N$) in all cases where $\mu + \rho$ (or $\nu + \rho$) would be larger than N. The third member in the equation is obtained by realizing that the sum must be extended over all possible positions of the closing link, so that it consists of two sums of the type $\sum \mu(N - \mu) \cong \tfrac{1}{6} N^3$ (compare Eq. (24)). If the distance between μ and ν is described by an angle $2\pi\tau = 2\pi|\mu - \nu|/N$, which would represent the angular distance between these links if the ring were a circle, one can also write

$$\bar{g}_{\mu\nu} \cong \tfrac{1}{6}N^2(1 - 3\tau + 3\tau^2), \qquad 0 \leqslant \tau \leqslant 1. \tag{39}$$

The identity of

$$\bar{T} = \frac{mL^2}{2N} \sum \bar{g}_{\mu\nu}(\mu\dot\nu) \tag{40}$$

with (38) results, of course, from the relations

$$\sum_{\nu=1}^{N}(\mu\dot\nu) = 0,$$

which in their turn are a result of the ring-condition

$$\sum_{1}^{N} \omega_\nu = 0. \tag{41}$$

From (41) it also follows that $\sum_{\mu,\nu}(\mu\dot\nu) = 0$, and we may therefore subtract a constant from $\bar{g}_{\mu\nu}$. In other words, instead of (39) one might also write

$$\bar{g}_{\mu\nu} = -\tfrac{1}{2}N^2\tau(1 - \tau). \tag{42}$$

The influence of a ring-shaped molecule on the friction will be given by

$$\delta\eta = \frac{\zeta L^2}{6N} \sum_{\mu,\nu}^{N-1} g_{\mu\nu}\langle(\mu\nu)\rangle_{\text{Av}} = \frac{\zeta L^2}{6N} \sum_{\mu,\nu}^{N} \bar{g}_{\mu\nu}\langle(\mu\nu)\rangle_{\text{Av}}. \tag{43}$$

To calculate the average values $\langle(\mu\nu)\rangle_{\text{Av}}$, one would have to know the distribution of the ω's. This distribution is determined in principle by the square root of the determinant A of the quadratic expression $T(\dot{q}_1 \cdots \dot{q}_s)$ after transformation into suitable generalized coordinates. It is easy to see that even our former $A^{1/2}$, in which $\vartheta_1\varphi_1, \cdots \vartheta_{N-1}\varphi_{N-1}$ served as generalized coordinates, will represent the distribution concerned, if we restrict ourselves to those values of ϑ and φ for which the distance between the Nth and the 1st particle is L. In fact, it would have been a sufficient approximation if the kinematic relation which expresses this condition is omitted and replaced by the introduction of a potential energy $U' = \tfrac{1}{2}K(r_{N1}^2 - L^2)^2$. Here r_{N1} represents the distance between the 1st and the Nth particle, while K is a constant of sufficient magnitude. The distribution of the ω's would then be accurately described by $A^{1/2}e^{-U'/\Theta}$, and this amounts exactly to the prescription given.

However, without going into detailed calculations, the sum in (43) can be determined for a pearl necklace with completely free rotation if we take into account that the correlation between two different links μ and ν which leads to a non-zero $\langle(\mu\nu)\rangle_{\text{Av}}$-value, will almost always be the same, independent of whether μ and ν are now lying close together or not. It thus follows from the ring-condition (41) that

$$0 = \sum_\nu \langle(\mu\nu)\rangle_{\mathrm{Av}} = \langle(\mu\mu)\rangle_{\mathrm{Av}} + \sum_{\nu(\neq\mu)} \langle(\mu\nu)\rangle_{\mathrm{Av}} = 1 + (N-1)\langle(\mu\nu)\rangle_{\mathrm{Av}(\mu\neq\nu)},$$

or

$$\langle(\mu\nu)\rangle_{\mathrm{Av}(\mu\neq\nu)} \cong -\frac{1}{N}. \tag{44}$$

Consequently, if we use expression (42) for $\bar{g}_{\mu\nu}$:

$$\begin{aligned}
\sum_{\mu\nu} \bar{g}_{\mu\nu}\langle(\mu\nu)\rangle_{\mathrm{Av}} &= \sum_\mu \bar{g}_{\mu\mu} - \frac{1}{N}\sum_{\mu\neq\nu} \bar{g}_{\mu\nu} \\
&= 0 + \frac{1}{N}\cdot N \int_0^1 \tfrac{1}{2}N^2\tau(1-\tau)\,dN\tau \\
&= \tfrac{1}{2}N^3(\tfrac{1}{2}-\tfrac{1}{3}) = \tfrac{1}{12}N^3.
\end{aligned} \tag{45}$$

Since the analogous sum was $\tfrac{1}{6}N^3$ for the free linear pearl necklace, it appears that *the ring formation has reduced the influence on the friction to $\tfrac{1}{2}$ of the original amount.*

For a linear pearl necklace with fixed angle α between two successive links $\sum_\nu \langle(\mu\nu)\rangle_{\mathrm{Av}}$ was equal to $\cot^2 \tfrac{1}{2}\alpha$, and the sum needed only to be taken over a small number of neighboring links (unless α would be too close to zero). It is clear that in the case of rings, the formula (44) must then be replaced by

$$\langle(\mu\nu)\rangle_{\mathrm{Av}(\mu\neq\nu)} \cong -\frac{1}{N}\cot^2\tfrac{1}{2}\alpha,$$

and that the sum (45) must also be multiplied by a factor $\cot^2 \tfrac{1}{2}\alpha$. Thus, here again, as was to be expected, it still holds good that ring closing reduces the effect to $\tfrac{1}{2}$ of the original figure.

There will now be no further difficulty in the problem of arbitrarily ramified molecules containing one or more separate rings. As soon, however, as the different rings are no longer separate, i.e., if they have one or more chain segments in common, the $\langle(\mu\nu)\rangle_{\mathrm{Av}}$'s can no longer be determined in this simple manner.

References

References to papers that have H. A. Kramers as author or coauthor can be found in his list of publications. It should be noted that in that list the authors are not listed in the order in which they appear on the paper in question.

Bird, R. B. (1989) *Physics Today*, October 1989, p. 14.
Bird, R. B., C. F. Curtiss, R. C. Armstrong, and O. Hassager (1987) *Dynamics of Polymeric Liquids*, Vol.2, Wiley-Interscience, New York.
Bohr, N. (1913a) *Phil. Mag.* **25**, 10.
Bohr, N. (1913b) *Phil. Mag.* **26**, 1, 476, 857.
Bohr, N. (1915) *Phil. Mag.* **30**, 381.
Bohr, N. (1918) *Proc. Dan. Acad. Sci.* (8) **4**, No. 1, parts I, II.
Bohr, N. (1920) *Zs. Phys.* **2**, 423.
Bohr, N. (1922) *Proc. Dan. Acad. Sci.* (8) **4**, No. 1, part III.
Bohr, N. (1923) *Zs. Phys.* **13**, 117.
Bohr, N. (1952) *Ned. Tijds. Natuurk.* **18**, 161.
Born, M. (1978) *My Life; Recollections of a Nobel Laureate*, Taylor & Francis, London.
Bothe, W. and H. Geiger, (1925) *Zs. Phys.* **32**, 639.
Brillouin, L. (1926) *Comptes Rendus* (Paris) **183**, 24.
Brinkman, H. C. (1932) *Zur Quantenmechanik der Multipolstrahlung*, Utrecht thesis, Noordhoff, Groningen, Netherlands; an English translation by E. R. Cohen and M. M. Mills can be found in Kramers 1947b.
Brinkman, H. C. (1956) *Applications of Spinor Invariants in Atomic Physics*, North-Holland, Amsterdam.
Burgers, J. M. (1917) *Ann. Physik* **52**, 195.
Casimir, H. B. G. (1952) *Ned. Tijds. Natuurk.* **18**, 167.
Casimir, H. B. G. (1963) *Zs. Phys.* **171**, 246.
Casimir, H. B. G. (1983) *Haphazard Reality*, Harper & Row, New York.
Casimir, H. B. G., A. D. Fokker, C.J. Gorter, S.R. de Groot, N. G. van Kampen, J. Korringa, and G. E. Uhlenbeck, (editors) (1956) *Collected Scientific Papers of H. A. Kramers*, North-Holland, Amsterdam.
Chapman, S. and T. G. Cowling (1939) *The Mathematical Theory of Non-Uniform Gases*, Cambridge University Press, Cambridge.
Compton, A. H. (1923) *Phys. Rev.* **21**, 207.
Compton, A. H. and A. W. Simon (1925) *Phys. Rev.* **26**, 289.
Debye, P. (1923) *Phys. Zs.* **24**, 161.
Debye, P. and E. Hückel (1923) *Phys. Zs.* **24**, 185.
Dirac, P. A. M. (1926) *Proc. Roy. Soc.* **A109**, 642.
Dirac, P. A. M. (1938) *Proc. Roy. Soc.* **A167**, 148.
van Dishoeck E. F. (1996) *Moleculen tussen de Sterren*, Inaugural Oration, Rijksuniversiteit, Leiden.

Dresden, M. (1987) *H. A. Kramers, Between Tradition and Revolution*, Springer, New York.
Dresden, M. (1988) *Physics Today*, September 1988, p. 26.
Einstein, A. (1917) *Phys. Zs.* **18**, 121; an English translation can be found in ter Haar 1967.
Gerritsen, A. N. (1946) *Physica* **12**, 311.
Gerritsen, A. N. (1948) *Physica* **14**, 381.
Gerritsen, A. N. (1949) *Physica* **14**, 407.
Gerritsen, A. N. and J. Koolhaas (1943) *Physica* **10**, 49.
Gibbs, J. W. (1902) *Elementary Principles in Statistical Mechanics*, Yale University Press, New Haven.
Goedkoop, J. A. (1996) *Ned. Tijds. Natuurk.* **62**, 123.
ter Haar, D. (1943) *Bull. Astron. Inst. Netherl.* **10**, 1.
ter Haar, D. (1955) *Proc. 1955 Paris Low Temp. Conf.*, p. 137.
ter Haar, D. (1966) *Elements of Thermostatistics*, Holt, Rinehart and Winston, New York.
ter Haar, D. (1967) *The Old Quantum Theory*, Pergamon, Oxford.
ter Haar, D. (1971) *Elements of Hamiltonian Mechanics*, Pergamon, Oxford.
ter Haar, D. (1995) *Elements of Statistical Mechanics*, Butterworth-Heinemann, Oxford.
Hänggi, P., P. Talkner, and M. Borkovec (1990) *Rev. Mod. Phys.* **62**, 251.
Heisenberg, W. (1925) *Zs. Phys.* **33**, 879.
Hermans, J. J. (1943) *Physica* **10**, 777.
Herzberg, G. (1950) *Molecular Spectra and Molecular Structure*, Van Nostrand, New York.
Hilgevoord, J. (1996) *Ned. Tijds. Natuurk.* **62**, 325.
Isihara, A. (1989) *Physics Today*, October 1989, p. 15.
Jaffé, G. (1913) *Ann. Physik* **42**, 303.
Jeffreys, H. (1924) *Proc. London Math. Soc.* **23**, 428.
van Kampen, N. G. (1951) *Proc. Dan. Acad. Sc.* **26**, Nr. 15.
van Kampen, N. G. (1954) *Physica* **20**, 603.
van Kampen, N. G. (1988) *Ned. Tijds. Natuurk.* **A54**, No. 1, 40.
van Kampen, N. G. (1991) *Ber. Bunsenges. Phys. Chem.* **95**, 225.
Keesom, W. H. (1942) *Helium*, Elsevier, Amsterdam, p. 174.
Kirchhoff, G. (1868) *Poggend. Ann.* **134**, 177.
Klein, M. J. (1952) *Am. J. Phys.* **20**, 65.
Klein, O. (1922) *Arkiv Mat. Astr. Fys.* **16**, No. 5.
Kronig, R. de L. (1926) *J. Opt. Soc. Am.* **12**, 547.
Kubo, R. (1943) *Busseiron Kenkyu* **1**, 1; an English translation of this paper was published in the *Selected Papers of Professor Ryogo Kubo*, published in 1980 on the occasion of his sixtieth birthday.
Kuhn, W. (1925) *Zs. Phys.* **33**, 408.
Lassettre, E. N. and J. P. Howe (1941) *J. Chem. Phys.* **9**, 747.
Lifshitz, E. M. and L. P. Pitaevskii (1980) *Statistical Physics*, Part 1, Pergamon, Oxford.
Lignac, W.P.J. (1949) Leiden Ph.D. thesis.
Margenau, H. and G. M. Murphy (1956) *The Mathematics of Physics and Chemistry*, Van Nostrand, New York, § 15.7.
Maxwell, J. C. (1879) *Phil. Trans.* **A170**, 231.
Mel'nikov, V. I. (1991) *Phys. Repts.* **209**, 1.
Montroll, E. (1941) *J. Chem. Phys.* **9**, 706.
Mulliken, R. S. (1932) *Rev. Mod. Phys.* **4**, 1.
Newell, G. F. and E. W. Montroll (1953) *Rev. Mod. Phys.* **25**, 353.

Onsager, L. (1944) *Phys. Rev.* **65**, 117.
Opechowski, W. (1941) *Physica* **8**, 161.
Opechowski, W. (1988) *Physics in Canada*, July 1988.
Pais, A. (1982) *Subtle is the Lord* ..., Oxford University Press, Oxford.
Pais, A. (1986) *Inward Bound*, Oxford University Press, Oxford.
Pais, A. (1991) *Niels Bohr's Times*, Oxford University Press, Oxford.
Pais, A. (1995) Unpublished text of a lecture delivered on September 14, 1995, at the Eindhoven University of Technology, on the occasion of the centennial of Kramers's birth.[1]
Pathria, R. K. (1972) *Statistical Mechanics*, Pergamon Press, Oxford.
Popov, V. S., B. M. Karnakov, and V. D. Mur (1996) *Phys. Lett.* **A210**, 402.
Schouten, J. A. (1918) *Proc. Kon. Ned. Akad. Wetensch.*, Amsterdam **21**, 533.
Serpe, J. (1940) *Physica* bf 7, 133.
Serpe, J. (1941) *Physica* bf 8, 226.
Shklovskii, I. S. (1946) *Astron. Zh.* **26**, 10.
Shklovskii, I. S. (1952) *Astron. Zh.* **29**, 144.
Shklovskii, I. S. (1953) *Doklad. Akad. Nauk SSSR* **92**, 25.
Slater, J. C. (1924) *Nature* **113**, 307.
Sommerfeld, A. (1915) *Münchener Ber.* **1915**, 425, 459.
Sommerfeld, A. (1916) *Münchener Ber.* **1916**, 131.
Thomas, L. H. (1926) *Nature* **117**, 514.
Thomas, W. (1925) *Naturwiss.* **13**, 627.
Townes, C. H. (1957) *IAU Sympos.* No. 4, p.92.
Van Vleck, J. H. (1932) *Theory of Electric and Magnetic Susceptibilities*, Oxford University Press, Oxford.
Wentzel, G. (1926) *Zs. Phys.* **38**, 518.
von Weizsäcker, C. F. (1933) *Ann. Physik* **17**, 869.
Weyl, H. (1931) *Theory of Groups and Quantum Mechanics*, Methuen, London.
Whitehead, A. N. (1974) *The Organization of Thought*, Greenwood Press, Westport CT, p. 127.
Wilson, W. (1915) *Phil. Mag.* **29**, 795.
Wycoff, D. and N. L. Balasz (1987) *Physica* **146A**, 175.
Zwaan, A. (1929) *Intensitäten im Ca-Funkenspektrum*, Utrecht thesis.

---

[1] I am indebted to Professor Pais for allowing me to quote from this lecture.

Publications of H. A. Kramers

1919*[1] Intensities of spectral lines, *D. Kgl. Danske Vidensk. Selsk. Skrifter* (8) **3**, 285. pp. 9, 11-15.
1920* Über den Einflusz eines elektrischen Feldes auf die Feinstruktur der Wasserstofflinien, *Zs. Phys.* **3**, 199. p. 15.
1921* On the application of Einstein's theory of gravitation to a stationary field of gravitation, *Amsterdam Proc.*[2] **23**, 1052. pp. 9, 109, 116, 117.
1922a Zonnestelsels en atomen, *Physica* **2**, 174. p. 9.
1922b Atomteoriens udvikling in de senere aar, *Fis. Tidsskr.* **20**, 1, 69. p. 9.
1922c Absorption af Røntgenstraaler, *Fis. Tidsskr.* **20**, 130. p. 9.
1922d Some main features of the modern theory of atomic structure, *Proc. Fifth Congress of Scandinavian Mathematicians, Helsinki.* p. 9.
1922e De bouw der atomen en de physische en chemische eigenschappen der elementen, *Physica* **2**, 269, 381. p. 9.
1922 (with H. Holst) *Bohrs Atomteori*, Gyldendal, Copenhagen.[3] pp. 8, 9, 15, 48.
1923a De bouw der atomen en de physische en chemische eigenschappen der elementen, *Physica* **3**, 12. p. 9.
1923b Korrespondenzprinzip und Schalenbau des Atoms, *Naturwiss.* **11**, 550. p. 9.
1923c* Über das Modell des Heliumatoms, *Zs. Phys.* **13**, 312. pp. 9, 15.
1923d* Über die Quantelung rotierender Moleküle, *Zs. Phys.* **13**, 343. pp. 9, 16.
1923e* On the theory of X-ray absorption and of the continuous X-ray spectrum, *Phil. Mag.* **46**, 836. pp. 9, 16-20.
1923* (with J. A. Christiansen) Über die Geschwindigkeit chemischer Reaktionen, *Zs. Phys. Chem.* **104**, 451. pp. 9, 93, 94.
1923* (with W. Pauli Jr.) Zur Theorie der Bandenspektren, *Zs. Phys.* **13**, 351. pp. 9, 16.
1924a* Law of dispersion and Bohr's theory of spectra, *Nature* **113**, 673. pp. 10, 23
1924b* Quantum theory of dispersion, *Nature* **114**, 310. pp. 10, 23.
1924c Chemische Eigenschaften der Atome nach der Bohrschen Theorie, *Verh. Ges. Dtsch. Naturf. Ärzte* **68**, 1924. p. 9.
1924d Chemische Eigenschaften der Atome nach der Bohrschen Theorie, *Naturwiss.* **12**, 1050. p. 9.
1924a* (with N. Bohr and J. C. Slater) Über die Quantentheorie der Strahlung, *Zs. Phys.* **24**, 69. pp. 8, 9, 21-23.

---

[1] Articles with an asterisk after the year are reprinted in H. A. Kramers's *Collected Scientific Papers*, North-Holland, Amsterdam (1956). The page numbers after the various publications refer to the places where the paper in question is mentioned in the first seven chapters of this book.
[2] *Amsterdam Proc.* ≡ *Proc. Kon. Ned. Akad. Wetensch.*, Amsterdam.
[3] An English translation was published in 1923, a German translation in 1925, a Spanish translation in 1925, and a Dutch translation in 1927.

1924b (with N. Bohr and J. C. Slater) The quantum theory of radiation, *Phil. Mag.* **47**, 785. pp. 8, 9, 21-23.
1925a Vekselwirkningen mellem lys og stof, *Fysisk Tidsskr.* **23**, 26. p. 9.
1925b De wisselwerking tussen stof en straling, *Hand. 20e Ned. Nat. Gen. Congres*, p.164. p. 9.
1925c* On the behaviour of atoms in an electromagnetic wave field, *6th Skand. Mat. Kongress*, p.143. pp. 10, 26.
1925d Enige opmerkingen over de quantenmechanica van Heisenberg, *Physica*, **5**, 369. Paper B; pp. 32, 33, 51.
1925* (with W. Heisenberg) Über die Streuung von Strahlung durch Atome, *Zs. Phys.* **31**, 681. Paper A; pp. 7, 10, 23-27.
1926a Vorm en Wezen, J. van Druten, Utrecht.
1926b Lysets Spredning i atomer, *Fys. Tidsskr.* **24**, 34.
1926c* Wellenmechanik und halbzahlige Quantisierung *Zs. Phys.* **39**, 828. Paper C; pp. 32, 35-39.
1926d Rapport sur de nouvelles experiences avec les supra-conducteurs[4] *Commun. Phys. Lab., Leiden* **15**, Suppl. 50a, 32.
1927a* Investigations on the free energy of a mixture of ions, *Amsterdam Proc.* **30**, 145. pp. 68, 71.
1927b* La diffusion de la lumière par les atomes, *Atti Congr. Intern. Fis., Como-Pavia-Roma*, II, 545. Paper D; pp. 10, 23, 28-30.
1927c Form og vaesen, *Fis. Tidsskr.* **25**, 128.
1927d De huidige opvattingen omtrent de quantumtheorie van het atoom, *Handb. 21e Ned. Nat. Gen. Congres*, p. 76.
1927* (with L. S. Ornstein) Zur kinetischen Herleitung der Fermischen Verteilungsgesetzes, *Zs. Phys.* **42**, 481. pp. 68, 71.
1928 De causaliteitswet in de moderne natuurkunde, *Voordracht, Diligentia* **7**, 33.
1928* (with R. de L. Kronig) Zur Theorie der Absorption und Dispersion in den Röntgenspektren, *Zs. Phys.* **48**, 174. p. 48.
1929a* Zur Struktur der Multiplett-S-Zustände in zweiatomigen Molekülen I, *Zs. Phys.* **53**, 422. p. 51, 52.
1929b* Zur Struktur der Multiplett-S-Zustände in zweiatomigen Molekülen II, *Zs. Phys.* **53**, 429. p. 51, 52.
1929c Moeilijkheden en successen in de moderne atoom mechanica, *Handb. 22e Ned. Nat. Gen. Congres*, p. 87.
1929d* Die Dispersion und Absorption von Röntgenstrahlen, *Phys. Zs.* **30**, 522. pp. 10, 28.
1929e Grepen uit de moderne atoomtheorie, *Versl. Prov. Utr. Genootschap*, p. 66.
1929f* La rotation paramagnétique du plan de la polarisation dans les cristaux uniaxes des terres rares, *Amsterdam Proc.* **32**, 1176. pp. 54, 68, 88, 89.
1929* (with J. Becquerel) La rotation paramagnétique du plan de polarisation dans les cristaux de tysonite et de xénotime, *Amsterdam Proc.* **32**, 1190. pp. 68, 88, 89.
1929* (with J. Becquerel and W. J. de Haas) Experimental verification of the theory of paramagnetic rotatory polarisation in the crystals of xenotime, *Amsterdam Proc.* **32**, 1206. pp. 68, 88, 89.
1929a* (with G. P. Ittmann) Zur Quantelung des asymmetrischen Kreisel, *Zs. Phys.* **53**, 553. pp. 52-54.
1929b* (with G. P. Ittmann) Zur Quantelung des asymmetrischen Kreisel II, *Zs. Phys.* **58**, 217. pp. 52-54.

---

[4] This is a note to a paper by H. Kamerlingh Onnes.

1930a De rotatie van meeratomige moleculen, *Amsterdam Versl.*[5] **39**, 11.
1930b Rotation paramagnétique dans les cristaux uniaxes de terres rares, *C. R. Paris* **191**, 784.
1930c* Zur Ableitung der quantenmechanischen Intensitätsformeln, *Amsterdam Proc.* **33**, 953. p. 45.
1930d* Théorie générale de la rotation paramagnétique dans les cristaux, *Amsterdam Proc.* **33**, 959. Paper E; pp. 32, 54, 55, 68, 69.
1930e Moderne molecuultheorieën, *Chem. Weekbl.* **27**, 406.
1930 (with J. Becquerel and W. J. de Haas) La loi de la rotation paramag-nétique dans le xénotime et sa vérification expérimentale, *C. R. Paris* **191**, 839.
1930* (with H. C. Brinkman) Zur Theorie der Einfangung von Elektronen durch α-Teilchen, *Amsterdam Proc.* **33**, 973. p. 54.
1930* (with G. P. Ittmann) Zur Quantelung des asymmetrischen Kreisel III, *Zs. Phys.* **60**, 663. pp. 52-54.
1931a* Die Multiplettaufspaltung bei Koppelung zweier Vektoren, *Amsterdam Proc.* **34**, 965. p. 45.
1931b Werkelijkheid en begrippenvorming, *Physica* **11**, 321.
1931c Virkelighed og Begrepsdannelse, *Fys. Tidsskr.* **30**, 103.
1932a* Propriétés paramagnétiques de cristaux de terres rares, *Amsterdam Proc.* **35**, 1272. pp. 68, 90.
1932b Oude en nieuwe natuurwetenschap, *Faraday* **3**, 33.
1932* (with A. R. Olson) The normal vibrations of acetylene, *J. Am. Chem. Soc.* **54**, 136. pp. 109, 117.
1933a Die Grundlagen der Quantentheorie, *Hand- und Jahrb. Chem. Phys.*, Akademie Verlag, Leipzig, Vol.I$_1$.[6] pp. 4, 8, 51-35.
1933b* Propriétés paramagnétiques de cristaux de terres rares II, *Amsterdam Proc.* **36**, 17. pp. 68, 90.
1933c Magnetische eigenschappen der zeldzame aarden, *Amsterdam Versl.* **42**, 60.
1933d Verleden en toekomst, *Handb. 24e Ned. Nat. Gen. Congres*, p. 76.
1933e Quantumtheorie en chemie, *Handb. 24e Ned. Nat. Gen. Congres*, p. 95.
1933f Prof. P. Ehrenfest, *Nature* **132**, 667.
1933g In memoriam P. Ehrenfest *Physica* **13**, 273.
1933h* Paramagnetische Eigenschaften der Kristallen seltener Erden, *Leipziger Vorträge, Magnetismus*, Hirzel, Leipzig, p. 43. pp. 68, 90.
1933* (with E. M. van Engers) Zur Anwendung der Methode der Phasenintegrale auf das Wasserstoffmolekülion, *Zs. Phys.* **82**, 328. p. 47.
1933a (with W. J. de Haas and E. C. Wiersma) Over het bereiken van lage temperaturen door middel van ontmagnetisatie, *Physica* **13**, 175. p. 8.
1933b (with W. J. de Haas and E. C. Wiersma) Das Erreichen niedriger Temperaturen mittels adiabatischer Demagnetsierung, *Naturwiss.* **21**, 467. p. 8.
1933c (with W. J. de Haas and E. C. Wiersma) Obtention d'une température extrêmement basse par démagnétisation adiabatique d'une sel d'une terre rare, *C. R. Paris* **196**, 1975. p. 8.
1933d (with W. J. de Haas and E. C. Wiersma) Temperature below 0.27° reached in Holland, *Nature* **131**, 719. p. 8.
1933* (with C. C. Jonker and T. Koopmans) Wigners Erweiterung des Thomas-Kuhnschen Summensatzes für ein Elektron in einem Zentralfelde, *Zs. Phys.* **80**, 178. pp. 49-51.

---

[5] *Amsterdam Versl.* ≡ *Verslagen Kon. Ned. Akad. Wetensch.*.
[6] The second part of this textbook was published in 1938 (Kramers 1938e); an English translation of the complete book was published in 1957 by North-Holland, Amsterdam (Kramers 1957).

1934a* L'interaction entre les atomes magnétogènes dans un cristal paramagnétique, *Physica* **1**, 182. pp. 69, 90, 91.
1934b De theorie van de electronenspin, *Amsterdam Versl.* **43**, 36.
1934c Natuurkunde en natuurkundigen, *Ned. Tijds. Natuurk.* **1**, 241.
1934d* On the classical theory of the spinning electron, *Physica* **1**, 825. p. 39.
1934* (with W. J. de Haas and E. C. Wiersma) Experiments on adiabatic cooling of paramagnetic salts in magnetic fields, *Physica* **1**, 1. p. 8, 68.
1934* (with G. Heller) Ein klassisches Modell des Ferromagnetikums und seine nachträgliche Quantisierung im Gebiete tiefer Temperaturen, *Amsterdam Proc.* **37**, 378. pp. 68, 72-75.
1935a Physiker als Stilisten, *Naturwiss.* **23**, 297.
1935b* Classical relativistic spin-theory and its quantization, *Verh. Zeeman Jubil.*, p. 403. Paper F; pp. 4, 32, 39-44.
1935c Mechanisch en magnetisch moment van het electron, *Ned. Tijds. Natuurk.* **2**, 25.
1935d Electronengeleiding in metalen *Ingenieur* **50**, E1.
1935e Atom- og kvanteteoriens udvikling i aarene 1913/25, *Fys. Tidsskr.* **33**, 82.
1935f* Das Eigenwertproblem im eindimensionalen periodischen Kraftfelde, *Physica* **2**, 483. Paper G; pp. 32, 39.
1935g De uitreiking der Lorentz medaille aan Prof. Dr P. Debije, *Chem. Weekbl.* **32**, 645.
1936a* Zur Theorie des Ferromagnetimus, *Commun. K. O. Lab.* **22**, Suppl. 83, 1. pp. 67, 68, 72, 75-81.
1936b Magnetisme by lage temperaturen, *Ned. Tijds. Natuurk.* **3**, 22.
1936c Over de theorie van het magnetisme, *Amsterdam Versl.* **45**, 61.
1936d De professor in en buiten de maatschappij, *Het Kouter* **1**, 365.
1936* (with P. A. Coenen) Zum Intensitätsverlauf in der diffusen Serie des Kaliums, *Physica* **3**, 341. p. 47.
1937a* The use of charge-conjugated wave-functions in the hole-theory of the electron, *Amsterdam Proc.* **40**, 814. Paper H; pp. 32, 54-57.
1937b Straling, *Handb. 26e Ned. Nat. Gen. Congres*, p.141.
1937c Kan de moderne atoomtheorie gepopulariseerd worden? *Chem. Weekbl.* **34**, 253.
1937d Negatieve en positieve electronen, *Amsterdam Versl.* **46**, 91.
1937e X-rays and wave mechanics, *Current Sc., Special Number*, p. 39.
1937f Wat is materie? *Ned. Tijds. Geneesk.* **81**, 5876.
1938a* Didaktisches zur Verwendung der grand Ensembles in der Statistik, *Amsterdam Proc.* **41**, 10. pp. 68, 71, 72.
1938b De dissertatie, het wetenschappelijke oeuvre en de publicaties van A.D.Fokker, *Ned. Tijds. Natuurk.* **5**, 204. p. 3.
1938c* Die Wesselwirkung zwischen geladenen Teilchen und Strahlungsfeld, *Nuovo Cim.* **15**, 108. p. 59.
1938d Dwalingen in de natuurwetenschap, *Het Kouter* **3**, 356.
1938e Die Grundlagen der Quantentheorie, *Hand- und Jahrb. Chem. Phys.*, Akademie Verlag, Leipzig, Vol.II.[7] pp. 4, 8, 31, 40, 44, 45, 57, 59, 60.
1939a Over de theorie van het ferromagnetisme, *Amsterdam Versl.* **48**, 37.
1939b Relativiteitstheorie en atoomtheorie, *Diligentia Voordracht* **17**, 45.
1939c Wetenschap, mens en maatschappij, *Het Kouter* **4**, 93.
1939d De studie aan de Philosophische Faculteit, from *De student aan de Leidse Academie*.

---

[7] See also Kramers 1933a.

1939e Uitreiking der Lorentz-medaille aan Prof. Dr Arnold Sommerfeld, *Amsterdam Versl.* **48**, 59.
1940a* Brownian motion in a field of force and the diffusion model of chemical reactions, *Physica* **7**, 284. Paper I; pp. 67, 93-102.
1940b De diffusie theorie van chemische reacties, *Amsterdam Versl.* **49**, 10.
1940c Over de voedzaamheid der filosofie, *Het Kouter* **5**, 264.
1941a Theoretische resultaten, *Handb. 28e Ned. Nat. Gen. Congres*, p. 81.
1941b J. J. Thomson † 1940, *Ned. Tijds. Natuurk.* **8**, 137.
1941c Levensbericht L. S. Ornstein, *Jaarb. Ned. Acad. Wetensch.*, 1940/1, 225.
1941d Kosmologische problemen en de moderne physica, from *Antieke en moderne Kosmologie*, van Loghum Slaterus, Arnhem, p. 153.
1941* (with F. J. Belinfante and J. K. Lubański) Über freie Teilchen mit nichtverschwindender Masse und beliebiger Spinquantenzahl, *Physica* **8**, 597. p. 32.
1941a* (with G. H. Wannier) Statistics of the two-dimensional ferromagnet I, *Phys. Rev.* **60**, 252. Paper J; pp. 6, 68, 81-88.
1941b* (with G. H. Wannier) Statistics of the two-dimensional ferromagnet II, *Phys. Rev.* **60**, 263. pp. 6, 68, 81, 88.
1942a Herdenking van Sir J. J. Thomson, *Jaarb. Ned. Acad. Wetensch.*, 1941/2, 211.
1942b Augustin-Jean Fresnel, *Ned. Tijds. Natuurk.* **9**, 311.
1942* (with D. ter Haar) Sur les tensions dans la cornée, *Physica* **9**, 234. pp. 109, 117.
1943a Psychologisch aanzicht der natuurkunde, *Ned. Tijds. Natuurk.* **10**, 228.
1943b* On multipole radiation, *Physica* **10**, 261. pp. 45, 47.
1943c Quantumtheorie der molecuulstructuur, *Handb. 29e Ned. Nat. Gen. Congres*, p. 99.
1943* (with J. Kistemaker) On the slip of a diffusing gas mixture along a wall *Physica* **10**, 699. pp. 67, 109-113.
1944a Het gedrag van macromoleculen in een stromende vloeistof, *Physica* **11**, 1. p. 92.
1944b, c Quantumtheorie der molecuulstructuur I, II *Ned. Tijds. Natuurk.* **11**, 41, 57.
1944d* Principiële moeilijkheden van een theorie der deeltjes, *Ned. Tijds. Natuurk.* **11**, 134. Paper K; pp. 59, 60, 64.
1945 E. C. Wiersma † 31-7-1944, *Ned. Tijds. Natuurk.* **11**, 145.
1946a Peter Debije, in *Nederlandsche Helden der Wetenschap*, p. 335.
1946b* The behavior of macromolecules in inhomogeneous flow, *J. Chem. Phys.* **14**, 415. Paper L; pp. 67, 93, 102-107.
1946* (with D. ter Haar) Condensation in interstellar space, *Bull. Astron. Inst. Netherl.* **10**, 137. pp. 49, 50, 109, 117, 118.
1947a* The stopping power of a metal for α-particles, *Physica* **13**, 401. pp. 109, 118.
1947b Topics in Theoretical Physics, unpublished notes of lectures given at the California Institute of Technology, May 22 to 29, 1947. p. 45.
1948* Remarks on the perturbation formulæ of Brillouin and Wigner, *Courant Anniv. Vol.*, p. 205. p. 39.
1949a Discours d'ouverture, *Nuovo Cim., Suppl.* **6**, 157.
1949b* On the behaviour of a gas near a wall, *Nuovo Cim., Suppl.* **6**, 297. pp. 67, 109, 113-115.
1949c* Vibrations of a gas column, *Physica* **15**, 971. pp. 109, 115.
1949d D. Coster 25 jaren hoogleraar, *Ned. Tijds. Natuurk.* **15**, 285.
1949e Wat "gebeurt" er eigenlijk in de natuur, *Diligentia Voordrachten* **28**, 55.

1950* Non-relativistic quantum-electrodynamics and correspondence principle, *Proc. Solvay Congr.* **8**, 241. pp. 59-61, 65.
1951* (with J. Becquerel and J. van den Handel) Sur l'aimantation et le pouvoir rotatoire paramagnétique du sulfate de nickel hexahydraté α, *Physica* **17**, 717. pp. 68, 90.
1952a* On the quantum theory of antiferromagnetism, *Physica* **18**, 101. pp. 67, 68, 91.
1952b* Some reflections on phonons and rotons, *Physica* **18**, 653. pp. 67, 68, 91, 92.
1952c* On a modification of Jaffé's theory of column-ionization, *Physica* **18**, 665. pp. 109, 118.
1957 *Quantum Mechanics*, North-Holland, Amsterdam.[8] pp. 31-35, 40, 44, 45, 57, 59, 60.

---

[8] This is an English translation of Kramers 1933a and Kramers 1938e and is referred to as QM. It was reissued in 1964 by Dover, New York.

Index

S-matrix   7, 254

Absorption   121, 123, 163
- negative   133
Absorption coefficient   29, 166
Absorption cross-section   166
Acetylene   109, 117
Action   10, 36, 100
Action and angle variables   12
Action integral   100
Action variables   10, 12
Activation energy   227
Adiabatic demagnetisation   8, 67, 90
Adiabatic invariants   10
Amplitudes   147
Angle variables   13, 150
Angular momentum   54
Anisotropy   176
Anisotropy constant   265
Antiferromagnetism   91, 245
Aperiodic damping   226, 230
Asymmetric top   52
Atomic magnetisation   173
Autocatalytic reactions   227
Axial symmetry   48

Bending vibrations   260
Birefringence   175, 176, 257, 264, 268
Birefringent crystal   173
BKS theory   8, 24, 121–123, 144
Black-body radiation   143
Body-centered cubic lattice   76
Bohr magneton   73
Bohr theory   165
Boltzmann distribution   103, 105, 216, 258, 262, 263, 265
Boltzmann equation   113
Boltzmann factor   89, 233, 238, 261
Boltzmann-Gibbs distribution   224
Born approximation   54
Branching points   257

Brownian forces   103, 217, 228, 229
Brownian motion   67, 93, 94, 213, 214, 229, 268

C-invariance   54, 56, 57
Ca spectrum   47
Canonical coordinates   190
Canonical equations   190
Canonical equations of motion   42
Canonical transformation   12, 61, 63
Canonical variables   126
Carbon bonds   264, 266
Cathode rays   249
Causality   22
Chain reactions   227
Chain-element
- statistical   259
Charge conjugation   32, 55, 203
Charge density operator   209
Charge-conjugate wavefunction   57
Chemical reactions   93, 94, 213, 219, 227
Circular polarisation   47
Coherent radiation   26
Coherent scattering   170
Column ionisation   109, 118
Combinatorial analysis   68, 77
Commutation relations   33, 51
Complementarity   21
Compton effect   4, 21, 22, 125
Conditionally periodic systems   11
Conduction electrons   118
Conservation
- of energy   21, 125
- of momentum   21, 125
Contact transformation   127
Continuous spectrum   31
Cooperative coupling   236
Coordinate transformation   116
Coordination number   79
Coriolis force   116

284   Index

Cornea   109, 117
- tangential tensions in   117
Correspondence principle   11, 14, 17, 19, 21, 26, 31, 36, 60, 121–123, 126, 128, 133, 135, 140, 143, 147, 150, 253, 254
Cosine-law reflection   113
Coulomb energy   57
Curie temperature   243, 245, 247
Curie transition   231

De Broglie reflection   110
Debye-Hückel theory   68, 71
Degeneracy   129
Demagnetisation   185
Density matrix   209
Detailed balancing   71
Diamagnetic rotation   181
Diatomic molecules   48, 118
- formation of   118
Dielectric coefficients   174
Dielectric permittivity   28, 163, 171, 181
Dielectric tensor   264
Diffuse reflection   110
Diffusion
- quasi-stationary   220, 224
Diffusion coefficient   111, 219
Diffusion current   217
Diffusion equation   94–96, 100, 213, 215, 216, 218, 268
Diffusion slip   109–111
Dipole   145
Dipole approximation   60, 62
Dipole moment   24, 122, 147
- induced   26–28, 47
Dirac equation   4, 32, 42, 44, 45, 50, 55, 188, 194, 203, 206, 251
Dirac theory   57, 60, 61, 65, 177, 212
Dispersion   24, 121, 123, 249
- negative   133
- positive   133
Dispersion formula   168
Dispersion theory   5, 93, 163, 170, 181
Double refraction   163, 257
Duane-Hunt limit   17, 19, 20

Eigenvalue spectrum   33
Einstein coefficients   11, 14, 18, 123, 129, 131, 166, 254
Einstein transition probabilities   17, 26
Electric dipole moment   47

Electric dipole transitions   48
Electric moment   126, 145
Electrolytes   71
Electromagnetic gauge   62
Electromagnetic mass   65, 250, 252
Electromagnetic momentum   254
Electron   249
Electron capture   118
Electron mass
- electromagnetic   65
- experimental   64
Electron radius   64, 250, 251
Electron scattering
- double   251
Electron spin   39, 177
Electron theory   60, 148, 164
- classical   64, 165, 169, 170, 249, 252, 253
- relativistic   252
Elementary particles   32
Elliptic polarisation   175
Energy bands   39
Energy conservation   21, 22, 33, 125
Energy level density   69
Energy operator   207
Ensemble theory   67, 71
Ensembles
- grand   68
Entropy   70, 77
Equilibrium statistical mechanics   68
Equipartition   258
Escape probability   99, 101, 220, 223, 225, 226, 229
Ether drag   249
Euler angles   52
Exchange forces   91
Exchange integrals   73
Excitons   92
Experimental mass   249

Face-centered cubic lattice   76
Fermi distribution   68
Fermi-Dirac distribution   71
Ferromagnet
- classical model   72
- Heisenberg model   72
Ferromagnetism   68, 73, 75
Field theory
- relativistic   252
Fine structure constant   251
First-passage time
- mean   97
Fission   213, 230

Fission probability   230
Flow birefringence   102
Fokker-Planck equation   95, 215
Free electrons   207
Free energy   76, 79, 227
Friction   216
Friction coefficient   261
Friction force   266

Gas constant   243
Gas reactions   228, 229
Gibbs equation   216
Gibbs-Helmholtz equation   77
Grand ensembles   68, 71
Grand partition function   71
Grand potential   71
Gravitational field   116
Gravitational potentials   116
Group theory   45
Gyromagnetic ratio   40, 41, 189

Hamilton-Jacobi equation   12, 36
Hamilton-Jacobi function   10, 12
Hamilton-Jacobi theory   12
Hamiltonian   146, 190
− renormalised   65
− structure-independent   61
Hamiltonian equation   190, 192, 194
Harmonic oscillator   22, 24, 60, 75, 97, 169
Heisenberg ferromagnet   72, 75
Helium atom   15
Helium ion   14
Hill's differential equation   195
Hole theory   56, 188, 205, 252
Huygens principle   22
Hydrogen atom   11, 44, 57, 210
Hydrogen molecule ion   47
Hydrogen-like atom   210
Hyperfine splitting   15

Incoherent radiation   26
Indirect exchange   90
Induced dipole moment   26, 28
Induced emission   11
Inertial mass   250
Infinitesimal Lorentz transformation   205
Infinitesimal rotation   205
Intensities
− of multiplets   45
Intensity
− of radiation   13

− of spectral lines   14
Internal friction   261
Interstellar chemistry   109
Interstellar grains   117
Irreducible representations   45
Irrotational flow   258, 261
Ising model   6, 68, 81, 93, 232, 233

JWKB approximation   35

K absorption   48
Kinetic theory   67
Kramers degeneracy   32, 68, 88, 173
Kramers equation   4, 93, 95
Kramers problem   93
Kramers-Kronig relations   10, 28, 30, 168

Lamé equation   54
Lamé functions   52, 198
Lamb shift   65
Laminar flow   261
Landé factor   89
Lifetime   141
Light quanta   124, 148, 252
Light scattering   60
Linear chain   73, 75, 232, 236
Liquid drop model   230
Liquid helium   91
Lorentz invariance   40, 44, 207
Lorentz transformation   56, 189, 193, 203, 205
− infinitesimal   205

Macromolecules   93, 102, 257
Magnetic moment   55, 180, 187
Magnetic rotatory power   173
Magnetisation   73, 75, 79, 81, 87, 185, 233, 236, 237, 242
− atomic   173, 185
− spontaneous   73
Magneto-optic effects   173
Magnetogyric ratio   189
Majorana calculus   205, 206
Majorana representation   205
Mass
− electromagnetic   65, 250, 252
− experimental   59, 249
− inertial   59, 250
Mass renormalisation   59
Mass velocity   111
Matching conditions   39
Mathieu functions   198

286   Index

Matrix algebra   146
Matrix elements   146
Matrix mechanics   32, 33
Maxwell distribution   97, 110, 217
Maxwell equations   61, 163, 174
Maxwell gas   111
Mean first-passage time   97
Mean free path   110, 113
Minkowski four-vector   41
Molecular spectra   16, 51
Molecular vibrations   16
Momentum   146
Momentum conservation   21, 125
Multiply periodic systems   10, 11, 16, 24, 25, 147, 150
Multipole radiation   45, 47

Negative energies   206
Nernst theorem   68
Neutron   253
Non-equilibrium statistical mechanics   68
Non-uniform flow   261
Normal density   92
Normalisation   34, 44
Nuclear fission   94
Nuclear matter   230
Nucleus
– liquid drop model of   230
Number operator   208

Occupation number representation   56
Old quantum theory   8, 155
Oscillator strength   27, 47, 131, 166
Oscillators
– classical   123
– virtual   24
Oxygen molecule   51

Pair annihilation   210
Pair formation   210
Paramagnetic rotation   68, 88, 173, 181, 185
Paramagnetic rotatory polarisation   54
Paramagnetic susceptibility   90
Partition function   67, 69, 72–74, 81, 83, 233, 235, 239
– grand   71
Pauli matrices   42, 44, 78, 177, 193, 204
Pauli principle   48, 90, 205
Pearl-necklace model   103, 259, 263, 266, 271

Periodic boundary conditions   74
Periodic potential   32, 39
Periodic system of elements   15
Perturbation theory   25
– singular   97
Phase space   95, 100
Phase transitions   7
Phonons   91
Photodissociation   118
Photoionisation   118
Photons   21, 254
– Slater's   21
Point electron   63–65
Point particles   251, 253
Poiseuille flow   112, 261
Poisson brackets   33, 43, 150, 190, 193
Polarisation   47, 129, 140, 163, 168
– elliptic   175
– of spectral lines   14
Polarisation coefficient   164, 167, 168, 170, 176
Polarization tensor   264, 265
Polymer molecules   67
Polymer physics   93
Polymers
– branched   106, 107
– ring   107
Positive energies   206
Potassium spectrum   47
Potential barrier   219, 228
Potential well   94, 101
Precession of earth axis   117
Pressure forces   266
Principal value   167
Probability density   95
Proton   188, 251–253

Quantisation rules   10, 13, 16, 26, 36, 128
Quantum conditions   147
Quantum defect   48
Quantum number   10
Quasi-classical approximation   35

Racemisation   228
Radiation
– absorption   11
– coherent   26
– emission   11
– incoherent   26
Radiation capture   118
Radiation field   61, 170
Radiation friction   170

Radiation intensity 13
Radiation reaction 249
Radiation theory 17, 59
Ramified molecules 262, 268
Random walk 80
Rare earths 88
Reaction kinetics 228
Reaction rates 93, 227
Reaction velocity 220, 223, 227
Reactions 227
- autocatalytic 227
- chain 227
- chemical 94
- gas 228
- unimolecular 94
Reactive vibration 230
Reflection
- De Broglie 110
- diffuse 110
- specular 110
Refractive index 28, 163, 168, 170, 264
Relativistic wave equation 4
Relativity 109, 117
- general 117
Relativity theory 249, 250
Renormalisation 59, 93
Renormalisation theory 6
Resonance radiation 122, 132, 140
Ring molecules 270
Rotation 116
- infinitesimal 205
Rotation group 45
Rotation vector 117, 173, 176, 181, 184
Rotatory power 181
Rotons 91
Runaway solution 64

S-matrix 59
S-matrix theory 59
Scattering 24, 123
Schrödinger equation 33, 39, 53, 177, 194, 195
Second quantisation 56
Second sound 92
Selection rules 48
Sellmeier's formula 24, 165
Semi-classical approximation 32, 35
Separation of variables 10, 12, 52
Simple cubic lattice 73, 75, 76
Singular perturbation theory 97
Six-vector 40
Skein theory 257
Slater's "photons" 21

Slip 109
- diffusion 109–111
- thermal 110
- viscosity 109, 110
Slip coefficient 113
Slip number 113
Slip velocity 110, 111
Small vibrations 117
Smekal-Raman effect 27
Smoluchowski diffusion equation 217
Smoluchowski equation 229
Solid-state physics 8
Specific heat 88, 242, 248
Spectral lines 11
- intensities of 11
- intensity of 14
- polarisation of 11, 14
Specular reflection 110
Spin
- electron 187
Spin conjugation 55
Spin Hamiltonian 43
Spin waves 75
Spin-conjugated states 90
Spinor 44, 45, 85, 203
- spin-conjugated 203
Spontaneous emission 11, 135
Spontaneous radiation 123, 144
Spontaneous transition 129
Spontaneous vibrations 115
Square lattice 73, 75
Square net 236
Stark effect 15
Stark splitting 15
Stationary states 148
Statistical chain-element 259
Statistical equilibrium 143
Statistical mechanics
- equilibrium 68
- non-equilibrium 68
Staudinger's rule 102, 105, 264
Steepest descent method 38
Stopping power 109, 118
Stress tensor 104, 261
Structure-independent Hamiltonian 61
Sum rules 47, 147
Sun 117
Susceptibility
- paramagnetic 236
Symbolic method 32

T-invariance 54

Temperature equilibrium   228
Thermal equilibrium   222
Thermal slip   110
Thermodynamic limit   4, 7, 72
Thermodynamic potential   71
Third law of thermodynamics   7, 68
Thomas factor   40, 42
Thomson scattering   17
Time reversal   54, 55
Transfer matrix   83, 85, 87
Transfer matrix method   81
Transformation
– coordinate   116
Transition state   227, 228, 230
Transition state method   98, 213, 223, 224, 226–229
Transition state model   230
Transition temperature   87
Tunnel effects   228, 229
Turning points   33, 37
Two-fluid theory   92
Tysonite   88

Unimolecular reactions   94

Valence bonds   260
Velocity potential   258
Vibration
– reactive   230

Vibrations
– of a gas column   109, 115
– of the acetylene molecule   109
– spontaneous   115
Virtual field   21
Viscosity   95, 102, 104, 113, 115, 213, 257, 261, 268, 269
Viscosity slip   109

Wall velocity   111
Wavefunction
– charge-conjugate   57
– spin-conjugate   55
Wigner-Jordan matrices   207
WKB approximation   32, 35, 54, 93
WKB method   47

X-ray dispersion   168
X-ray refractibility   167
X-ray spectrum   16
– continuous   17
X-rays
– absorption of   16
Xenotime   88, 90, 185

Zeeman effect   45, 181, 187, 249
– anomalous   187, 249
Zeeman splitting   89
Zero-point energy   207